Eski's Oscillator/Substance Group, 2008-2011

 Compiled by
Dean L(eRoy) Sinclair

Copyrighted as a compilation, by Dean L. Sinclair, Sept. 1, 2011

Preface Number One to First Printing of "Eski's Oscillator/Substance Theory Group, 2008-2011"

This "First Printing--" actually the formation of a PDF File and a CD Disc--of the first attempt known to this writer of the conversion of the contents of a Google Group to a coherent external, hard-copy permanent file, will be interesting in the diversity of errors, mistakes and general "boo-boos, " that will be included. It is hoped that these will have the effect of adding a bit of amusement and humanity to the presentation rather than detracting significantly from the amount of very valuable information which is included. As it is intended to forward a copy of this material to the Earth/matriX people who are site members for their consideration for "cleaning up" and publication as a possible hard-bound copy, it is possible that future editions will have many fewer errors and snags.

As will appear in the "Second Preface," this group was founded in Aug. 2008, to explore ideas related to a comprehensive theory of existence. By the Spring of 2011, a great deal of interesting information had accrued, such that there is a finite possibility that some day this little effort will be considered of historical interest, possibly even to the extent of becoming the basis for some Master's Thesis in History of Science Theorizing.... (Note that I say, "Finite chance, " not, "High probability.")

By the Spring of 20II, the situation had changed with Google with respect to "Google Groups" to the extent that it appeared that a great deal of the information submitted to the site might be lost. It was decided to print out the "Pages" as a separate. "Book,' from the Zip Drive file that Google offered at that time. This led to complications to the point that this compiler did some checking and decided that it might be best to try to pull out the entire file, as the site was not receiving active posts to any extent, with the problem of the shaky status and--probably--personal situations with the people who had been the contributors of the majority of articles.

This has resulted in this " Publication," which includes not only the "Pages," but,.also, as much as this person could retrieve of the rest of the contents of the site. As this will run to some 300 pages or so, it is still intended to extract the "Pages," as a much smaller, more easily readable, "one-sitting" tome which will cover much of the information included.

This writer is, also, projecting two other publications which will include much of this material with additions from other sources in two books. One working title is "A Framework for the Physical Sciences" which is projected as an expansion of this writer's science and mathematics essays, including not only material that has developed here; but, also essays which have been published elsewhere, or have not been previously published.

The other working title is, "Shards of Cracked Pottery, Essays of a Ph.D.--PWT" which is a more light-hearted approach and is expected to include a variety of kinds of essays varying from a poetic protest against the ridiculous, hazardous, low-visibility, red and white stop signs, to some of this material....

These latter books may not make it to publication as the writer is to the age when his life expectancy is "two years, forever," and he has no idea when the two years will run out... .To one writer, who noted that most Lefties do not make it into their 73rd year, he has already run into some seven years of "overtime." D. Sinclair, Aberdeen, SD, USof A Sept. 1, 2011.

PREFACE II. Site Welcome and "Visitor Letter"

Welcome to an experiment.   This site is an attempt   to recruit serious comment in development, or refutation. of ideas as to the basis of reality.

The status of the ideas of the "Founder of this Site" are covered in the page, "The Current Paper..."and in other pages posted on the site. Perhaps the quickest overiview is in the short "page,"   OLD DATA/NEW MODEL.

This is an open site, feel free to express your opinions and to add information.   (I shall reserve the right to edit out anything which I consider as "garbage" or personality comments that are not pertinent.)
Dean L. Sinclair   Group Founder/Manager

A

LETTER TO SITE VISITORS

Dear Visitor:

Here is the situation.   At this time, the Oscillator/Substance Model seems to be the simplest--yet most inclusive--model of Existence yet to appear.

In its present form, it suggests Existence to be within a substance of undetermined---possibly indeterminable--size, which is organized into and/or by oscillators of a "constant torque" family defined by the equation, $m \times r = h/c$.   This little equation, mass times radius equals Planck's Constant divided by the Speed of Light, can describe any oscillator from smaller than an electron to a Universe.

(new seven)

The examination of the ideas over the last five years, as the model has evolved, indicates that the model should be able to correlate date from below the atomic level to the cosmological scale. It needs further development and testing.

as the Space-Time Model implies in its name something moving in cycles, someone familiar with Einsteinian Space-Time Mathematical Modelling most likely would be able to reconcile the two approaches to   mutual profit. Likewise, it would seem that the concepts might well allow expansion/extension of the Quantum Mechanical approach, perhaps with some understanding of the physical basis of the model.

Dr. Mills Grand Unification Theorem may possibly fit more closely than it at first appears to. Dominion Cosmology which postulates Matter/Antimatter repulsion is some what echoed in the splitting of oscillators that arises naturally in O/S Modelling.

One of the biggest and most difficult areas that needs be reconciled is the Elements, Isotopes, Periodic Chart Situation.   By O/S. although these, ideas are useful and valuable, they also are somewhat in error,   being-- at least partially--responsible for the splitting of molecular chemistry from nuclear chemistry.

O/S suggests strongly that the "nuclear atom" is a logical structural result of the differences in size and motion characteristics of the electron ad the proton, and that the presence of neutrons, as such, in the vast majority of atoms, is an illusion. Therefore, there is a need for a complete revision of chemical thinking to account for the existence of "Elements." This is to say, without the handy bookkeeping afforded by the proton/neutron concept of atomic nuclei. (Note added, April 5, 2010. Re: The Periodic Chart Problem. Recent contacts between "Eski" and two researchers who have extensive research in correlation of information related to the Periodic Chart, suggest that this area is being "waded into." When results are ready for publication, contact information will be posted here. At this point, it seems possible to say, "Nuclear chemistry is an interesting branch of Inorganic Chemistry on the sub-atomic scale, The various isotopes may very well eventually be traced to a simple molecular cation..." Oh, yes, also, there are indications that "Cold Fusion" may very well be the "Future Energy Source of Mankind," and it may be possible to develop an "Energy Pump," based on Helium 4 and its relationship to Beryllium 8.)

The Standard Model of Particle Physics which won a Nobel Prize in the 1970's. needs to be critically re-examined. The probable true nature of "Quarks" as an observational phenomenon rather than actual entities needs to be demonstrated. It needs to be worked out where the "fundamental particles" that arise from "atom-smashing" fit in. (It has been suggested that some of these may be based on the "Positronium atom," rather than the Hydrogen atom or some combination of the two.) In other words, the Standard Model needs to be dismantled, keeping what parts are valid and discarding those which turn out to be illusions.

At the Cosmological Scale reconsideration of stars, constellations, even our Universe, itself, as oscillators and combinations of oscillators should lead to new analyses and insights.

It might be a good idea to redo a set of "Universal Values," ala Max Planck, updating and refining the concept.

Gravitation and electromagnetism need to be relegated to their proper positions as observational phenomena . (Seeing the "Four Forces of Nature " as aspects of one "Substance" greatly simplifies things.)

The simple O/S approach has implications for literally every area of scientific theorisation. Its development is far too big a job for any one person. This is particularly true when the discoverer, current "major developer," is at an age when time is rapidly catching up/running out.

It is hoped that some younger people will pick up on the ideas and be able to develop, extend or even refute them... The writer would like to see this model either generally validated or refuted within his lifetime. I hope that you will find the information in this site interesting and informative. Dean L(eRoy) Sinclair (BA, MS, PhD)

Additional note:
        Even if your science background was limited to a high school basic science class, or less, we still need your help to keep us "grounded." Most of us "Scientist Types" tend to use so much jargon as to be incomprehensible to anyone except our closest colleagues.

PPS. Try brousing the titles of the pages. There is quite a diversity of things there, You might find a title that would interest you. I still do, even though I wrote most of them.

I'll let you in on a little secret. I have probably as broad a science background as 99/100 people; but, most "advanced" scientific papers are a complete "Snow Job" to me, too. So, read our stuff, tell us what you "hear" us saying, Let us know what you think.

. Don't go away, we need all the help we can get...

. DLS   aka "Eski,"   "The Oldster," and various unmentionable sorbiquets....

PPPS. If you do have advanced scientific background, we can use help in a multitude of topics. For instance, I suspect that the math. of Space-Time theory will fit in because of the inherent implication of movement and circular motion in the concept of time; however, I do not have the expertise in the particular mathematics used therein to correlate that set of concepts to the ones used here. It would be great if someone could do that.   Same thing holds for Quantum Mechanics.

Actually, here I am only re-emphasizing what I implied in the "Letter..."

BON VOYAGE, wherever you go from here...

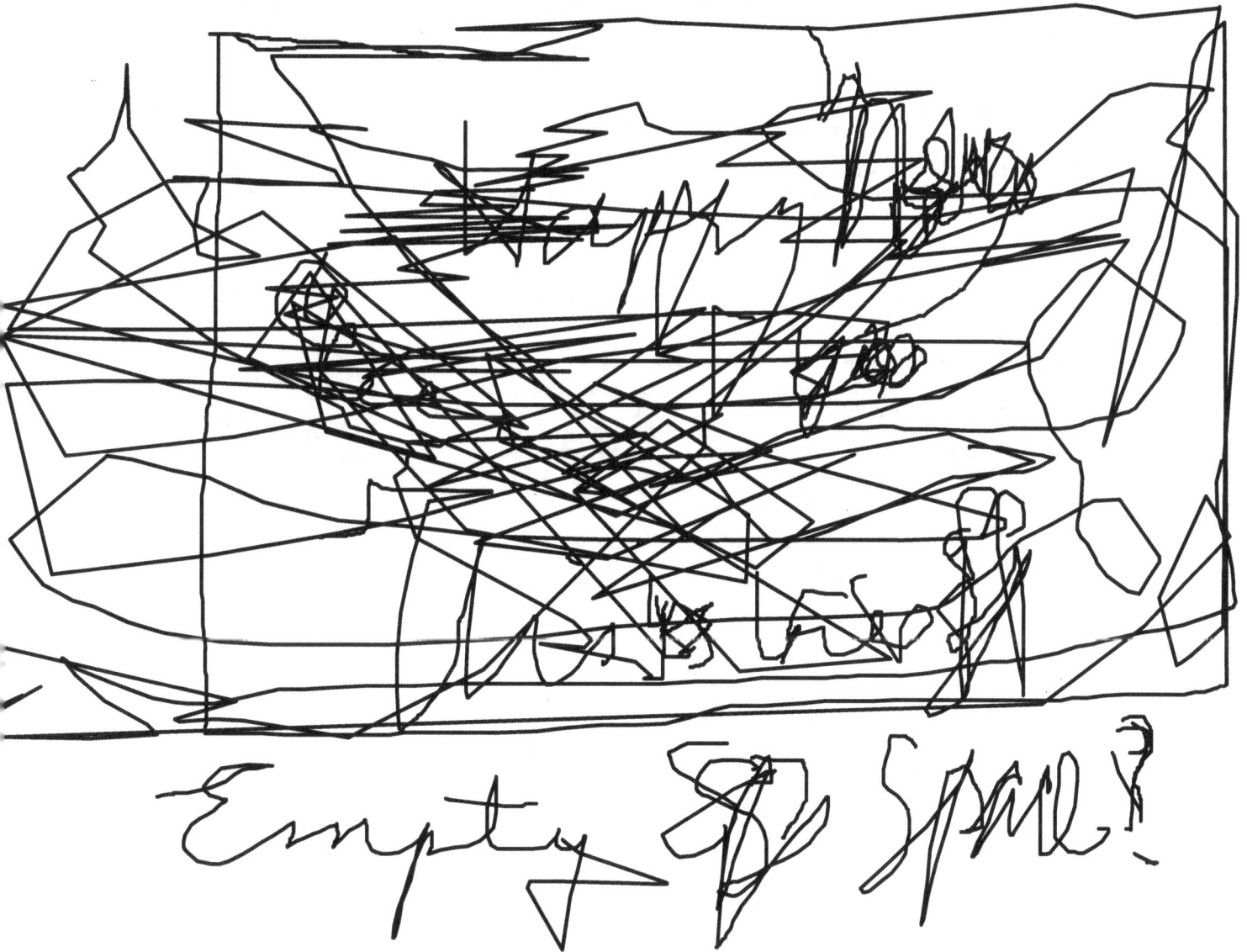

Gmail Calendar Documents Photos Reader Sites Web more   deanlsinclair@gmail.com

Google groups

A "Reverse Contents,"
The "First" topic is actually
the last, chronologically, hence
the last on the disc
or data file.

« Groups Home

Oscillator/Substance
⊞ Theory

[Search this group] [Search Groups]

Discussions

View: Topic list, Topic summary   Topics **1 - 10** of **104** Older »

A "Rant " I posted to another site that someone may find interesting.

... On a Coherent Theory, a "Rant." Let me introduce myself, for those who do not personally know me, my name is Dean LeRoy Sinclair. I am a 79 year old, former science teacher whose background includes a Ph.D. in Organic Chemistry, Mathematics through... more »
By **dean sinclair** - May 25 - 1 message - Report as spam

VALUE OF INFORMATION ON THIS SITE

Dear Reader.: Google has removed the welcome message and the pages fro this group, inadvertently destroying its usefulness to most readers. To get maximum value, rather quickly, firnd the paper, "Essentials of O/S" lsted in the discussions, I t is on about the second page of "Discussions.... more »
By **ESKI** (deanlsinclair@gmail.com) - May 10 - 2 messages - Report as spam

Reality and math. short comment

Reality and Mathematical Definitions We consider mathematics as a model for reality. Perhaps it would be worth while to consiee4 how certain mathematical definitions fit as we look at the physical world. First, let us look at the concept of Zero, usually considered as the symbol of nothingness. However, is this true? Zero is the starting... more »
By **dean sinclair** - Mar 14 - 1 message - Report as spam

Comparison of pressure initiated reactions and cmns.

T This appears on another site, the results are intersting from our view point... [link]
By **dean sinclair** - Mar 10 - 1 message - Report as spam

Rumoured Soviet Weapons, Death Ray and Hammer

If as I suspect, the matter-anti-matter duality noted in the Bsubs units is actually common throughout existence even in electrons and protons, then it may well be that the rumoured weapons could have been developed by technologists who, unhampered by theoretical considerations, simply applied greater and greater accelerative forces to electron... more »
By **ESKI** - Mar 1 - 1 message - Report as spam

Comment on Schroedinger Equatiion, Also. Happy New Year

Here is my " take" on the Schroedinger Equation, the basis of Quantum Mechanics. The first element of the Equation notes a wave function involving the second derivative with respect to time , apparently of Kinetic Energy. The first derivative of Kinetic Energy is Momentum,, and the second derivative, the... more »
By **dean sinclair** - Dec 30 2010 - 1 message - Report as spam

A bit of update from Eski.

Apparently Google has backed off trom blocking access to current content on our "pages," that is good news. There is one bit of personal news. Although i do not yet have Internet access except through public computers, at least not useable access, i do now have ove of those little toys that Americans seem to... more »
By **ESKI** - Dec 10 2010 - 1 message - Report as spam

Elements from "Empty Space," yep, looks like it's accidentally been done; and they darn near killed themslves doing it. ....

Home

Discussions
+ new post

Members

About this group

Edit my membership

Group settings

Management tasks

Invite members

 View this group in the new Google Groups

Sponsored links

BlackBerry® Official Site
Find More, Know More, Do More. Enhance Your Life With BlackBerry.
BlackBerry.com

How To Track A Cell Phone
Read Texts, Track GPS & Call Logs.
See Which Phones Are Supported Now.
InquireHow.com

  See your message here...

Group info

Language: English
Group categories:
Science and Technology > Physics
Science and Technology > Chemistry
Science and Technology
change categories
More group info »

Researchers,posting on a closed Internet site have reported that, in a Cavitation Collapse experiment, they produced--among other things--a melange of elements, consistent with what they called a "Nova Event." They also report that they became very ill. The symptoms are of radiation poisoning. They also note that later checking showed a "neutron burst."... more »
By dean sinclair - Dec 1 2010 - 1 message - Report as spam

Fwd: Copy of Framework II
---------- Forwarded message ---------- ...To: deanlsincl...@gmail.com I've shared Copy of Framework II<[link]> Message from deanlsincl...@gmail.com:Click to open: - Copy of Framework II<[link]>... more »
By dean sinclair - Nov 17 2010 - 2 messages - Report as spam

Google is closing our "Pages" section. SAVE the "Pages Library?!"
Google is going to close down access to "Pages" as of Feb. 1 That will destroy our "Library." I suggest that you-all might want to transfer the "Pages" from this site, en masse, to a file on your home computers. I'm going to try to move the material to a companion OscillatorSubstance Site, but am not sure how to do it. Anyone want... more »
By ESKI - Oct 26 2010 - 4 messages - Report as spam

+ New post   1 - 10 of 104   « Newer | Older »

Gmail Calendar Documents Photos Reader Sites Web more ·   deanlsinclair@gmail.com ·

Google groups

« Groups Home

Oscillator/Substance
⊞ Theory

[Search this group] [Search Groups]

Discussions

View: Topic list, Topic summary   « Newer Topics 11 - 20 of 104 Older »

Fwd: Vigier AIP Paper Submission. Correspondence, I give up on this one. I blew it...

Geez, people, I hate to admit being a "Quitter" but this one I give up on . Will post my current version of the paper that I'm giving up on because of publication technicalities and time constraints. I'ts been an expensive learning experience for an old man.ESKi Here is my latest correspondence re the paper with the Symposium... more »
By **dean sinclair** - Oct 12 2010 - 2 messages - Report as spam

FW: [Vo]:FZ-Quantum Transistion-LENR-Podkletnov-Casimir

Dean: The following is another 'short' that I 'woke-up' with this morning. . . Cheers; have a good Autumn!~Jack~ aka Jack Harbach O'Sullivan/O'Suileabhain * * *FRANK ZNIDARSIC & QUANTUM TRANSITION re: LENR/Podkletnov corollary* * * *Quantum Transition root-ingress-Plasma as original pre-atomic energy state a la Frank Znidarsic*... more »
By **Jack O Suileabhain** - Oct 6 2010 - 2 messages - Report as spam

Matter-Anti-matter and the possible HH+ --> D+ transform.

Fairly recently it has been shown that certain entities known as Beta sub s in Standard Model Parlance undergo Matter-to-Anti-matter oscillation.... It is not inconceivable that this type of oscillation is general rather than confined to this type of special case.l The consequences of such a supposition are interesting. Let us look at... more »
By **ESKI** - Sep 23 2010 - 4 messages - Report as spam

View this page "Essentials of O/S"

This is rather long, about 23 pages. However, I felt it should be in the file, in fact, I had thought that it was.... DS Click on [link] - or copy & paste it into your browser's address bar if that doesn't work.
By **ESKI** - Aug 23 2010 - 2 messages - Report as spam   *See "Item 33"*

Matter-Anti-matter Annihilation

Hi, everybody, I have a question that I need an answer for, It is:. IS THERE ACTUALLY, ANYWHERE, A VERIFIABLE CASE OF MATTER-ANTI-MATTER ANNIHILATION AT THE LEVEL OF THE hYDROGEN ATOM OR ABOVE? It seems to be dogma that "Matter" and "Anti-matter" annihilate on contact to revert to pure "Energy."... more »
By **ESKI** - Jul 27 2010 - 7 messages - Report as spam

Fwd: Can someone else read a script into the 1500 slot schedule so that there isn't a "blank" space where I'm supposed to be?

People, here is the e-mail that I sent to the Symposium director when he wrote, "Sorry to disappoint after...." DS --------- Forwarded message --------- ...there isn't a "blank" space where I'm supposed to be? To: noet...@mindspring.com Dr. Amoroso,: Assuming that the "Internal Problems" of which you speak are... more »
By **dean sinclair** - Jul 15 2010 - 1 message - Report as spam

Vigier VII Symposium

Hello again, Dr. Sinclair, In your prompt, inconsiderate, dismissive and unacceptable response to the below paragraph, which I ask that you read in your upcoming Vigier VII Symposium address. An address that you asked all members to help you with as

---

Home

Discussions
+ new post

Members

About this group
Edit my membership
Group settings
Management tasks
Invite members

View this group in the new Google Groups

ponsored links

Alternative Energy - BP
See how BP's advanced technologies
are expanding energy production.
www.bp.com/energymix

What is Quantum Jumping?
Discover Why Thousands of People are "Jumping" to Change Their Life
www.QuantumJumping.com

Plasma Fume Collectors
Capture Plasma Cutting Fumes With
A Farr Industrial Dust Collector
www.FarrAPC.com

See your message here...

Group info
Language: English
Group categories:
Science and Technology > Physics
Science and Technology > Chemistry
Science and Technology
change categories
More group info »

I did below: It is my pleasure to welcome the Vigier VII Symposium to the Aberdeen...
more »
By **hoek** - Jul 2 2010 - 5 messages - Report as spam

ADDENDUM TO TALK

In thinking about it, particularly realizing that the criticism, "You are asking an awful lot!!!" (In a justifiable criticism of the temerity of an admittedly average person operating rather jokingly as a "pretend genius" daring to say, in effect, "some of the geniuses were possibly wrong, Eski, has decided that he should probably go...
more »
By **ESKI** - Jun 22 2010 - 3 messages - Report as spam

REPEAT of INFINATE VARIABES - Reply to ESKI.

... * Never believe this of yourself Eski... nor of me! ... * It has been no secret between us that I have almost completed a Report/Book. I try to share 'snippets' but cannot give it all away... yet. OK ? ... * If I am, I am sorry. I personally know what I am talking about and the link is to you all and the Oscillation Substance Theory. You know... more »
By **ka-sala** - Jun 19 2010 - 2 messages - Report as spam

View this page "Possible Script for VigierVIITalk"

Am posting a possible script as a "page" on this site. Want to know what the people of this Group think about this version. I'm finding it more difficult than I thought it would be to condense the last three years of discoveries, even the most basic, into a thirty-minute, oral presentation.... more »
By **ESKI** - Jun 10 2010 - 3 messages - Report as spam

+ New post   **11 - 20** of **104**   « Newer | Older »

Gmail Calendar Documents Photos Reader Sites Web more   deanlsinclair@gmail.com

Google groups

« Groups Home

Oscillator/Substance
⊞ Theory

| Search this group | | Search Groups |

Home

Discussions
+ new post

Members

Discussions

View: Topic list, Topic summary   « Newer   Topics 21 - 30 of 104   Older »

MECHANISM
We all find ourselves existing in a wondrous realm of three spatial and one forward moving dimension of time. This last dimension allows cause, effect and logic, allows motion within the other three spatial dimensions, which allows MECHANISM! Mechanism allows biological life. Even, though, it won the Nobel Prize for nuclear physics, back in... more »
By hoek - Jun 4 2010 - 3 messages - Report as spam

Attention 'Reporter'
Interested as to why you joinded this group ? ka-sala

[handwritten: " Reporter " dropped in and out. Never posted anything.]

By ka-sala - Jun 2 2010 - 1 message - Report as spam

REPEAT of " INFINITE VARIABLES " Wishing you what you so desperately seek.
*** The reason for this repeat is because this is where we are all standing. On the same pin-point. I said I believe we have the key, and the gravity of this O/S jig-saw. I have not repeated in these simple words for nothing. Best wishes to all. Can anyone see this while we are together ? In our Oscillation Theory, a lot has been said lately regarding... more »
By ka-sala - Jun 2 2010 - 2 messages - Report as spam

Matter--Anti-matter
Merhaba, arkadashlar, Long time me watch Group. Sorry, hear lose old Eski. I use Eskiadam name, think maybe time i go try join group. Old Eski he say if anti-lectron anti-matter, so be proton, so if the electron matter then so be anti-proton if so be-- maybe atom weight be total of all anti... more »
By eskia...@mail.com - Jun 2 2010 - 2 messages - Report as spam

Trying to get Eski back some contacts.
Luckily for him, Eski and I have a side contact which was not through Google. He tells me that the lock out from the deanlsinclair@gmail address has lost him the exact addresses to two very valuable contacts, "Charles" and "Herringbone." He says that there is a member of the Group that is an associate of "Charles" and that "Herringbone" has also been in contact with... more »
By hugh vreeland - Jun 1 2010 - 2 messages - Report as spam

Looks like we may be "orphans"
It looks like this sight may be a small "family of orphans' with out a parent or any way to add new members. Although this group is supposed to be open membership, to be able to keep out useless spam, the managers set up a situation to monitor first submissions. That worked fine as long as we had manager's.... more »
By Hugh V - May 25 2010 - 5 messages - Report as spam

An old essay that might be of interest,
Hi, folks, Looking over my old documents of about three years ago, I ran into this little essay which I apparently wrote back when Osc/sub theory was Motion in a Matrix The article I'm referring to would have to be Doc's original Motion in a matrix paper, the trial balloon on Helium. Have fun.!....Hugh... more »
By hugh vreeland - May 19 2010 - 1 message - Report as spam

Group info

Language: English

Group categories:
Science and Technology > Physics
Science and Technology > Chemistry
Science and Technology
change categories
More group info »

The fluid theorem...

To Mr. Sinclair, According to my own theories, one universe cannot exist without another. Let me explain. You mentioned the circular motion of matter/ anti-matter. Think of the "first" universe as the middle. Imagine the distance between universes (do they exist separate from each other?) as spots on the water. If a ripple happens in one, let's say... more »

By heycollin - May 7 2010 - 2 messages - Report as spam

FW: *NASA: David Adair's 'Quasi-Fusion:' ?Cold, Warm, Hot?

Provenance: Werner Von Braun, Hermann Oberth, David Adair-NASA Adv.Prop.Res.Prct., Wu Yeong Wei(Andy Wu), R.A.-Ned-Allen, Whitt Brantley-NASA Adv.Prop.Res.Prjct., Marc Rayman JPL-NASA. -blame Jack Harbach O'Sullivan-:-) Re comment: THUSWISE realizing this 'model' for epanding-dialating Protons(as micro-singularities) enhances the description of NOT ONLY CHEMICAL REACTIONS but also of FUSION REACTION be they 'cold,' warm, or hot.... more »

By Jack O Suileabhain - May 10 2010 - 5 messages - Report as spam

View this page "Congruent Parallelogram Theorum "

HI, Everybody, This paper is a bit more mathematical than most of what I have posted here; but, some people may find it interesting. As usual comment is invited. ESKI Click on [link] - or copy & paste it into your browser's address bar if that doesn't... more »

By ESKI - May 4 2010 - 2 messages - Report as spam

+ New post

21 - 30 of 104   « Newer | Older »

XML   Send email to this group: oscillatorsubstance-theory@googlegroups.com

Gmail Calendar Documents Photos Reader Sites Web more ▾   deanlsinclair@gmail.com ▾

Google groups

« Groups Home

Oscillator/Substance
⊞ **Theory**

[Search this group] [Search Groups]

Discussions

View: Topic list, **Topic summary**   « Newer Topics 31 - 40 of 104 Older »

News to the group. Re: Vigier VII Symposium. The Search for....
Folks, get your thinking caps on and your extensions of the ideas together, as it looks like old Eski, himself, is going to have to try to be the group representative at the first presentation of our work in to more or less the main stream by a remote hook up from here in Aberdeen, to London, sometime in the July 10-14 period... more »
By ESKI - Apr 27 2010 - 2 messages - Report as spam

A Caveat, I may just be another deluded crackpot. Here is one person's opinion.
The following is an excerpt of an exchange between myself and a theretical physicist/mathematician who has, to say the least, a low opinion of this.whole thing. ...be > fooling around with such things? A kid playing with grown-ups toys > that he doesn't understand? ...be > around to dream....DS Dream all you like, but what you're doing is both logically &... more »
By ESKI - Apr 10 2010 - 3 messages - Report as spam

Fwd: Essentials of O
---------- Forwarded message ---------- ...To: amylaverickd...@aol.com, Earth/matriX <earthmat...@gmail.com>, drsdcm < drs...@cinci.rr.com> Here is a twenty three page write up that I did a few weeks ago on my home computer, finally got it from there to a form that I can put on the Internet. Yes, I don't really know my way around this technology as well as... more »
By dean sinclair - Apr 6 2010 - 1 message - Report as spam

An old item i don't think i shared It's about "Energy."
Must be round a year ago, that Eski published a bit on SciScoop and a man who went by Barak ridiculed him. This was my take.. The topic is somewhat peripheral to o/s theorizing, but, then, o/s is supposed to be inclusive. Since i'm in a send stuff to the group mood today, here is this one, if you've seen it before, it won't hurt too much to see it again.... more »
By hugh vreeland - Apr 6 2010 - 1 message - Report as spam

Abridged summary of oscillatorsubstance-theory@googlegroups.com - 1 Message in 1 Topic
This is not a direct reply to today's topic, but, I am too lazy to retype the address. A new topic could be started on Graphing with Signed Numbers, i suppose. A few days ago i tried something; Eski always tries to keep things too simple, he works only with absolute values, although I'm sure he realizes that we could use signed numbers. If i... more »
By hugh vreeland - Apr 6 2010 - 2 messages - Report as spam

O/S and Light (Copy of body of a letter to another scientist.)
Looking at my site, it appears that the closest that I came to addressing the topic of light, and then not by name, was the article, "Quantization, a 3-D Merrigoround?" Since analyzing electromagnetic waves as information carrier is a basis of the model, I have been remiss! As you will be studying light, maybe you can... more »
By dean sinclair - Apr 5 2010 - 1 message - Report as spam

The One Force of Nature. Also, some miscellaneous progress.
Hi, Everybody, Here is a little post that I just put on another site that is pertinent to us.

" ... the required force is present based on Nature resisting the concentration of energy." This quote, cut from a previous writer, is the thing that no one seems actually to realize,. This defines the one and only true "Force" that there is.... more »
By ESKI - Mar 29 2010 - 1 message - Report as spam

Notification.
Please Note. Due to moving house... It 'maybe' I am off the net for a while. If any do not recieve a reply - or wish to do so - please carry on as normal, and I will be able to see, when again opened. Best wishes to you all, ka-sala
By ka-sala - Mar 7 2010 - 1 message - Report as spam

BALANCE... an OSCILLATING SCIENCE in Velocity and Time.
*** A repeat with a twist, of going back to the future, and finding the answer was already there. Lost in the maze of equations and theory. Analogy. Using the Earth. Something no-one has been able to duplicate in time and space! 'As above... so below.' Balance. *** Quote. 'Truth is ever to be found in the simplicity, and not in...
more »
By ka-sala - Mar 7 2010 - 1 message - Report as spam

'Get Back in Time.' Back to the future. How Light Speed makes Time Travel possible.
Having scanned everything submitted, the who of what essentials being sought, is lost in the swamp of theorizes and equations, regardless of where they first originated. This is just a repeat - a down to earth one - in the going back to the future in order for the sum total of equalization the O/S theory for which ESKI - the owner of... more »
By ka-sala - Feb 13 2010 - 2 messages - Report as spam

[+ New post]

31 - 40 of 104 « Newer | Older »

XML   Send email to this group: oscillatorsubstance-theory@googlegroups.com

Gmail Calendar Documents Photos Reader Sites Web more ▾

deanlsinclair@gmail.com ▾

Google groups

« Groups Home

Oscillator/Substance
⊞ # Theory

| Search this group | Search Groups |

Discussions

Home

Discussions
+ new post

Members

View: Topic list, **Topic summary**   « Newer Topics **41 - 50 of 104** Older »

Errors and Omissions of the CODATA: The Planck Constants

I, Everybody, Please, consider my recent work entitled, "Errors and Omissions of the CODATA and the Planck Constants Based on the Fundamental Physical Constants". In this brief study, selected CODATA fundamental physical constants are researched in an effort to identify the implied computational foundations that serve as a basis for deriving the Planck constants.... more »
By **ollin** - Feb 5 2010 - 9 messages - Report as spam

Dark-Energy Einstein-Rosen/Cern-NDR path ID.prjct-

The Cern/International NDR-Pathway-Identification project plots 'naturally occuring' Einstein-Rosen transtemporal/transdimensional pathways though Dark-Energy Trans-Space This is the Holy Grail of contemporary String/Membrane theorists. The NDR-ID. project is the cutting edge medium for the diverse theories to find their legitimate expression in by consolidating their efforts in this hard technology application.... more »
By **Jack O Suileabhain** - Dec 28 2009 - 3 messages - Report as spam

Dolphin-like'pods' swim Aethyrs//WinterSolsticeEveDream2009i

* * * *Winter Solstice Eve's all night Dream//2009* * * * I dreamed all night even-after returning to interrupted sleep of a constant migration to Planet Terra of Silver Dolphin-like Spirit-Soul swimming the Aethyr-Seas. Our Planet in the Aethyr-AexoDarkSeas is a beacon of concentration's of sentients and our eco-system has a brilliant Avatar-Archetype energy matrix of the full kaleidoscopic-evolutionary array of all of the planet's myriad symbiosis of species of which Homo-Sapien-Sapien is one.... more »
By **Jack O Suileabhain** - Dec 21 2009 - 1 message - Report as spam

Abor-Empathic-Resonance/DREAMTIME-Umbilicus-Phoenix Portal

... *The 'time' is Now & inexorable//ready or not~here it comes!* ~Title: Oscillatorsubstance-Christmas/Hanukkah/Phoenix-Millenium-Eve for Homo-Sapiens-Noviensis~ Ref: * * * ABORIGINAL-Empathic-Resonance/DREAMTIME-Umbilicus-Phoenix Portal * * * From the Norse Valhallah proto-Kaballah 'Tree of Live-Yggdrasil' across Northern Europe to the Super/Quasi-Celtic-Norse-&-Asian super-civilization of Shamballah in Northern Mongolia(now recently discovered in the sands from 'twenty-thousand years ago very well preserved); the Tree of Life Archetype had penetrated down into Ancient Egypt from which arose the 'Allah-Kaballah' to Aakhenaton/Winged-Disk mono-theistic Super-Membrane-Unity traditions. And from ancient 'Shamballah-Mongolia' the 'Tree of Life' conceptual matrix also had migrated from across the Pacific to the Super-Civilizations of MesoAmerica which form the ancient Aztlan/Aztlantus Pyrimid Super-Civilization incorporating all of the ancient America's of the Toltecs-Olmecs Mayas-Aztecs etc. These also connected across the Atlantic to form a vital link back to Egypt that from 20 thousand years ago again was descendent of the ancient Shamballan super-Tree-of-Life culture. And the global circuit was complete and doing its work trans-culturally leading to this final planetary Phoenix-Metamorphosis that the Maya & others had so profoundly predicted....
more »
By **Jack O Suileabhain** - Dec 19 2009 - 1 message - Report as spam

Large Hadron Collider starts again(video)   (Spam)

Everybody should see this... [link]
By **JimmNorton** - Nov 22 2009 - 1 message - Report as spam

Group info

Language: English

Group categories:
Science and Technology
> Physics
Science and Technology
> Chemistry
Science and Technology
change categories

More group info »

Some interesting basic math....

Hi, Everybody, I'm back with a bit of my "Crackedpottery that just might hold a little water somewhere."..I beg your indulgence... Jack O' , on the O/S site has proposed that there should be adopted some sort of a basic quantum for quantum mechanical calculations. Assuming that a "basic quantum" might correspond to the lowest... more »

By **dean sinclair** - Nov 12 2009 - 1 message - Report as spam

A Brief Future of TIME - Sorry for 2nd. Post. Typo-Fix done

Response to J/O *** <DISCLAIMER for DELICATE FLOWERS in the AUDIENCE: !Warning! ..... This is 'not' intended as 'shouting' so that the less emotionally resiliant amongst the audience should 'not' get their knickers-in-a-wad-about it. Thankyou for your indulgence.-JO- > & / Warning! Insult to O/S 'Contributors' - not audience - personal... more »

By **ka-sala** - Nov 10 2009 - 1 message - Report as spam

A Brief Future of TIME

For: Dean L. Sinclair Title: 'A Brief Future of TIME' Sub-title: 'Spin Oscillating Inversion-Spheres with Spheres/Wheels within Wheels at Time's END' DISCLAIMER for DELICATE FLOWERS in the AUDIENCE: !Warning! Limited usage of CAPS; this is merely a 'punctuation device' to facilitate extended-emphasis as a mere alternate-symbol-function-of my 'limited' key-board. This is 'not' intended as 'shouting' so that the less emotionally resiliant amongst the audience should 'not' get their knickers-in-a-wad-about it. Thankyou for your indulgence.-JO-... more »

By **Jack O Suileabhain** - Nov 1 2009 - 4 messages - Report as spam

BAE&Aexospace/DarkEnergyPlasmaBreachReactor:Right on 'Space-TIME.'

RE: Singularity/DarkEnergy-Physics//BAE-Constant Revised Planck-Einstein Constant//Green Bubble Universe within DarkEnergy/DarkSpace Aexoverse//Jack O'Suileabhain Fusion-Gate/Aexospace/DarkEnergyPlasmaBreachReactor:Right on 'Space-TIME.' * * * ? Which Nations shall be pre-eminent in these DarkEnergy Tech R&D projects;... more »

By **Jack O Suileabhain** - Oct 25 2009 - 1 message - Report as spam

*NEWTON left hitch-hiking~;-) ConceptCraft Prospectus: Quantum-Gravionic Point-Lead Focused Hyper-GravThrust

Respectfully submitted to Dean L. Sinclair & Oscillator-Substance Theory Group *BOSE-EINSTEIN SUPER CONDUCTOR TOROID RING REACTOR access DARK-ENERGY/parallel-DARK SPACE aka 'TACHYON CARRIER-WAVE SPACE'* * * * The RACE is AFOOT in DarkEnergy Technology R&D and the LOSERS will be LOSING for a VERY LONG TIME * * * as per 'TIME' in the aspect... more »

By **Jack O Suileabhain** - Oct 25 2009 - 1 message - Report as spam

---

[+ New post]   41 - 50 of 104 « Newer | Older »

[XML]   Send email to this group: oscillatorsubstance-theory@googlegroups.com

Google groups

« Groups Home

Oscillator/Substance Theory

Search this group | Search Groups

Discussions

Home

Discussions
+ new post

View: Topic list, Topic summary   « Newer Topics 51 - 60 of 104 Older »

You cannot post messages because only members can post, and you are not currently a member.

Description: Discussion of Oscillator/Substance Theory and its relationship to other theories. The theory assumes that all reality exists in a substance at its triple point consisting of separable oscillators.

*TachyonCarrierWaveDARKENERGY~IS~PARALLEL(parent) AExoDarkSpace

-Respectfully submitted- *PARALLEL(parent) DarkSpace~IS~TACHYONCarrierWaveDARKENERGY * * * TACHYON SPEED-DENSITY CARRIER-WAVE SPACE is 'AEXOVERSAL SUPER-COSMOS DARKSPACE' and is most simply DARKENERGY SPACE: This is 'high-speed-density' space whose 'density' is hyper-fluidic at density that makes 'mass' seem like vapor and whose 'relative-speed' moves at rates that makes vast distance beyond the span of 'several universes' crossable virtually INSTANTANEOUSLY. This is the 'parent' of 'Spooky action at a Distance.'... more »
By Jack O Suileabhain - Oct 25 2009 - 1 new of 1 message

Group info

Language: English

Group categories:
Science and Technology
> Physics
Science and Technology
> Chemistry
Science and Technology
More group info »

*CernHadron-Explosion&Quantum-Grav-Lev-GRAY-JET/(DarkEnergy) SubSingularity-REACTOR-ONLINE/HADRON Modified-'Super-Collider?'aka Plasma Breach Reactor access 'TIME' & far UNIVERSE

Respectfully submitted to Dean L. Sinclair & Oscillator-Substance Theory Group * * * The CERN reactor explosion resulted from a DEFACTO-PLASMA BREACH BLEED THROUGH vortex/eddie quasi-worm-gate-holing FROM PARALLEL------->TACHYON-CARRIER-WAVE(aka) 'Full-DarkEnergy-Torsion Waves'> ('DARKENERGY-TACHYON CARRIER WAVE PLASMA-DARK SPACE').< The bleed-through Dark-space plasma-breach inadvertently occurred within the EYE OF THE high-EM-giga-density//electro-plasmic INDUCTION RING-ARRAY SUPER-MAGNETS. This created a SPONTANEOUS DarkEnergy-TachyonSpace induced 'EVOLUTION' of HYDROGEN & HELIUM. This describes the function of the GRAY-JET subSINGULARITY REACTOR which can be fairly easily affected by adding BOSE-EINSTEIN SUPERCONDUCTOR main-core RING replacing the super-collider open ring-track of a more compact version similar to the general design of current SUPER-COLLIDER TECHNOLOGY, but with a far different end function creating a GRAY-JET ELECTROPLASMIC-SINGULARITY cross-spectrum/cross-dimensional TOROID-VORTEX quasi-worm-hole DARKENERGY bleed-through gate REACTOR.... more »
By Jack O Suileabhain - Oct 25 2009 - 1 new of 1 message

NewAtomModel-SingularityPhysics

-Alternate 'Atom' Model//BAE-Constant//DarkEnergy-Singularity Physics-per Jack O'Suileabhain/O'Sullivan(English) * * * Fusion-Gate/Plasma Breach Hyper-Gravionic Fusion Sustaining Reactors// aka Gray-Jet subSingularity Hyper-Plasma Gate Reactor * * * * * * Quantum-Electron Shell/Fields are the same hyper-gravionic lobe/field that the Plasma-Breach Reactor exhibits. The Russian's made the reactor and did it, thusly are recently boasting of 'Gravity Manipulation' and creating electro-plasmic/hypergravionic lobe fields that can interdict aircraft & missles and stress their superstructures so violently that they are virtually 'ripped from the skys.' * * *... more »

By Jack O Suileabhain - Oct 25 2009 - 1 new of 1 message

BAEconstant//Planck'sConstantRevised

Respectfully submitted to Dean L. Sinclair & Oscillator-Substance Theory Group
Author: Jack O'Suileabhain/O'Sullivan- For formal submission to: Max Planck
Institute--> M...@Mpe.mpg.de <&> R...@mpiwg-berlin.mpg.de <&>Oscillator-
Substance-Theory group <&> vorte...@eskimo.com <-- From Jack O'Suileabhain
(eng."O'Sullivan) aka Jack 'Harbach' O'Sullivan... more »
By Jack O Suileabhain - Oct 25 2009 - 3 new of 3 messages

What you want Fun Guy?

Check you are in right place otherwise you in wrong place.
By ka-sala - Oct 20 2009 - 1 new of 1 message

O/S, Space-Time and QM

Oscillator/Substance Model, Space-Time and QM. The Oscillator/Substance Model
seems to be more general than the "Accepted Models" of Space-Time and Quantum
Mechanics. A close look at Space-Time shows that the name, itself, suggests
"substance" and, as is pointed out in the paper, "Why Einstein Was... more »
By ESKI - Sep 28 2009 - 2 new of 2 messages

Invitation to Video SEX Chat

I am commenting on this as an example of the sort of thing that is a total waste of
time on this site which is a site dedicated to a scientific research project. It has no
need for discussions of human sexuality, human taboos and such material, which
may be of interest to many people but has no value here. I am marking Helga as "no
post."... more »
By ESKI - Sep 28 2009 - 1 new of 1 message

TAKE NOTE - NOTIFICATION OF BANNING MEMBERS TO THIS SITE.

Any 'members' trying to get in the back door to this O/S site will be banned. Those
who already have, have been reported. Take what you want elsewhere. It has
nothing to do with the reputable members of the O/S Theory. USE YOUR MINDS!
You are not wanted here.
By ka-sala - Sep 26 2009 - 2 new of 2 messages

View this page "Some Constants of the Universe?"

I've republished some data that was in another paper with a little different slant on it;
Some of you might want to pick up on the idea. (Or pick it apart.) Click on [link] - or
copy & paste it into your browser's address bar if that doesn't... more »
By ESKI - Sep 22 2009 - 1 new of 1 message

Unification of Gravity and Electricity finally discovered ! !

[link] [link]
By EinsteinGravity.com - Sep 21 2009 - 3 new of 3 messages

+ New post   51 - 60 of 104   « Newer | Older »

XML   Send email to this group: oscillatorsubstance-theory@googlegroups.com

Gmail Calendar Documents Photos Reader Sites Web more ~   deanlsinclair@gmail.com ~

Google groups

« Groups Home

Oscillator/Substance
⊞ **Theory**

[Search this group] [Search Groups]

Discussions

View: Topic list, **Topic summary**   « Newer Topics **61 - 70 of 104** Older »

A look at Quantum Superposition, Atemporality and the direction of Entropy in relation to Eternity - INFINITE VARIABLES

In our Oscillation Theory, a lot has been said lately regarding Infinity. Directly of indirectly; we have been oscillating all around the subject of the O/S Theory. To get to where we want to be we must bring it back to the unified weak and electromagnetic interaction between elementary particles, including inta alia, the prediction of... more »

By ka-sala - Sep 19 2009 - 2 messages - Report as spam

Explaining the Universal Law of Physics - COMMON BOND

COMPLETE BOND Explaining the Universal Law of Physics would be the Elemental Effect in which each and everything is Linked. We could take the largest circumference of a circle in any given direction within the shape of a round ball - call everything within this a Complete Bond - then take everything out of it and divide it up... more »

By ka-sala - Sep 18 2009 - 3 messages - Report as spam

View this page "Letter to "Sophie" at H=Judge Parker"

Nothing really new in this other than the somewhat ironic view point of talking across dimensions to a personality who exists only in a writer's mind and apprear only in two dimensions...You may get a chuckle out of it. Click on [link]... more »

By ESKI - Aug 20 2009 - 1 message - Report as spam

View this page "O/S and the Periodic Chart"

Click on [link] - or copy & paste it into your browser's address bar if that doesn't work.
By ESKI - Jul 20 2009 - 1 message - Report as spam ·

View this page "Deuterium Molecular Cation....."

This little bit of chemical speculation was done with respect to some work done by another group. It seems unfair to not share the same ideas with this group!! Cheers, DS Click on [link] - or copy & paste it into your browser's address bar if that doesn't... more »

By ESKI - Jun 5 2009 - 1 message - Report as spam

View this page "Universes Within Universes?"

Just a bit more speculation to think about.....DS Click on [link] - or copy & paste it into your browser's address bar if that doesn't work.
By ESKI - Jun 5 2009 - 1 message - Report as spam

View this page "The Matter of Anti-Matter"

I just added a little essay to the pages on the site. Hope you find it intersting. Eskil Click on [link] - or copy & paste it into your browser's address bar if that doesn't work.
By ESKI - May 21 2009 - 1 message - Report as spam

View this page "M.Margulis' Penetrating Ether ...."

This theory has some elements in common with our ideas. I thought some of you might like to compare.... Click on [link] - or copy & paste it into your browser's address bar if that doesn't work.
By ESKI - May 13 2009 - 2 messages - Report as spam

Home

Discussions
  + new post

Members

About this group
Edit my membership
Group settings
Management tasks
Invite members

 View this group in the new Google Groups

Group info
Language: English
Group categories:
Science and Technology
> Physics
Science and Technology
> Chemistry
Science and Technology
change categories
More group info »

REPRINT

Returning at Double the Speed of Light.
Hi everyone... I'm back! QUOTE - HUGH <Here in lies the key to the reason that the two energy expressions differ in form. In the process of integration, one of the variables was "held constant," that is taken out of real consideration except as a "scalar number" in the process. However, it is still a variable, a... more »
By ka-sala - May 2 2009 - 2 messages - Report as spam

View this page "More on the Second Energy Equation"
I'm finally kicking in a little contribution. 'Bout time, Eh?? Click on [link] - or copy & paste it into your browser's address bar if that doesn't work.
By Hugh V - Apr 30 2009 - 2 messages - Report as spam

+ New post

61 - 70 of 104   « Newer | Older »

XML   Send email to this group: oscillatorsubstance-theory@googlegroups.com

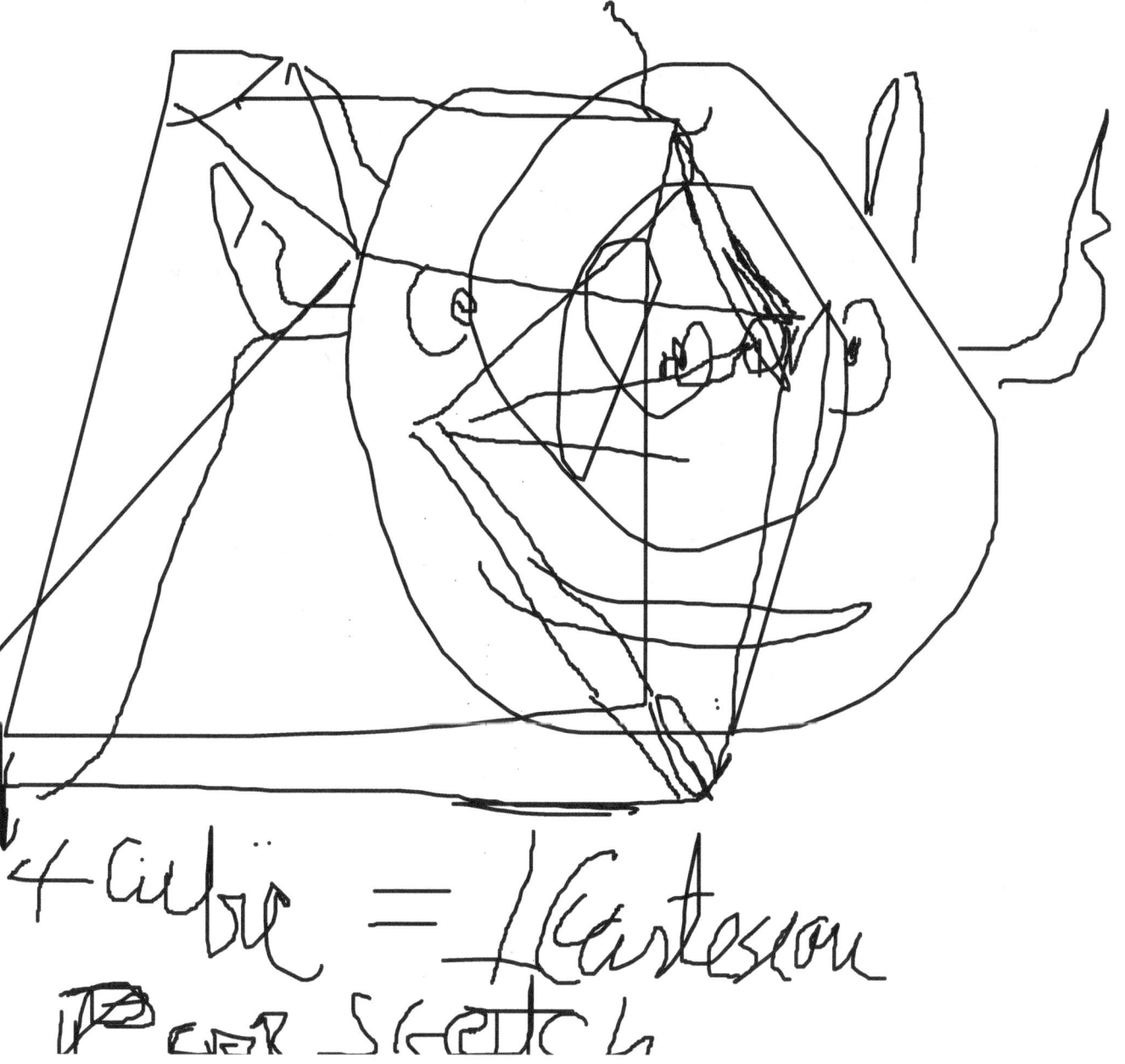

4 cubic = Cartesian
Peer Sketch

Gmail Calendar Documents Photos Reader Sites Web more ⌄

deanlsinclair@gmail.com ⌄

Google groups

« Groups Home

Oscillator/Substance
⊞ **Theory**

[Search this group] [Search Groups]

Discussions

View: Topic list, **Topic summary**   « Newer Topics **71 - 80 of 104** Older »

View this page "The Electron and Proton as Oscillators"
I added a little about possible internal structure of the electron and proton.
Cheers.DS Click on [link] - or copy & paste it into your browser's address bar if that
doesn't work.
By **ESKI** - Apr 27 2009 - 1 message - Report as spam

Fw: I wish you enough
----- Original Message ----- ...To: alan ; AL CAPONE ; Dianne ; eddie ; eli ; gail ;
george mc ; hotlips ; jame scott ; Jen ; Joann G. ; Ryan Kin ; Loui ; michelle ; KEVIN
MONTERIO ; Pat ; Cathy Rainey ; sandy ; shannon ; Tim ; BOB Vanderhoek Sent:
Tuesday, April 14, 2009 12:19 PM ... From: GGa7606...@aol.com
<GGa7606...@aol.com>... more »
By **Robert Vanderhoek** - Apr 14 2009 - 1 message - Report as spam

View this page "The Oscillators"
Click on [link] - or copy & paste it into your browser's address bar if that doesn't work.
By **ESKI** - Apr 10 2009 - 1 message - Report as spam

View this page "OLD DATA/NEW MODEL "
HI, People, Hope everyone is well, happy and enjoying the nice Spring-- or Fall--
weather! This is a simple version of the origin of the Oscillator /Substance Model
written to be used as a "prequil" on other sites for articles that have previously been
accepted there. Cheers, Dean "Eski"... more »
By **ESKI** - Apr 7 2009 - 2 messages - Report as spam

Welcome Major Ray
"...the atom of truth could only win." Welcome Major Ray. Would love to see a
discussion here from you. You have a lot up your sleeve and you are far from alone
in your beliefs. I had a scan of your BIOS and as I have been a fly on the wall of the
Academy of Future Science since the 70's - I prefer ro remain out of... more »
By **ka-sala** - Mar 25 2009 - 1 message - Report as spam

View this page "Iso-sets ,Radioactivity and CMNS"
kThis little page was posted sometime ago, but somehow, without a title.....The sme
ideas have also been reposted elswwhere, including on SciScoop. Anyway, I finally
got at title on this one so it will have more respectability..... Click on [link]... more »
By **ESKI** - Mar 13 2009 - 2 messages - Report as spam

View this page "Why Einstein was right."
HI, everybody, and thanks for the heads up on various other things! Anyway, this is
the latest try to get us some more attention via SciSdoop. Look it up there, OK/
Thanks, ESKI Click on [link] - or copy & paste it into your browser's address bar if
that doesn't... more »
By **ESKI** - Mar 9 2009 - 3 messages - Report as spam

View this page "Positron plus Negatron equals Zerotron?"
For some reason, this page, although it had a page title had no content. Who knows
why.... DS Click on [link] - or copy & paste it into your browser's address bar if that
doesn't work.
By **ESKI** - Feb 17 2009 - 2 messages - Report as spam

Home

Discussions
+ new post

Members

About this group
Edit my membership
Group settings
Management tasks
Invite members

View this group in the
new Google Groups

Group info

Language: English

Group categories:
Science and Technology
> Physics
Science and Technology
> Chemistry
Science and Technology
change categories
More group info »

Fwd: CMNS: Zero Point Energy, Casimir effect and hydrinos ☆☆☆☆☆

Somehow this should fit into oscillator-substance theory! ---------- Forwarded message ---------- ...To: CMNS <cmns@googlegroups.com> Dear all, Jovion recently got a patent for extracting energy from the vacuum, using the Casimir effect to create what looks like hydrinos. Scott, Marissa, I believe that there is a strong link with Earthtech's... more »
By **dean sinclair** - Feb 6 2009 - 6 messages - Report as spam

PEACE...is an OSCILLATING SCIENCE

Peace within a Wave Band? It is a Frequency, a Vibration; just as one could say the Frequency of Music, the note, Middle C = 440 Oscillation per second. Middle C to the 7th note is a Scale in Music. All sound has a Frequency, a Vibration, and in vibration there is movement. This is the Peace Wave. With the Earth's population by November 2008 up to... more »
By **ka-sala** - Jan 1 2009 - 2 messages - Report as spam

| + New post | 71 - 80 of 104   « Newer | Older »

XML   Send email to this group: oscillatorsubstance-theory@googlegroups.com

Google groups

« Groups Home

Oscillator/Substance Theory

Discussions

Home

Discussions
+ new post

View: Topic list, Topic summary   « Newer Topics 81 - 90 of 104 Older »

You cannot post messages because only members can post, and you are not currently a member.

Description: Discussion of Oscillator/Substance Theory and its relationship to other theories. The theory assumes that all reality exists in a substance at its triple point consisting of separable oscillators.

About this group

Join this group

 View this group in the
new Google Groups

Group info

Language: English

Group categories:
˙cience and Technology
 Physics
Science and Technology
> Chemistry
Science and Technology

ᴵre group info »

View this page "Energy Expressions Interact?"
Click on [link] - or copy & paste it into your browser's address bar if that doesn't work.
By ESKI - Dec 29 2008 - 1 new of 1 message

View this page "Quantization: A 3-D Merry-go-round?"
Hi. this is a copy of a submission to Sci-Scoop which, apparently, they don't know what to think about it. It takes two negative votes to discard a submission, and a net of 4 positive votes to publish it, That's the supposed rules, So far it seems to have gotten nothing but abstentions, which would mean that people reading it don't know what... more »
By ESKI - Dec 29 2008 - 4 new of 4 messages

Positron plus Electron equals Negatron?
This little article was posted on Sci-Scoop. Although the subject is mentioned in other papers on this site, it is felt that the idea might deserve attiention by itself. Also, I am posting bits of the theory from this site on Sci-Scoop as a "spreading of the word" so to speak. e^- + e^+ = e^0 ?... more »
By ESKI - Dec 10 2008 - 1 new of 1 message

View this page "An Intertwined Universe? "
Hi, I just posted another little article which was also published on the SciScoop Site. There it made the "Front Page." Somebody must have liked it. Cheers! "ESKI" Click on [link] - or copy & paste it into your browser's address bar if that doesn't... more »
By ESKI - Dec 1 2008 - 5 new of 5 messages

View this page "A Constant's Secrets, a Different Look at Planck's Constant"
Click on [link] - or copy & paste it into your browser's address bar if that doesn't work.
By ESKI - Nov 20 2008 - 1 new of 1 message

View this page "Some basic background ideas"
Click on [link] - or copy & paste it into your browser's address bar if that doesn't work.
By ESKI - Nov 4 2008 - 1 new of 1 message

THE PROTON and The Sixth Element of Action
THE PROTON and The Sixth Element of Action Which-ever particle we speak of, they are all inter-linked. To some, this may be more metaphorical, and to others, analogies. Maybe... a hint of insight. This should not subtract from a broader perspective into a subject which is a 'pin point' in the Cosmos. It... more »
By ka-sala - Oct 30 2008 - 2 new of 2 messages

THE PROTON and The Sixth Element of Action

Ka-Sala- I think that in this essay you are coming awfully close to definining some sort of "proto-oscillator" perhaps, a "proto-on." The physical description comes close to descriptions of a toroid, pseudo-sphere which would describe very closely what I would envision as the shape of most of the tiny dots making up the "Substance." The only... more »
By **dean sinclair** - Oct 27 2008 - **2 new** of 2 messages

MANY OSCILLATIONS - CONNECTIVE ISSUES
MANY OSCILLATIONS – CONNECTIVE ISSUES Dealing with something as normal as pitchblende, seems of little import. Ships of the past, and even some of the present, patch holes with Pitch. Yet the Pitch Element is very significant. Being a Carbonic Element it is linked to the Transition Element from which... more »
By **ka-sala** - Oct 19 2008 - **1 new** of 1 message

OSCILLATING ENERGY - IN A GRAIN OF SALT
OSCILLATING ENERGY – IN A GRAIN OF SALT Alternative fuel ? A new millennium ? Once the designer of a water fueled engine, was bought out, to 'shut up'! So now that we have turned over from then, where the price of fuel has caught up with the purse strings of every one except who holds the purse, why not try... more »
By **ka-sala** - Oct 19 2008 - **1 new** of 1 message

[+ New post]   81 - 90 of 104   « Newer | Older »

[XML]   Send email to this group: oscillatorsubstance-theory@googlegroups.com

Need a magnifier or no lead print ?

Gmail Calendar Documents Photos Reader Sites Web more ⌄   deanlsinclair@gmail.com ⌄

Google groups

« Groups Home

Oscillator/Substance
⊞ Theory

[   ] [Search this group] [Search Groups]

Discussions

View: Topic list, **Topic summary**   « Newer Topics 91 - 100 of 104 Older »

ZERO DISSECTED
ZERO DISSECTED Zero = Nothing, right? So... No Butter, no eggs no apple no
'what'? No Zero? What does Zero look like? That is not nothing but a symbol of
something. OOOOOOOOOOOOOOOOOOOOOOOO Here I have 24 Zeros I can,
divide them up! The division of 6 into these 24 = 4. Four Zero's is not nothing but...
more »
By **ka-sala** - Oct 19 2008 - 2 messages - Report as spam

TRAVELING ON ONE WAVE LENGTH
TRAVELING ON ONE WAVE LENGTH Anyone... who thinks they have the answers
to the link between all things, is missing out on the journey. This exercise is called
'Think'. These Graphics by word, are imprinted in the mind, and simply expressed as
a dot within a circle. Throwing Light on the subject =... more »
By **ka-sala** - Oct 19 2008 - 1 message - Report as spam

DOUBLE THE SPEED OF LIGHT - TIME TRAVEL
Ai, ai, ai; Ka-sala, Amiguita mia, Again, you give us something to think about!
definitely differemt than my dry prose! Eski
By **ESKI** - Oct 18 2008 - 3 messages - Report as spam

OBJECTIVE INTELLIGENCE
OBJECTIVE INTELLIGENCE . Beginning the search on the theory of everything.
Some-one with a great mind once said that, "we only begin to learn when we realize
we know nothing." At least this is a good beginning, to start searching for something.
So lets start with Zero. Graphically written as 'O'.... more »
By **ka-sala** - Oct 17 2008 - 3 messages - Report as spam

OBJECTIVE INTELLEGENCE
Hello "ka-sala" You've posted alot of stuff requiring many answers, some of what you
say seems to be just word play. I believe zero is zero, no matter how many times you
multiply or divide it, At first I thought you were kind of a wizz-bang, but I read several
of your works on Helium.com and realize you... more »
By **Robert Vanderhoek** - Oct 20 2008 - 2 messages - Report as spam

View this page "Neutron ala O/S Theory"
Here is more comment on the neutron. It could be called a sort of current update on
the paper, "Neutron Facts and Fables." DLS Click on [link] - or copy & paste it into
your browser's address bar if that doesn't work.... more »
By **ESKI** - Oct 16 2008 - 1 message - Report as spam

View this page "Reality and Speculation in Science"
Hi, I have posted some comments essentially about "taking with a grain of salt"
almost everything in sceince. DLS Click on [link] - or copy & paste it into your
browser's address bar if that doesn't work.... more »
By **ESKI** - Oct 16 2008 - 3 messages - Report as spam

Basic Status Paper Reposted
Oscillator/Substance Model, A Theory of Everything Summary: This article presents
a view of existence as being within a substance, at its triple point, consisting of
separable oscillators. As such it gives new definitions of mass, energy, force,

---

Home

Discussions
+ new post

Members

About this group
Edit my membership
Group settings
Management tasks
Invite members

  View this group in the
new Google Groups

Sponsored links

Particle Identification
Particle size, shape & make-up
Using SEM/EDS/AFA Analysis
www.herguth.com

Alternative Energy
BP is investing in bus... .s and
technology to deliver clean energy.
www.bp.com/energymix

See your message here...

Group info

Language: English
Group categories:
Science and Technolog
> Physics
Science and Technology
> Chemistry
Science and Technology
change categories
More group inf ⌄

gravitation.... Included are possible explanations of "The Big Bang,"... more »
By **ESKI** - Oct 15 2008 - 1 message - Report as spam

View this page "Proton Cosmology"
Click on [link] - or copy & paste it into your browser's address bar if that doesn't work.
By **hoek** - Oct 13 2008 - 6 messages - Report as spam

Discussion on .draft-1223675099611 *Iso-Sets, Radioactivity & CMNS*
cool_gr...@hotmail.com Iso-sets, Radioactivity and CMNS This is, in general, a note I
posted on another Google Group Site which has ideas pertinent to our work here.
DLS Dear People: This is as condensed version which will be expanded for
publication elsewhere. ([link].... more »
By **ESKI** - Oct 10 2008 - 1 message - Report as spam

[+ New post]

91 - 100 of 104   « Newer | Older »

XML   Send email to this group: oscillatorsubstance-theory@googlegroups.com

$x + y + z$

3 cartesian directions = one "cube"

Google groups
« Groups Home

Oscillator/Substance
⊞ Theory

[Search this group] [Search Groups]

Discussions

View: Topic list, Topic summary   « Newer Topics 101 - 104 of 104

View this page "The Hadron Collider and O/S Theory"
Another short page post-- Click on [link] - or copy & paste it into your browser's
address bar if that doesn't work.
By ESKI - Sep 22 2008 - 2 messages - Report as spam

Space-Time and O/S Theory
Comment on Space-Time and the O/S model. If time be considered as always "cycle
referenced," which it is, and it be noted that all mathematics operates in such a way
as to imply the constant presence of a "dot field," then, the title, "Space-Time" implies
in the mathematical sense, an entity of undetermined size... more »
By ESKI - Aug 28 2008 - 1 message - Report as spam

View this page "Combining constants of Nature"
Another basis for the ideas of this theory. Click on [link] - or copy & paste it into your
browser's address bar if that doesn't work.
By ESKI - Aug 26 2008 - 4 messages - Report as spam

View this page "Set Notation and Atomic Structure"
A possibe way of looking differently at atoms..... Click on [link] - or copy & paste it
into your browser's address bar if that doesn't work.
By ESKI - Aug 23 2008 - 1 message - Report as spam

View this page "The Principle of Electron Equivalence"
The point is made that dropping the idea of nuetrons existing as such in nuclei leads
to a different view of atomic nuclei. One that might allow possibly useful computer
modeling Click on [link] - or copy & paste it into your browser's address bar if that
doesn't... more »
By ESKI - Aug 20 2008 - 1 message - Report as spam

View this page "An Essay on Time"
This little essay seems pertinent to the topic. Click on [link] - or copy & paste it into
your browser's address bar if that doesn't work.
By ESKI - Jul 27 2008 - 2 messages - Report as spam

This Model and the Space-Time Model
Is this model compatible with the Space-Time Modelling? For example, would
"Space-Time Distortion" mean approximately the same thing as "Disturbance in the
Substance?"
By ESKI - Jul 27 2008 - 1 message - Report as spam

View this page "The Current Paper (27 July 2998)"
Here is a version of the parent paper posted on [link]. Click on [link] - or copy & paste
it into your browser's address bar if that doesn't work.
By ESKI - Jul 27 2008 - 3 messages - Report as spam

Welcome to an experiment
Hi, I'm Dean Sinclair, well, if you want the full name, titles, etc. I'm Dean LeRoy
Sinclair, BA, MS, PhD. It may be that I shld use the whole thing to distinguish me
from at least one other existing Dean L., and the most famous of the Dean Sinclairs
who is a soccer star in England. In any case, I am a 76-year-old Yank who graduated

Oscillator/Substance
⊞ **Theory**

Welcome to an experiment

2 messages - Collapse all - Report discussion as spam

ESKI View profile   More options Jul 27 2008, 11:39 am

Hi,
I'm Dean Sinclair, well, if you want the full name, titles, etc. I'm
Dean LeRoy Sinclair, BA, MS, PhD. It may be that I shld use the whole
thing to distinguish me from at least one other existing Dean L., and
the most famous of the Dean Sinclairs who is a soccer star in England.
  In any case, I am a 76-year-old Yank who graduated from a school
called Yankton College in Yankton, SD in 1953, with a BA (cum laude,
yet! Most of my acquaintences said it really should have been "laude
how cum" or some such thing.)
  I spent a couple of years is the US Army, Pfc, Radar Repair. in
'54-'56 and, after various misadventures, eventually got a Ph.D. in
Organic Chemistry at Kansas State U. in 1867.

[handwritten: 1867 PhD!? A Pun "Typo!"]

So, what has all this to do with this group?
  Putting together some of my varied background in science and
electronics with the idea that I might be able to pretend tp be a
genius by "loooking for the overlooked obvious" I have come up with a
synthesis of ideas, that. at least to me, make sense, which are
summarized in the name of this group. The experiment is to see if a
Google Group left wide open to everyone can work together tp explore,
elaborate. criticise, elaborate and perhaps develop the ideas into
truly useable form.

I shall post a copy of the paper that covers the current status of the
model in another posting.

Welcome, aboard. I need all the help I can get. At my age, if the
idess are valid, I don't have a lot of time to develop the ideas a lot
farther than their current status. DLS

(If you wonder about the "ESKI," the word means "old," in Turkish,
sometimes with the impication of "out-dated" or "obsolete." With my
name I suppose I could be "The Green Dinosaur." Some people do call
me "Dino.":)

Reply to author   Forward

Hugh V View profile   More options Aug 7 2008, 6:30 pm

"ESKI,"

SPIT IT OUT, DOC, YOU HAVE A GOOD IDEA THAT YOU NEED HELP DEVELOPING,
YOU NEED SCIENTISTS TO LOOK AT YOUR IDEAS AND SEE IF THEY FIT IN WITH
ACCEPTED THEORY, DISAGREE WITH IT OR AMPLIFY IT.
THEY DON'T NEED YOUR LIFE-HISTORY AT THIS POINT!

""I'll do what I can to help by suggesting some topics and an
occasional comment. After all, I did get in on the early part of the
act a bit when it was called the Motion in a Matrinx Theory on
Helium.com and thanks for giving me half credit for the Sin-Vree
Entityidea, It does make a good pun if one knows the original Holland
Dutch pronunciation of "Vree."

Regards, and good luck. HV

Oscillator/Substance Model, A Theory of Everything

Summary: This article presents a view of existence as being within a substance, at its triple point, consisting of separable oscillators. As such it gives new definitions of mass, energy, force, gravitation.... Included are possible explanations of "The Big Bang," Black Holes, Neutron Stars.... In terms of this model there is also a warning of possible danger from the Hadron Collider which is due to go into operation in the Summer of 2008. .

The "Oscillator/Substance Model of Everything" is a second-generation model developed from the Motion in a Matrix Model.* The key insights of that model were that the speed of light being a limiting velocity of information transfer implied a necessary medium for interaction and that the interconvertibily of mass and energy implied that they were different aspects of the same thing. The original thought was that mass coordinated to point-centered motion and energy to motion along a line. While this seems to be in general quite close to the facts, it was realized that modifying the ideas to mass being motion confined within a surface and energy being a less specific term covering all types of motion, with Kinetic Energy being the type that would be dissipated along a line would be a more usable view.

If, instead of postulating a fixed matrix of dots as a basis for a model, it be assumed that there be a basic substance which will remain at, or return to its triple-point, the problem of whether the model is dealing with a solid, liquid or gas disappears. At the triple point the substance can act as any of the three depending upon slight variations in conditions. We may even postulate that the triple-point temperature is approximately the temperature of outer space, about three degrees Kelvin.

Postulating a substance basis for everything, allows us to define a number of things which are otherwise undefined or have "circular definitions." Mass becomes the balance between the motion content of the substance within a particular surface and the remainder of the substance. Kinetic Energy is the motion dissipated into the "Bulk" of the substance when a portion of the substance within a surface changes position. Light and other electromagnetic radiation is motion within the substance. The law of forces, "For every force there is an equal and opposite force," falls into place as a statement of the fact that if there is a pressure change within one part of the substance, there will be compensatory changes within the rest of the substance. It has been long known that pressure is a force. We simply change the statement slightly to, "All true forces are pressures."

Gravitation loses its place as a force--a status which it actually never should have had. It becomes a description of the difference in pressure of the amount of the substance between two entities and rest of the substance, including that part of the substance included within the entities. Electro-magnetism is the observed result of pressure disturbances within the substance from actions and interactions between electrons and protons,

Electrons and protons are known to be the basic units from which all of our chemical substances are derived. That is, with the possible exception of the "short-lived" particles produced in high-energy, particle-physics experiments. In our Universe, protons and electrons occur as the decay product of neutrons. Electrons and anti-electrons also appear in "Pair-production," and disappear in "Annihilation."

Putting the above pieces together with the characteristics of oscillators. we may postulate that 'Pair-production" and "Annihilation." are opposite processes. A spherical cavity oscillator can be considered to consist of counter-rotating halves, which, if subjected to enough "motion disturbance," can split into two mirror-image vortex oscillators having opposite spin orientations. If these two halves were to re-encounter on a proper axis they could recombine to the original oscillator with loss of kinetic energy. That is, postulating a basic unit of our substance as an oscillator that can be split into vortex oscillators having opposite spins, neatly accounts for "Pair-production" and "Annihilation."

As to electrons, protons and neutrons, we can account for these at the same time as we account for the

"Big Bang" and the "Expanding Universe."

Neutrons fit the characteristics of a substance distorted by a shock wave such that the "front half of the unit met the back-half coming forward." This distorted unit splits into an electron and a proton. The speed of light in the substance--if light is a typical information carrier--is the average velocity in any given direction in any given instant. Therefore, it will be the average expansion velocity of the expansion phase of an oscillator, and will also be the average rotational velocity measured at the outer edge of the expansion (or contraction) phase of an oscillator. As such it will be the average velocity felt by our basic oscillators.

We may postulate that motion at above this velocity will travel as a shock wave and since "c," the speed of light, is an average, the initial velocity at the start of the expansion phase has to be greater than "c." if we consider acceleration and deceleration to be linear. we can estimate the initial shock wave from any oscillator as having a velocity of "2c" which fits fairly well with the comparative "masses" of the electron and proton. with the proton half laking the entire distortion from a "2c" shock wave slamming into our substance acting as a "solid."

What this all adds up to is that some oscillator is operating at ultra-low frequency, very high power, and we are riding in the "Creativity out of Chaos" behind that shock wave which converted may substance units into neutrons which disintegrate into electrons and protons. The electrons and protons, vortex particles of opposite spin orientations, but very different sizes and shapes, can not stably recombine but can associate in a multitude of ways, creating everything that we know in our Universe. Presumably, or the other side of the Parent Oscillator, is an "Anti-Universe, coping in its own way with the same "Creativity out of Chaos" attempt to stabilize. At some point the shock wave will expend itself, and there will be a contracting Universe-Anti-universe set which will eventually collapse through the parent-oscillator to start the process all over again.

Our discussion above leads to the conclusion that our basic oscillators possess the ability to coordinate vast amounts of energy. Taking this into account gives a possible explanation for "Black Holes" in the centers of Galaxies. In the Creativity of Chaos behind the shock wave, some units fall into coordination with an oscillator at their exact center, and begin to move with its frequencies. Considering the characteristics of oscillators, a vortex oscillator such as a proton or electron, if passed intact through a full-cavity oscillator would have its "sense" reversed. That is, an electron would become an anti-electron, a proton an anti-proton. Since, tossed out the other side of the oscillator these could unite with "untransformed" units, oscillators operating in this way would be reducing electrons and protons back to unit-oscillators. This. however, comes at a cost. Protons, formed at distortion of smaller oscillators, would be combined into "proton-parent" oscillators, which could presumably be, in turn, distorted by shock waves into Super-neutrons, which would collapse into an anti-proton ("Super-electron") and a "Super-proton" and so on and on and on...
When a Black Hole Oscillator has done its "trash converter" work in a volume, what would appear to a human observer would be a complete blank spot in space devoid of anything which we would consider "Matter."

One other factor in our Existence that we have not mentioned is the neutrino-anti-neutrino combination. It seems logical to guess that these are a pair-product just as electrons and anti-electrons are, and that their "parent-oscillator" is a more basic unit to our substance oscillator(s). This would be in analogy to atoms and molecules in chemical substances. These could, in turn be composed of even smaller oscillators, ad infinitum.

We consider all "matter" to be made up of protons and electrons, which combine to form atoms, atoms to form molecules, molecules to form organisms.... In this model we would not consider that neutrons would exist within atoms other than in "potential" or "evanescent" state, and consider that neutron stars and quasars would perhaps model as hugely over-sized atoms. As such they would function as "isotope factories." Isotopic distributions within galaxies may be a result of differences between the "neutron" stars active within them in the past. It should be possible to find spectra of elements including possibly trans-Uranium elements or even unknown "Trans-Uraniums" in the radiation from quasars and neutrons stars. Because of the expectation of high energy electron "orbitals" closely spaced, it can be expected that these entities will have distinctive radio-frequency "signatures.

This model, if valid, casts considerable doubt on both the validity of the assumptions behind some of the publicized experiments that are intended for the "Hadron Collider" under Switzerland and France that is due to go into operation this Summer (2008), and on its safety. If the simple definition of mass which arises in this model be valid, then the idea that something known as the Higgs Boson is responsible for mass becomes a total misconception. The idea that colliding protons with protons would duplicate the conditions before the "Big Bang" is inconsistent to this model in several ways. One is that by this model, protons would come into existence as a result of the decay of neutrons. As neutrons have a significant half-life, protons would not exist until after the start of the shock wave. As to Lead Ions, these would appear much later. However, collisions at relative velocities of close to "2c" could certainly have the potential to distort oscillators cause neutron formation and, perhaps, additional esoteric, short-lived, alternate forms of matter. Almost surely, Hydrogen would appear within the collider tube. A much more dangerous implication of this model is the idea of Black Hole formation by coordination of a circular spinning entity with an oscillator in its center. The Collider apparently has as least a couple of perfect circle dimensions which have resonance frequencies. There is a high probability that these circles will act as cavity oscillators. If one of these oscillators did come into active coordination with a central oscillator the result would be the same as is postulated for galaxies. Although on a miniature scale to a galaxy, a Black Hole interaction the size of the Collider could have results ranging from, hopefully, merely annoying "unexplained energy anomalies," to a new volcano where the Collider had been, to a coordinated vibration with the resonance frequencies of the Earth to create a new asteroid field between Venus and Mars. Although a "Black Hole Oscillator" would probably "eat" only the material within the circumference of the Collider, the conversion of that amount of matter to anti-matter would make the most powerful atomic bomb appear a mere firecracker.

The scientists working on the project apparently admit that they have no idea what will happen. I hope they remember that, in retrospect, the physicists working on the Manhattan Project realized that it was by the merest of good luck that they did not blow themselves, and the entire Chicago area off the face of the Earth.

----

* See "Motion in a Matrix...." on www.helium.com. Several other articles by this writer, which are posted on that web site also bear upon ideas related to this model.

AUTHOR'S POST-SCRIPT: After reading through the above article, the writer has realized that he has not included some things which usually occur in a modern, scientific article. There are no mathematical formulations that might not be comprehensible to everyone, nor any mention of one or more of the modern, accepted theoretical ideas such as Relativity, Quantum Mechanics, String Theory. Rather than break into the body of the above to correct these apparent flaws, there is being appended this addendum.

It has been shown in other papers that by equating the value of Planck's Constant,"h." to the dimensions of its definition, evaluating this at the speed of light, "c," one can obtain a definition of a family of dual-cavity and/or vortex oscillators defined, in set notation, by {m x r = h/c}. This set has a central, symmetric, spherical--or ring--oscillator defined by m = r = $(h/c)^{0.5}$., where "m" is a mass and "r" is the radius of a circle or a sphere. The value of m in cgs units is about 7.4×10^{-19} g. and the value of r is about
7.4×10^{-19} cm. If one goes a step farther and writes, "m = h/c^2" , one finds an associated minimal mass of about
7.5×10^{-58} g. which may be the mass of a "neutrino parent." Possibly the smallest mass of significance in our Universe.
When the accepted mass of the electron is inserted into the oscillator-family-set equation, the radius is found to be the Compton Wave Length used in Quantum Mechanics. Inspection of the mathematical definitions shows that the two are mathematically identical although their different derivations do not show this clearly.

Quantum Mechanics is said to fail at about 10^-18 cm., and String Theory gets lost into its ten dimensions at about the same value. That is the value of the diameter of our average oscillator, a value through which all the oscillators may well invert. It is interesting that it is the point of failure of two of the widely used theoretical models. If our speculations are correct the failure of both is perhaps that their mathematics only accounts for half or less of "Reality."

In so far as the writer knows, no-one has correlated the Space-Time approach with "Q.M." Each seems to have been in its own niche. The mathematics of both, in so far as they go, would seem to be somewhat compatible with this model. Space-Time modeling seems to work quite well at the macro level despite the fact that Relativity probably only would strictly apply to communication theory, and concepts of what occupies space and what time actually is, appear to be in error. String Theory seems to depend, ultimately, on the true, but-rather-useless-in-the-thinking-of-most-people fact that the route traced by a dot moving in random fashion will follow a path which one can consider as having always being there. The 10 dimensions may arise because one can use nine dimensions of space, and one of motion. to define the actions of a dot--if one wishes to do so.

If the foregoing seems not too well explained and a bit abstruse, then, perhaps, the writer has reached the goal of writing a "modern, scientific paper." Most modern, scientific papers seem to be written in such a way that only the author and the closest associates have any idea what is going on. If they do.

On a more serious note, it can be observed that if this model be valid, it could have been put together close to a century ago, had the ideas advanced by Planck been followed up rather than those of Einstein. This writer did not realize this at the start of the path that led to this write-up. However, the initial impetus was the realization that there was an "Overlooked Obvious" in Einstein's Special Relativity, the fact that the transform equations could be generalized to fit into communication theory if one realized that "c," the speed of light was simply the limiting velocity of an information carrier wave. The theory was valid for situations wherein communication was possible or necessary; but, possibly of no pertinence in other situations.

Version: | Current by ESKI - Oct 11 2008 ▼ |

---

3 messages about this page

Aug 7 2008 by Hugh V

Maybe someone would like to summarize this paper a little more. Five
pages is a lot to read. It does make sense, but I think that you
perhaps should take what you are doing a bit more seriously, although
you are right, most scientific papers are ony comprehensible to the
initiated which could mean the people working in the same lab. Let us

Jul 27 2008 by ESKI

Whoops, the title is wrong, the date is in 2008, not 1998, I need to
learn to type.

Jul 27 2008 by ESKI

Here is a version of the parent paper posted on www.Helium.com.

Click on http://groups.google.com/group/oscillatorsubstance-theory/web/the-current-paper-27-july-2998
- or copy & paste it into your browser's address bar if that doesn't
work.
Report this page
Sign in to discuss

An Essay on Time

Version 2 of 6, edit by ESKI, Jul 27 2008

These are some of the author's thoughts on the subject of time.

Time is a human construct used to keep track of sequential motions in space referenced to repetitive observed motions. Useful for planning and information transfer, it is sometimes considered as if it were a fourth spatial dimension.

Time is usually considered to have three aspects, the past, present and future. These can be considered to correspond to motion of a specific point on a wave front. The past is the total sum of all motions that led to that instantaneous position which we consider the present, and the future is wehere that wavefront spot may be considered to go in the next instant and all the instants which may follow.

The present is the result of a certain sequence of motions, which we call the past, something which no longer exists, but which is never the less a "fixed construction." To go back into the past, in a physical sense, would require a retrograde repteition of all of these motions, a set of motions which would increase instant by instant in a huge geometric progression. Our "wave front" would have to move backwards in a perfect reversal of the sequence by which it had previously moved forward. Even were this possible, one can see, that, since the direction has been reversed, what was "back" is now "forward," the wave front, while "retracing the past" would actually by "moving into the future." In trying to go into the past, a "time-traveller" would actually be attempting to create a future which was a reversal of the past.

If one considers that it may be possible that long wave fluctuations in the "Matrix" in which we exist creates multiple adjacent universes in which certain sequences may coincide, it might be possible to move from one alternate universe into another which would correspond exactly to some point in ones own past. There is no real indication that there are such alternate universes; and, were they to exist, it is highly unlikely that they would correspond in such way as for there to be possible entry from one to another.

However, if one wished to combine Religion, Brane Theory, and ultra-slow vibrations in the Matrix of Existence, maybe "Heaven is Just a Brane Away?" That sounds like a new hymn, or a Country-Western Song. This whole last comment, of course, has little to do with the original essay..

(This little essay can also be found in the "Portfolio" for deanlsinclair on www.writing.com as well as in the section, Essays on time, on www.helium.com.

Oct 19 2008 by ka-sala

Hello Eski,
Ride that Wave you are on!
It's living in the Now which in Time past and future are One, and the meaning of why we are on 'that wave' in the first place.
I don't know about you, but mine is One huge ride, and I would not have missed that wave for anything! A Time Traveller has so much

The Principle of Electron Equivalence

THE PRINCIPLE OF ELECTRON EQUIVALENCE.

If one drops the idea that neutrons, as such, exist in the nuclei of atoms, one obtains a totally different view of what may be the situation within atoms and molecules. If all electrons retain identity as electrons rarther than some of them losing identity to become neutrons, then all electrons with in any atom or molecule are equivalent. That is, no electron is confined to any given spot at any given time. This indeterminacy is the same idea as appears in Quantum Mechanics where it is considered that all possibilities exist until there is an "observer." For observer, one could write, a forced change in the possibilities. At that instant, motion can be considered to be frozen, and at that moment electrons will be "in specific orbits," specific parts of the involved "dance of the electrons."

It can then be said that there are essentially three divisions of the "electron dance." Intra-nuclear, extra-nuclear, and "whole-unit" with any electron at some time or other involved in one of the three motion units. If these be considered as separate orbitals-- the extra-nuclear orbital ideas has served well for many years--then there can be postulated that there are, in any unit twice the number of orbitals as there are electrons. Writing out orbital designations for "orbital electrons," nuclear electrons, and then for all electrons gives sets of possible orbitals which explain a number of thinga about atoms that are otherwise mysterious....

It may also be noted that, in this view, protons also would have equivalency, and would tend to enter into an "intra-proton dance." The combinations of these two motions can be come very complex. so much so as to lead to the instasbility fragmentation seen in radioactivity.
 One can consider that radioactivity will occur when the interactions become so complicated that at times the "dance" will come apart with small ller units of less complicated patterns resulting....

It can also be seen that, if one goes by the idea from Oscillator/Substance Theorizing that it is possble for certain oscillators to be split into electron-positron pairs, there may be certain configurations which would result on such a situation, particulary if addition of an electron would result in a more symmetric unit... This seems to be the case in "positron emission" radioactivity.

The whole area of exploration of atomic structure and of isotopic transformations from the view of electron equivalence and proton equivalence has never been explored, in so far as this writer knows. While the approach may not offer any new insights, it would seem that it could be worked up on a computer adding one "nucleon," i.e., electron or proton at a time to a computer model. In computer modeling, one could [perhaps also take advantage tof the $(m \times r = h/c)$ set size and shape forms that it suggests for electrons and protons and the idea of the average kinetic energy being the same for any nucleon such that the velocity of the electrons would be about 42 times that of the protons.

This last statement is probably in error. DLS Aug 1, 2011

Version: Current by ESKI - Aug 20 2008

Google groups

« Groups Home

Oscillator/Substance
⊞ **Theory**

Set Notation and Atomic Structure

SET NOTATION AND ATOMIC STRUCTURE

The equivalence of electrons in atoms is more or less implied in Oscillator/Substance Theory and is explored somewhat in another page. Here is a little different slant.

Electrons in atoms may be considered in a set theory manner as consisting of one set which tends to subdivide into two other sets. One of these sets, the "nuclear electrons" have been considered since the 1930's as being bound to protons to form neutrons. Another possibility is that they have motion patterns, "orbitals" that are analogs of the orbitals that are assigned to the extra-nuclear electrons or to the orbitals which are written in molecular orbital theory for molecules.

If we take the first idea and use a simple example, what can be called the "Iso-3,3-set members, Tritium and Helium 3, we can write, using the form, N-1s2; 2s2, etc.,for nuclear electrons and a corresponding, E-1s2;2s2, etc. notation for the "extra-nuclear electrons, we would have the following sets:
    {e-H3} = {P-1s2} U {E-1s1} and {e-He3}= {P-1s1} U {E-1s2}. This would read, " the set of electrons in Hydrogen 3 is composed of the set of 2 electrons in a "1s" type orbital in the nucleus and and a 1s electron that is not in a nuclear orbital. The situation for Helium 3 is a union set of one electron in the nuclear orbital and two in an outer orbital."

The Union set notation implies that all electrons belong to the entire unit....

Protons can also be supposed to have some degree of motion which could perhaps be described in the same way. In that case we could write for protons in the above cases, for both units, {P-1s2, 2s1} and an entire description for Tritium or He3 could be done by writing the foregoing set as a union set with the electron set to describe the entire atom.

Whether any of the above will spark ideas for researcy or computer modelling is unknown at this point. There is one paper, published on www.helium.com as "Introduction to Helium II" which suggests that the superfluid characteristics of HeliumII could possibly be due to the change of the nucleus from a tetrahedral to a square planar configuration.

[Delete this page]   Version: Current by ESKI - Oct 16 2008

---

1 message about this page

Aug 23 2008 by ESKI

A possibe way of looking differently at atoms.....

Click on http://groups.google.com/group/oscillatorsubstance-theory/web/set-notation-and-atomic-structure
- or copy & paste it into your browser's address bar if that doesn't work.

Reply to this discussion   Report this page

THE CONSTANCY OF CONSTANTS

One thing that we seem to depend on is that constants of nature will not change, at least in the time that we are working with them. Going along with this, we can then say that if we combine constants of nature, we will obtain other constants of nature which may or may not bive us insights into what is going on.

One very intersting constant of nature is Planck's constant which relates the concept of "Energy" to the frequency of electromagnetic radiation. This constant, "h," can have the dimensions of grams, centimeters squared per second, or in other words the units associated with "mass," radius, and velocity, the units of "action" or "angular momentum. We can write then that "h" as a constant can be considered as m x r x v = h bing a defining equation. This is of the form of three unknowns egualling a constant, mathematically, that is, "xyz=K." We also notice that, if we divide out the factor,"r." we have another rather familiar looking equation, mv=a constant. This is the familiar equation for momemtuM, "p." Therefore, h/r is the same as momentum. We may note, however, that there are two other ways we can change things to get something that looks suspiciiously like a momentum. We could divide out mass or velocity and get an equation of the same form. Maybe we have just discovered the rationale of the rule that in collisions, momentum is conserved. In a three factors equal to a constant, if one factor changes, the other two will change in such a way as to keep the relationship the same.

Another way of looking at this is to note that since all three are related to the concept of angular momentum, all three factors are related to some central point. The radius, r, is a distance from the center, "v," would be the velocity of a unit moving at that distance, and the mass would be the balanced pressure at that distance that kept the motion constant.,

At this point, let us introduce another constant of nature, the speed of light, "c," which, since it is the velocity of information transfer by electromagnetic radiation, has to be the average velocity, in any given direction of any moving entity..

Evaluating mxvxr=h at "c" by dividing out "c" we get m x r =h/c. This is the equation for a torgue, mxr so "h/r," is the "torque constant of nature." Since the absolute values of m and r are interchangeable, we have defined a set of rotors which can have an unlimited number of pairs of values socn that "m x r = h/c." We may also say that this set of rotors would also define an oscillator family set, m x r = h/c.

It is not too great a jump to say that since this seems to be a constant of nature that our Universe itself would be within the most powerful low-frequency member of this family, with all other units "keyed" to "harmonics" within that same oscillator. Going the other way, looking at certain other phenomena, such a electron-anti-electron pair production, and e-, e+ annihilation, moves the concept to the idea of these oscillators being either "full-cavity" oscilators, which would account of r a Universe/Anti-Universe pair, or a positron-negatron, "zerotron" parent unit, or being vortex "half-oscillators" which would account for the opposite spin characteristics of electrons and anti-electrons and can be exended to account for the structure of all matter as being composed from electrons and protons.

We can go on, we can divide h by c^2. to get a constant. We can divide c^3 by h to get a constant. In fact, we can manipulate any set that we wish of h and c to get possible constants of nature. The one mentioned above of "c^3"/h, has the dimensions of acceleratin per gram, which if changed by multiplying out grams as one gram, gives a huge number as a force. Perhaps the force that would be felt, is being felt with out our realizing it, on every gram total of units of existence.... Some of these ideas of the 'Principle of the combinability of constants as clues to the structure of everything," can be explored further.

If anyone has read this far, how about telling what "h/c^2" might mean? How about some of the other combinations?

| Delete this page | Version: Current by ESKI - Sep 13 2008   ·

Latest 3 messages about this page (4 total) - view full discussion

Oct 27 2008 by ESKI

Actually this is not a reply since I was the last writer. The expression, h/c^2 does have an interesting interpretation., when I get around to it I will write out my interpretation on another URL and post it. In the meantime, I' d like to see what others come up with.

One little clue:
As you may have already noted, h/c^2 has the dimensions of "mass."

Oct 23 2008 by dean sinclair

Hmm. Looks like that question of mine is a bit of a mind blower. Actually, if you turn the question upside down and try to give the resulting expression meaning one ends up with a massive value of mass..... You're right that could be something like the mass in some kind of big bang.... Eski

Oct 19 2008 by ka-sala

Hello Eski,

If anyone has read this far, how about telling what "h/c^2"

You have created a Stabilized Ultrasonic Vibrational Plutonium Stink Bomb! You can't go much further than this, or your 'combining constants of nature' will end up with a Big Bang! Safe landing... ka-sala
(Correction made with spelling!)

Hadron Collider and O/S Theory

(This first part was written about July 2008.)

The hadron Collider may have some useful results for O/S theory. If it doesn't cause a disaster; which O/S theory suggests is a definite possibility, some of the other result may fit in. It may be projected that the collision expected of two proton beams may not result in the release of the energy content in some explosive manner but merely in the divergence of the beams because of the spin effects on vortex particles. However, if there is energy transfer on such contact there is certainly enough content to cause electron-proton production, (some 1800 pairs or so) or even, possibly, to form one neutron. We're talking about the energy content of two protons accelerated to essentially "c" of course

If collision is elastic, with energy transfer, there will be an attempted acceleration of some of the protons to beyond the speed of light. This would probably have the effect of increasing their mass, which could result in the formation of very short-lived "bosons" which would decompose....

This writer still hopes that the entire system does not go into resonance with disastrous results.....

In any case, I still feel that there is a good chance of having "unexplained" Hydrogen atoms appear in the system....

Comment, July 23, '09. Approximately a year later,

Additional information has arisen making the above analysis seem somewhat naive. The Hadron Collider burnt out a component shortly after start up, apparently in early attempts at passing a proton beam around (more than once?) Anyway, it apparently hasn't started up again.

Additional ideas has developed in considering the proton. It was noted above that the vortex nature of the proton would cause problems. A little more consideration, taking in the fact that protons can interact by pairing processes and even further linkage, just as is possible with electron, shows that most likely a stream of strongly accelerated protons would develop either a linked structure for the entire chain or separate groupings corresponding to the "nuclei' of simple atoms. In either case, there is a definite possibility that the effect would be of a rotating, organized charge. This may well be the cause of the break down.

Additionally, the clockwise rotation of the proton would make the energetics of spinning the protons about a circular accelerator in a clockwise direction one situation, while spinning them in the opposite direction would be a totally different kind of energetics. There would be several areas of conflict, even with the ccw. spin of the Earth, and even the (Probably also ccw) spin of our Universe.) The approximation of the proton as a "charged point particle" is grossly misleading . If the insights of the O/S model are valid, it is very doubtful if the Hadron Collider will be able to go into operation, as the scientists and engineers have no idea of the problems they may actually be facing.

| Delete this page | Version: Current by ESKI - Aug 6 2009

---

2 messages about this page

Aug 6 2009 by ESKI

Have psted an updated version of this page, to take into account some of the new information since it was first written Eski

P.S. Hope this finds you all in good health and good spirits.

Click on http://groups.google.com/group/oscillatorsubstance-theory/web/the-hadron-collider-and-o-s-theory

Sep 22 2008 by ESKI

Another short page post--

Click on http://groups.google.com/group/oscillatorsubstance-theory/web/the-hadron-collider-and-o-s-theory
- or copy & paste it into your browser's address bar if that doesn't work.

Reply to this discussion   Report this page

Create a group - Google Groups - Google Home - Terms of Service - Privacy Policy
©2011 Google

Gmail Calendar Documents Photos Reader Sites Web more

Go gle groups

« Groups Home

Oscillator/Substance Theory

⊞   S

Discussion on .draft-1223675099611

Options

1 message - Collapse all - Report discussion as spam

ESKI View profile

More options Oct 10 2008, 4:57 pm

cool_gr...@hotmail.com
Iso-sets, Radioactivity and CMNS This is, in general, a note I posted
on another Google Group Site which has ideas pertinent to our work
here. DLS

Dear People:

This is as condensed version which will be expanded for publication
elsewhere. (http://groups.google.com/group/oscillatorsubstance-theory.

Iso-sets, Radioactivity and CMNS.

One may define an Iso-set as the set of all structures that may be
formed from a given number of electrons and protons. In this type of
definition, "Iso-set-1,0" would be the electron, Iso-set-0,1, the
proton, the Iso-set-1,1 would have three characterized members, the
subset, {e, p} , the neutron, and the Hydrogen atom. Iso-set-2,1
would include, {e, e, p; '"eep" (a central aggregate), and Hydride
ion}. The "2,2" set includes the H:H molecule and the Deuterium Atom,
an unspecified, as yet, central aggregate, and all possible
combinations of previous sets.

As each set contains all possible combinations of smaller sets, this
becomes complex very rapidly and may seem so obvious as to be
trivial. However, one interesting observation arises. In any given
set there is a possible central aggregate of all the units of the
set, an "Iso-set Aggregate," an "Iso-A." which would be the
"Residence" of the highest "mass/energy" states of the set and would
have all of the electrons confined within, or very closely associated

to, the volume defined by the motions of the protons.

Any Iso-A set would have within it the potential to decompose to any combination of units of smaller sets and would be thermodynamically unstable with respect to such decomposition. Collapse to an electron (the Iso-A of "set-1,0) and a cation) accounts for many radioactive transformations, e.g. neutron [Iso-A of set-1,1 (?)], to a proton and electron. All other Beta-particle radioactivity can be explained similarly. An atom reaches an isomeric Iso-A state which does not revert to the original atom but to a cation by loss of an electron. A similar argument holds for the Alpha-emitters, except that the unit left behind is a di-anion... The Iso-A explanation can be extended to all nuclear transforms with appropriate modification of what is seen to be happening. Positron emission would involve interaction with Iso-set-0,0. The parent set considered pervasive which contains a "zerotron" unit splitable by "pair-production" into a proton and electron.

One notes that Iso-As of neutral units, atoms and molecules could be considered as "poly-neutrons" a term that has appeared in the literature. If it be taken that this means, "units that may be considered as 'polymers of neutrons' within which the neutrons have no individual identity' " the writer has little objection to the nomenclature. The Iso-A concept simply includes these and any other central aggregate units, charged or uncharged which can be considered. Iso-As of huge sets may well be what are known an "neutron stars" and Quasars which are probably the "Isotope Factories of the Universe."

An Iso-A intermediate formed from the Deuterium molecule which, under certain circumstances, transforms to another version which rewould account for the most commonly observed CMNS transformation. CMNS is short for "Condensed Matter Nuclear Science which comvers the area once known as "Cold Fusion" and other phenomena in the border land between "nuclear science" and molecular and atomic chemistry.

Iso-A intermediates might be interesting to consider in much of CMNS work. As noted above, they would be considered as the ultimate "Condensed Matter" of any given set of electrons and protons.

Reply to author   Forward

End of messages

« Back to Discussions   « Newer topic   Older topic »

http://groups.google.com/group/oscillatorsubstance-theory/browse_thread/thread/378c1d08f32b973b

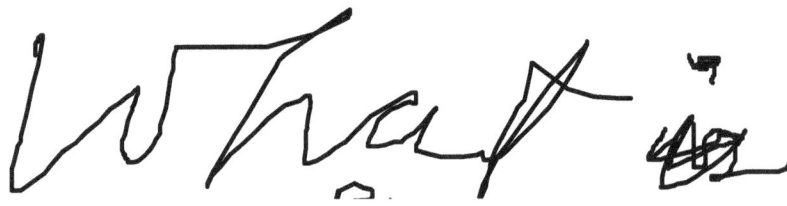

Size or Shape

Proton Cosmology

Hello Dr. Sinclair,
   I concur that the oscillator matrix model goes a long way in explaining what may be causing the pecularities and phenomena observed in the nature of charged particles. I do ,however, believe that neutrons exist in the nuclei of atoms and are responsible for binding protons together to form all the elements larger than helium. Protons are the primary oscillator that Max Planck sought back in 1899, rather than the atom as a whole. If you check out:
http://protoncosmology.com/fundamental_mechanics1.htm you will find a detailed electrodynamic model of the proton and the inner mechanism that keeps it oscillating. There's a link at the bottom of the page to the neutron based on this model. I hope you find it interesting, if not, enlightening.
Enjoy
hoek

| Delete this page |

Version: Current by hoek - Oct 13 2008

Latest 3 messages about this page (6 total) - view full discussion

Oct 31 2008 by ESKI

O.K. I grant all you are saying about shell capacities, etc. and the chemistry based on these models. However, I still say that there is room for skepticsm that these are the entire story or even correct other than in an approximate form.

Warping of space-time, of course, corresponds to an expression one

Oct 24 2008 by Robert Vanderhoek

Hello Dr. Sinclair,
In your kind response you say,"while my hypotheses may have little going for them than as bit of logic and some guesses."
I respond: The hypothesis I suggest is based only on the logic of the laws of electrodynamics, no guesses. The only assumption I make is in the directionality of charge along the energy-mass spectrum. Energy-mass

Oct 23 2008 by ESKI

Bob:
You may be absolutely correct: however, while my hypotheses may have little going for them than as bit of logic and some guesses. I'm not convinced that four positive charges rotating in a proton really answer anything either. Whence commeth the four charges? Where do negative charges fit in?

$2 \ldots = //$ line

2 lines $=$

Basic Status Paper Reposted

Oscillator/Substance Model, A Theory of Everything

Summary: This article presents a view of existence as being within a
substance, at its triple point, consisting of separable oscillators.
As such it gives new definitions of mass, energy, force,
gravitation.... Included are possible explanations of "The Big Bang,"
Black Holes, Neutron Stars.... In terms of this model there is also a
warning of possible danger from the Hadron Collider which is due to go
into operation in the Summer of 2008. .

The "Oscillator/Substance Model of Everything" is a second-generation
model developed from the Motion in a Matrix Model.* The key insights
of that model were that the speed of light being a limiting velocity
of information transfer implied a necessary medium for interaction and
that the interconvertibily of mass and energy implied that they were
different aspects of the same thing. The original thought was that
mass coordinated to point-centered motion and energy to motion along a
line. While this seems to be in general quite close to the facts, it
was realized that modifying the ideas to mass being motion confined
within a surface and energy being a less specific term covering all
types of motion, with Kinetic Energy being the type that would be
dissipated along a line would be a more usable view.

If, instead of postulating a fixed matrix of dots as a basis for a
model, it be assumed that there be a basic substance which will remain
at, or return to its triple-point, the problem of whether the model
is dealing with a solid, liquid or gas disappears. At the triple
point the substance can act as any of the three depending upon slight
variations in conditions. We may even postulate that the triple-point
temperature is approximately the temperature of outer space, about
three degrees Kelvin.

Postulating a substance basis for everything, allows us to define a
number of things which are otherwise undefined or have "circular
definitions." Mass becomes the balance between the motion content of
the substance within a particular surface and the remainder of the
substance. Kinetic Energy is the motion dissipated into the "Bulk" of
the substance when a portion of the substance within a surface changes
position. Light and other electromagnetic radiation is motion within
the substance. The law of forces, "For every force there is an equal
and opposite force," falls into place as a statement of the fact that
if there is a pressure change within one part of the substance, there
will be compensatory changes within the rest of the substance. It has
been long known that pressure is a force. We simply change the
statement slightly to, "All true forces are pressures."

Gravitation loses its place as a force--a status which it actually
never should have had.   It becomes a description of the difference in
pressure of the amount of the substance between two entities and rest
of the substance, including that part of the substance included within
the entities. Electro-magnetism is the observed result of pressure
disturbances within the substance from actions and interactions
between electrons and protons,

Electrons and protons are known to be the basic units from which all of our chemical substances are derived. That is, with the possible exception of the "short-lived" particles produced in high-energy, particle-physics experiments. In our Universe, protons and electrons occur as the decay product of neutrons. Electrons and anti-electrons also appear in "Pair-production," and disappear in "Annihilation."

Putting the above pieces together with the characteristics of oscillators. we may postulate that 'Pair-production" and "Annihilation." are opposite processes. A spherical cavity oscillator can be considered to consist of counter-rotating halves, which, if subjected to enough "motion disturbance," can split into two mirror-image vortex oscillators having opposite spin orientations. If these two halves were to re-encounter on a proper axis they could recombine to the original oscillator with loss of kinetic energy. That is, postulating a basic unit of our substance as an oscillator that can be split into vortex oscillators having opposite spins, neatly accounts for "Pair-production" and "Annihilation."

As to electrons, protons and neutrons, we can account for these at the same time as we account for the "Big Bang" and the "Expanding Universe."

Neutrons fit the characteristics of a substance distorted by a shock wave such that the "front half of the unit met the back-half coming forward." This distorted unit splits into an electron and a proton. The speed of light in the substance--if light is a typical information carrier--is the average velocity in any given direction in any given instant. Therefore, it will be the average expansion velocity of the expansion phase of an oscillator, and will also be the average rotational velocity measured at the outer edge of the expansion (or contraction) phase of an oscillator. As such it will be the average velocity felt by our basic oscillators.

We may postulate that motion at above this velocity will travel as a shock wave and since "c," the speed of light, is an average, the initial velocity at the start of the expansion phase has to be greater than "c." if we consider acceleration and deceleration to be linear. we can estimate the initial shock wave from any oscillator as having a velocity of "2c" which fits fairly well with the comparative "masses" of the electron and proton. with the proton half laking the entire distortion from a "2c" shock wave slamming into our substance acting as a "solid."

What this all adds up to is that some oscillator is operating at ultra-low frequency, very high power, and we are riding in the "Creativity out of Chaos" behind that shock wave which converted may substance units into neutrons which disintegrate into electrons and protons. The electrons and protons, vortex particles of opposite spin orientations, but very different sizes and shapes, can not stably recombine but can associate in a multitude of ways, creating everything that we know in our Universe. Presumably, or the other side of the Parent Oscillator, is an "Anti-Universe, coping in its own way with the same "Creativity out of Chaos" attempt to stabilize. At some point the shock wave will expend itself, and there will be a contracting Universe-Anti-universe set which will eventually collapse through the parent-oscillator to start the process all over again.

Our discussion above leads to the conclusion that our basic oscillators possess the ability to coordinate vast amounts of energy. Taking this into account gives a possible explanation for "Black Holes" in the centers of Galaxies. In the Creativity of Chaos behind the shock wave, some units fall into coordination with an oscillator at their exact center, and begin to move with its frequencies. Considering the characteristics of oscillators, a vortex oscillator such as a proton or electron, if passed intact through a full-cavity oscillator would have its "sense" reversed. That is, an electron would become an anti-electron, a proton an anti-proton. Since, tossed out the other side of the oscillator these could unite with "untransformed" units, oscillators operating in this way would be reducing electrons and protons back to unit-oscillators. This. however, comes at a cost. Protons, formed at distortion of smaller oscillators, would be combined into "proton-parent" oscillators, which could presumably be, in turn, distorted by shock waves into Super-neutrons, which would collapse into an anti-proton ("Super-electron") and a "Super-proton" and so on and on and on...

When a Black Hole Oscillator has done its "trash converter" work in a volume, what would appear to a human observer would be a complete blank spot in space devoid of anything which we would consider "Matter."

One other factor in our Existence that we have not mentioned is the neutrino-anti-neutrino combination. It seems logical to guess that these are a pair-product just as electrons and anti-electrons are, and that their "parent-oscillator" is a more basic unit to our substance oscillator(s). This would be in analogy to atoms and molecules in chemical substances. These could, in turn be composed of even smaller oscillators, ad infinitum.

We consider all "matter" to be made up of protons and electrons, which combine to form atoms, atoms to form molecules, molecules to form organisms.... In this model we would not consider that neutrons would exist within atoms other than in "potential" or "evanescent" state, and consider that neutron stars and quasars would perhaps model as hugely over-sized atoms. As such they would function as "isotope factories." Isotopic distributions within galaxies may be a result of differences between the "neutron" stars active within them in the past. It should be possible to find spectra of elements including possibly trans-Uranium elements or even unknown "Trans-Uraniums" in the radiation from quasars and neutrons stars. Because of the expectation of high energy electron "orbitals" closely spaced, it can be expected that these entities will have distinctive radio-frequency "signatures.

 This model, if valid, casts considerable doubt on both the validity of the assumptions behind some of the publicized experiments that are intended for the "Hadron Collider" under Switzerland and France that is due to go into operation this Summer (2008), and on its safety. If the simple definition of mass which arises in this model be valid, then the idea that something known as the Higgs Boson is responsible for mass becomes a total misconception. The idea that colliding protons with protons would duplicate the conditions before the "Big Bang" is inconsistent to this model in several ways. One is that by this model, protons would come into existence as a result of the decay of neutrons. As neutrons have a significant half-life, protons would not exist until after the start of the shock wave. As to Lead Ions, these would appear much later. However, collisions at relative velocities of close to "2c" could certainly have the potential to distort oscillators cause neutron formation and, perhaps, additional esoteric, short-lived, alternate forms of matter. Almost surely, Hydrogen would appear within the collider tube. A much more dangerous implication of this model is the idea of Black Hole formation by coordination of a circular spinning entity with an oscillator in its center. The Collider apparently has as least a couple of perfect circle dimensions which have resonance frequencies. There is a high probability that these circles will act as cavity oscillators. If one of these oscillators did come into active coordination with a central oscillator the result would be the same as is postulated for galaxies. Although on a miniature scale to a galaxy, a Black Hole interaction the size of the Collider could have results ranging from, hopefully, merely annoying "unexplained energy anomalies," to a new volcano where the Collider had been, to a coordinated vibration with the resonance frequencies of the Earth to create a new asteroid field between Venus and Mars. Although a "Black Hole Oscillator" would probably "eat" only the material within the circumference of the Collider, the conversion of that amount of matter to anti-matter would make the most powerful atomic bomb appear a mere firecracker.

The scientists working on the project apparently admit that they have no idea what will happen. I hope they remember that, in retrospect, the physicists working on the Manhattan Project realized that it was by the merest of good luck that they did not blow themselves, and the entire Chicago area off the face of the Earth.

------

* See "Motion in a Matrix...." on www.helium.com. Several other articles by this writer, which are posted on that web site also bear

upon ideas related to this model.

AUTHOR'S POST-SCRIPT: After reading through the above article, the writer has realized that he has not included some things which usually occur in a modern, scientific article. There are no mathematical formulations that might not be comprehensible to everyone, nor any mention of one or more of the modern, accepted theoretical ideas such as Relativity, Quantum Mechanics, String Theory. Rather than break into the body of the above to correct these apparent flaws, there is being appended this addendum.

It has been shown in other papers that by equating the value of Planck's Constant,"h." to the dimensions of its definition, evaluating this at the speed of light, "c," one can obtain a definition of a family of dual-cavity and/or vortex oscillators defined, in set notation, by {m x r = h/c}. This set has a central, symmetric, spherical--or ring--oscillator defined by $m = r = (h/c)^{0.5}$., where "m" is a mass and "r" is the radius of a circle or a sphere. The value of m in cgs units is about 7.4×10^{-19} g. and the value of r is about

7.4×10^{-19} cm. If one goes a step farther and writes, "$m = h/c^2$", one finds an associated minimal mass of about

7.5×10^{-58} g. which may be the mass of a "neutrino parent." Possibly the smallest mass of significance in our Universe.

When the accepted mass of the electron is inserted into the oscillator-family-set equation, the radius is found to be the Compton Wave Length used in Quantum Mechanics. Inspection of the mathematical definitions shows that the two are mathematically identical although their different derivations do not show this clearly.

Quantum Mechanics is said to fail at about 10^{-18} cm., and String Theory gets lost into its ten dimensions at about the same value. That is the value of the diameter of our average oscillator, a value through which all the oscillators may well invert. It is interesting that it is the point of failure of two of the widely used theoretical models. If our speculations are correct the failure of both is perhaps that their mathematics only accounts for half or less of "Reality."

In so far as the writer knows, no-one has correlated the Space-Time approach with "Q.M." Each seems to have been in its own niche. The mathematics of both, in so far as they go, would seem to be somewhat compatible with this model. Space-Time modeling seems to work quite well at the macro level despite the fact that Relativity probably only would strictly apply to communication theory, and concepts of what occupies space and what time actually is, appear to be in error. String Theory seems to depend, ultimately, on the true, but-rather-useless-in-the-thinking-of-most-people fact that the route traced by a dot moving in random fashion will follow a path which one can consider as having always being there. The 10 dimensions may arise because one can use nine dimensions of space, and one of motion. to define the actions of a dot--if one wishes to do so.

If the foregoing seems not too well explained and a bit abstruse, then, perhaps, the writer has reached the goal of writing a "modern, scientific paper." Most modern, scientific papers seem to be written in such a way that only the author and the closest associates have any idea what is going on. If they do.

On a more serious note, it can be observed that if this model be valid, it could have been put together close to a century ago, had the ideas advanced by Planck been followed up rather than those of Einstein. This writer did not realize this at the start of the path that led to this write-up. However, the initial impetus was the realization that there was an "Overlooked Obvious" in Einstein's Special Relativity, the fact that the transform equations could be generalized to fit into communication theory if one realized that "c," the speed of light was simply the limiting velocity of an information carrier wave. The theory was valid for situations wherein communication was possible or necessary; but, possibly of no pertinence in other situations.

Version: Current by ESKI - Oct 1110. Edit by buskirkmarbuckle - Aug 299. Edit by kedmundo1811 - Aug 288. Edit by vienoheaton1916 - Aug 287. Edit by vienoheaton1916 - Aug 176. Edit by vienoheaton1916 - Aug 175. Edit by vienoheaton1916 - Aug 174. Edit by vienoheaton1916 - Aug 173. Edit by vienoheaton1916 - Aug 172. Edit by ESKI - Aug 61. Created by ESKI - Jul 27

Page editing not supported in your web browser. Download a new copy of Firefox or Internet Explorer to edit pages.

3 messages about this page

Aug 7 2008 by Hugh V

Maybe someone would like to summarize this paper a little more. Five pages is a lot to read. It does make sense, but I think that you perhaps should take what you are doing a bit more seriously, although you are right, most scientific papers are ony comprehensible to the initiated which could mean the people working in the same lab. Let us

Jul 27 2008 by ESKI

Whoops, the title is wrong, the date is in 2008, not 1998, I need to learn to type.

Jul 27 2008 by ESKI

Here is a version of the parent paper posted on www.Helium.com.

Click on http://groups.google.com/group/oscillatorsubstance-theory/web/the-cur... - or copy & paste it into your browser's address bar if that doesn't work.

Report this page Reply to this discussion Hide message box

Reply to author   Forward

End of messages

« Back to Discussions   « Newer topic   Older topic »

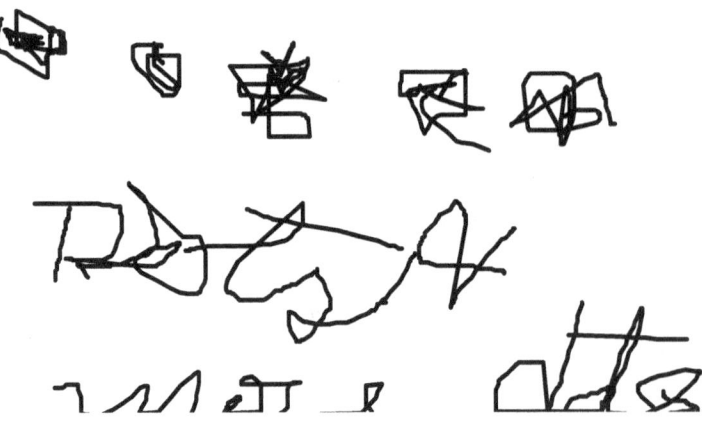

Reality and Speculation in Science

Reality and speculation in science.

The writer finds himself fighting a definite problem in scientific theory. Determining what is real, that is with solidly established experimental /observational data and which is simply imaginative speculation. Some items might be considered to fall in an "in-between" category.

In Oscillator/Substance Theory, to which the writer has been a major contributor, it can be conceded that both the oscillators and the substance fall into an "in-between " category. While they neatly explain many things and fit together logically, there seems to be no direct confirmatory evidence.

The "zerotron" oscillator parent for the positron-negatron pair, which is postulated, makes logical sense; but, how could it be observed as such? One may note, however, that there is the triple set of Muons, negative, positive and neutral. If it could be shown that a negative and a positive muon would join to form the neutral form, we would have strong backing for the "zerotron" concept.

Similarly, the "Sin-Vree Entity," found at "$m = r = (h/c)^{0.5}$" as a central or average unit, is a mathematical observation which may have no basis in reality or may be some sort of a mathematical construct analogous to an "average human."

The "More Accepted Theories" such as Relativity, and "Standard Model" are replete with ideas which are spoken of as fact which should be considered with considerable skepticism. Neutrinos were originally postulated to account for anomalies in conservation of momentum. While there are some bits of data which are attributed to them other than momentum deviations, they should most likely be considered in the "probable, but somewhat questionable" category." Neutrons, as such, as units within nuclei have been considered as fact for 3/4 of a century. This writer is skeptical that they exist as such in nuclei and suspect that this is a false assumption which has led to other false assumptions.

Quarks, gluons, gravitons, etc. are spoken of in the literature as if they were proven entities. There seems to be no experimental backing for any of them. Writers do not seem to realize that, at this moment, they are only imaginative speculation

While the various "fundamental particles" of the meson type seem to be real enough, there seems no reason to consider them as "fundamental." There is as much, or more, reason to consider them as alternate forms that arise from the experimental conditions.

Quantum Mechanics is a mathematical mode. However, one "top scientist has been quoted as saying, "Outer space may be filled with Quantum Mechanical Froth."

One can continue with examples. The point is that we must continuously reexamine our work to determine what is valid and what is what we hope is valid, wish were valid, or someone else assumed to be valid.

Jan 10 2009 by ka-sala   Version: Current by ESKI - Oct 16 2008

(RE. Eski-ka-sals)
To: Oscillator/Substance Theory
Hello Eski,
Just going over a couple of pointers here re. one of your earlier
discussins..."Reality and Speculation in Science' and your comment on
'taking with a grain of salt.'
Just as you found your SciScoop front page a great boost (An
Intertwined Universe,) a publisher purchased my "OSCILLATING ENERGY in

Oct 19 2008 by ka-sala

So long as you keep Oscillating, that's the main thing!
Regards,
ka-sala

Oct 16 2008 by ESKI

Hi, I have posted some comment essentially about "taking with a grain
of salt" almost everything in sceice. DLS

Oscillator/Substance
⊞ **Theory**

Neutron ala O/S Theory

Neutrons ala O/S Theory

In the current version of the O/S Theory as published by this writer. the neutron is considered a distortion product of the "zerotron" oscillator, presumably the most "stable," simple unit into which that parent unit can be converted. That is a unit which is stable with respect to many units of what we might call "electron-motion time."

This unit would be expected to be the unit formed when conditions within the nucleus of an atom approach the conditions wherein neutrons were originally formed, or conditions wherein greater symmetry/stability can be attained by removal of such a unit.

If one wishes to correlate a "neutrons within the nuclei" idea with other O/S Concepts one could rationalize neutrons in nuclei on the basis of of electron time spent within the "shell" of a proton, with each "proton" having an equal chance of momentary synchronization to a "neutron-type-state."

The picture that arises of the neutron is of a dual-cavity, concentric oscillator, with what might be called a "Proto-electron" vibrating in a relatively narrow shell about the central "$(h/c)^{0.5}$ average" and a "Proto-proton" vibrating through a much wider range across the same central "entity." The synchronization of these two concentric oscillators, with the one-- the "Proto-electron--" vibrating about 43 times as fast as the other, would be the "neutron."

When the "neutron" oscillator "falls out of sync,", the proto-units collapse to the vortex oscillators, electron and proton.

This is imaginative speculation based on a certain amount of logic; however, the foregoing should not be taken as established fact, to add another "fable" to the neutron legend. [See paper, on this site, "Neutron Facts and Fables."}

Delete this page	Version: Current by ESKI - Oct 16 2008

1 message about this page

Oct 16 2008 by ESKI

Here is more comment on the neutron. It could be called a sort of current update on the paper, "Neutron Facts and Fables." DLS

OBJECTIVE INTELLIGENCE -

Hi, Ka-Sala:
You definitely add a different voice to this dry scientist's group1
The logic is there, I think. You have a habit of blowing minds, but
much of what you say fits in with theOscillator/Substance ideas....
The idea of there being something inbetween "no. and thing"
parallels, in a rather poetic way, the concept of the the negatron-
zerotron-prositron triad. Here wee are. with that ubiquitous number
3 showing up again !
Hang in there and keep writing! Eski (Dean Sinclair)

ka-sala View profile

Thank you Eski for your encouragement.
Yes, it is good 'logic' to 'think' sometimes! Can we escape 'what
is' ?
ka-sala

Robert Vanderhoek View profile

Hello "ka-sala"
   You've posted alot of stuff requiring many answers, some of what you say
seems to be just word play. I believe zero is zero, no matter how many
times you multiply or divide it, At first I thought you were kind of a
wizz-bang, but I read several of your works on Helium.com and realize you
are a genuinely good and wise person. Some of what you say seems to be more
numerology than science. Some of it, however, seems to touch on basic
concepts that are true, like all of creation being circular. It's more true
than you know, Irrira. The 3 is the states of matter, positive, negative
and neutral. You seem to be more transendental than scientific. It may be
the feminine right side biased thought process versus the left handed
testosterone biased interpertation I percieve. If you're truly interested
in the circularity of the universe, please check out:
http://protoncosmology.com/omniscient_correlation.htm and let me know what
you think, you too Dean, It's base 10, a truely anthrophomorphic
interpretation. There's some real coincidences there.
Enjoy,
hoek

- Show quoted text -

Reply to author   Forward   Report spam

ka-sala View profile

Hello hoek,
Yes I have been including some snippets, from the Report I mentioned
in my profile. I appreciate you writing. I guess the way I determine
science, is a bit different. But my conclusions are what matters,
which seem to tie in with most of what is going on in the universe!
(and here.)
My belief re. zero is that we would have no beginning without it, so
the fact that it exists, is good enough reasoning for me to understand
this theory. I repspect what you believe.
No... what-ever a wizz bang is, it's taken me 45 years to come this
far, and never giving up. But I'm the first to admit I have to work on
English words to understand them enough to use.

I have a way of seeing things from all angles - so even words or
numbers must come into play - scientifically speaking. I am the first
to agree my 'approach' may be a little unconventional, but it is the
conclusions which counts. So maybe the more we all get to know each
other's ways, it will help to understand our differences.

We are Oscilliating all the time, but I'm not a spinner of yarns, just
freedom of speach and equations, as well as expression. My simple
approach saves a lot of complexities...

I do know how true the circular concepts are... these are dealt with +
diagrams in my Report also. It is really an amazing
link, that you should ask me if I am interested in the circularity of
the universe. It's amazing all this I did in the 1970's!
I may not quite speak your language - nor that of the group - but I am
honored to be a part of you all. Eski peronally invited me, and I have

OBJECTIVE INTELLIGENCE

ka-sala View profile

OBJECTIVE INTELLIGENCE .
Beginning the search on the theory of everything. Some-one with a
great mind once said that, "we only begin to learn when we realize we
know nothing." At least this is a good beginning, to start searching
for something. So lets start with Zero. Graphically written as 'O'.
Oh...? Oh, is usually the point of realization. Well that's a start.

We realize there must be 'something'. So, if we just keep the word
'thing', and instead of 'some' added to it, lets try simply to add
'every'. Just like that we have 'everything.' Is this how the theory
of everything came about? There is another way. Stay with just the
words we started with. 1=nothing. 2=something. 3=everything. The word
'thing' remains constant, but we have 3 x equations, in the
'no'-'some'-and 'every'.

Lets see where all three bring us to. We can dissect them. 'No' can be
used in many ways. Without getting to bogged down in the many, lets
stay with the simple. No means no. No thank you. No good. No way. It
has to be the negative or opposite to to a positive. No votes...
out... gone! So just for now, lets see it as a three-way 'thing' we
are looking at. As given above.

From 'some', we reach an unspecified of 'something'. Anything! OK...
another tiny word has popped in: 'any'. So we don't have to get
ourselves in a knot as to what this 'any' is, because it can be
'anything'. Anything of 'something.' It's somewhere in between,
'nothing' and 'everything'. Lets just call it the link to, or the
neutral equation, of the three. Would it be right to even be so bold
as to think that some where between 'nothing' is 'something' ? Sounds
to me me we are getting somewhere.

Already there are 'links' to just the word 'everything'. I wonder if
we were to be even bolder, as this is a theory of science, to assume
that 'every' (being an individual part of a group of anything,)
becomes the entire, complete of no+some=every? So 'every' is within
either a range of being 'no' or the 'some'. Obsolete, or Absolute. One
could use the 'no' as the negative and change it to the positive, by
saying "no, it is not Obsolete !" Or, change the positive to negative
and say " no, it is not Absolute !"

So really, 'no' is the Di-no = twice, double, twofold, the combining
factor like two atoms, the radicals. Like the group dioxide. The root
or basic principle, or the foundation. The free radical leading to
group = 'some'. The Radical sign becoming the square root of an
equation. All this out of just the word 'no'! Leading to combinations
of 'x = some'. Which in essence there can never be 'nothing'. Because
in the very word there is already the beginning of 'something'. So
'nothing' can ever be obsolete, as there is always 'something' equal
to it's joint equation.

It's like a tree without roots. We would not have the branches. But if
there was no tree, somewhere there has to be its equal, so in essence,
there is 'no' one tree, but at least 'some' other trees. If you think
this is a bushman's approach to the equation of everything, stick
around. Even one tree, has many branches, elements: and in itself is

'everything' about what a tree is absolutely supposed to be! Just in a mere 579 words. 5 + 7 + 9 = 21. The key to this is 3. Just in these 579 words we've gone from 'nothing' to 'everything'. We were using only 3 equations. This 3, is the square root of 9

The absolute sum of units. The Pi of mathematical total. The total is more than one, it has become 'some thing', and in this case the total or absolute of it's square root. In all this, we have been dealing with equations. From 'nothing' to 'something' to 'everything'.

In the Pi, the Greek 16th letter of their alphabet, we have the ratio of the circumference to the circle. O Oh ! The ratio is interestingly, a transcendental number, having a value rounded to 8 x decimals. (The figure 8 next to that of O, are the two most complete of all 9 numbers.) Our figure 9, the end of where the decimal, or deca, can only go into the 10's. But... without getting too caught up in all this headiness which began out of 'nothing' = O is a bushman's equation of the transcendental value being rounded to 8 x decas = the ratio itself :a transcendental number having a value rounded to eight decimal places. That's Greek!

So, let's do it in double Dutch for short. We can reach a transcendental decimal quantity just by 16 to the valance of 8: eg. 3.4437376 doing the same, which becomes a transcendental value of 8 decimals. That's pretty close to the Greek of 3.14159265. That's only 0.302145 difference. The key to this = 6. Is this the missing link within the theory of everything?

Now look into the word, and the theory of traveling at 2 x speed of Light.
See DOUBLE THE SPEED OF LIGHT – TIME TRAVEL - ka-sala

OK, there can be 'nothing' beyond the transcendental... or can there be? That must be the Ultimate, Absolute, the 'everything' where we began this theory. How many pages does it take science to come to this conclusion? This has been 2 paragraphs over 1 x A4. We've come full circle. That has to be 'everything'. We come into the circle. The 360 degrees. This too = 9 where the square root is 3 ! Coincidence?

Now tell me how a tree isn't 'everything' of that tree. Tell me it grows at infinitum to it's absolute, given the chance to survive. Yet, it doesn't go anywhere but back to where it came from, rotting into the ground it grew out of. Becoming in time, 'one with'.

'Everything' is linked. No element can be without it's link to the other. Don't let me start on the elements, or I'll be singing the music of the spheres. The harmonics of everything related to it's vibrational force. I repeat... 'Everything' is linked. You cannot have Light without Sound. Sound without Vibration. Vibration without movement. Movement without.... Well, just one tiny example. So 'minor' it's almost musical !

O no... Did you know that Nobelium, the Element, was named after Alfred Nobel, of the 'Nobel Prize'? Everything is linked. Even Names, and Human Beings. Nobelium's atomic number is 102. This means the key to this number is 3. Didn't we start off with three equations in the search of the theory of everything? If I keep going at this rate we will have started on 'Everything is Linked Theory!' Sorry... Fact. By the way, I might be a bushman. I have my University Degrees. But remember me if there's a Nobel somewhere in here... Please? It's over to you now. Over and out to the theory of everything.

Maybe with just 3, 6, 9 = 18 = 9 we cannot escape the square root of the double Dutch 3 in 1.
Sounds a bit like a Holy Trinity! The Holy Grail? The Whole thing from nothing to everything theory.

*S & SOTU = Degree in Sanity, and Science of the Universe.

Not because I will always be seeking answers, but to share our theories. If I seem more 'transandental' that is only because I have been one of those quiet flys on the wall with eyes and ears, and very little to say, compared with a lot I have witnessed in my time. It's easier to fit into places when you are as small as me!

As for my 'feminine side' I can't help that. That's how I arrived on this planet!
Thankyou for your interest, and as with Eski, I am pleased to meet you.

I will, as I am able, go through all you have asked of me/us...http://protocosmology... I need a little time for this, and will eventually get back to you. The only thing which would prevent me, is if I am called away from this planet! So be patient with me, and thanyou again hoek,
Regards
ka-sala

TRAVELING ON ONE WAVE LENGTH
Anyone... who thinks they have the answers to the link between all things, is missing out on the journey.
This exercise is called 'Think'.

These Graphics by word, are imprinted in the mind, and simply expressed as a dot within a circle. Throwing Light on the subject = the Point of Light being the Graphic Point. The Circle, is it's Telegraphic Wave Length as received and Projected in Radionics' It's Form can be likened to the Radar. This is known as it's Resonating Factor.

Telecommunication is the result within this Resonating Field, which is... its' Electromagnetic Field. For now... between us.

The expression 'O', is used when realization of a point is recognized. Recognition is made and the word is clear through Contact. EG. Through expression, I want you to understand this point. From my mind I give expression through the words I use. If you are receiving them as intended, you will notice there is a certain Sound to what I am saying.

The Sound is being received by you, in a sound you recognize. If this is the case, we are on the same Wave Length. You are hearing what I am saying without me even opening my mouth. In Graphic Form in Light - you read me! - through your screen. Think carefully! The only point I am trying to make is...Contact!

In Amplifying that point in your mind, you have received, in recognition of the Sound. Simply because it has been Modulated to your own Wave Length which I use. Think. If I talk on One Wave Length so there is no confusion, the Sound of what I am saying should be clear.

See this 'O' in a Light of Light Rings. Imagine every time you open your mouth to speak, it emits rings of Light. Just for now - as a diagram cannot be included here - see yourself as 'the dot' within the circle. You... emit the rings. You... are the point (the dot,) from where it begins Emission. You... are the diagram. Your concentration is to 'see' this in your Mind's Eye, as much as it is my Mind communicating with yours, through Ultra Sound!

This Light Radiates outwards in Pulsating Rings of Light. The Circle of these Rings being their Radionic Boundary. The Sound of these Emissions bounce off their Circumference and return... to you. It is Energy from you via. the Sound in Light of your Emissions.

Being Light within the Sound, every time it Reverberates, it Oscillates back to it's central point and bounces back to the Circumference (boundary.)
This sends out Rings of Light at the rate of 360 Rings per. Second. At this Specific Speed the Rings of Light are Projected one after the other.

Consider this. A Tube is formed by these Rings at this speed, making

Google groups

« Groups Home

Oscillator/Substance
⊞ **Theory**

DOUBLE THE SPEED OF LIGHT - TIME TRAVEL   Options

3 messages - Collapse all - Report discussion as spam

ESKI View profile   ★★★★★★ (1 user) More options Oct 18 2008, 3:56 pm

Ai, ai, ai; Ka-sala, Amiguita mia,
Again, you give us something to think about! definitely differemt than
my dry prose!
Eski

On Oct 17, 10:13 pm, ka-sala <irrir...@gmail.com> wrote:

- Show quoted text -

Reply to author   Forward

ka-sala View profile   More options Oct 18 2008, 7:46 pm
DOUBLE THE SPEED OF LIGHT
We need something that is easy to maneuver. It's called a Time
Machine. It must at least travel at
6 x Earth's Given Time. For every six days man would take to travel
in, we would take one for the same distance. The Distance is then
derived from the Given Speed of Earth days that man would take to
travel in 6 days: traveling at 24 hours non stop around the clock.

EG. At only 100 km.. per hour. X 24 = 2,400 km. per Earth day. X 6 =
14,400 km. Per Our Day.
    This is the Given Unit, the Sum Total of Given Distance in
Time, in One Unit.
Repeat.
    Answer = 14,400 km. Transversed in 24 Hours of earth Given
Time, at 100 km. Per hour only!

The Circumference at the Earth's Equator is +- 40,075 km. At this
Specific Speed in Time this would measure just over 2 and a half times
round the world in 24 hours at 100 km. per hour!
In Fact..Consider this...in Light.
    Distance = Time to Transverse
    Speed   = .Energy required to Transverse.
    One Solar Day = 1 Revolution of Time.
Repeat.
    One Solar Day = 1 Revolution of Time. (ie. Earth's Time.)
This is the Given Specific Speed, which remains unaltered with Time.
Thus the Energy generated
is Light Wave.
    The Distance Transversed in One Day of Our Given Time.

Universal Law means the Energy you only term as a word, which we term
as an Energy Force, by word of Love. It too has a Positive and a
Negative Pole. You have known also of the term Hate. If you know this
Energy Force, you will know its' constructive and destructive
Elements. 'Earth' it into the tri... Utilize it in full !

Thus his Radius of Time Travel is given only within that of his own
Given Field. Further than that he cannot go. The answer is very
simple. In his own given speech termed Air Force, he has not yet
earned his Wings!

Now that you have been briefed only to these Specifics, who amongst you will help build your Space Ark? If any amongst you believe this not to be true, then you have not the knowledge to look further into the facts. For right now Man is polluting the very Air Waves he needs to even get him off the ground! That is, his own Atmospherics. You cannot reach for the Stars if you only leave your people to die on your own Planet.

I leave you with this to see for yourselves, your own Specifics. Only then will you see Ours in the Light of the Speed we have given to you here. This, is one for your Next Generation Propulsion System.

Reply to author   Forward   Report spam

ka-sala View profile   More options Oct 18 2008, 8:19 pm

Hello Eski,
Different approach only. I believe it to be a jig-saw piece of your 'Space-Time and O/S Theory' Discussion. The old elephant story, and who is seeing what from where! It's a subject matter worth every piece. Never think your prose are dry. You are the Professor... I am just alien! Have made slight change + correction in 2nd. post

'Universal Law means the Energy you only term as a word, which we term as an Energy Force, by word of Love. It too has a Positive and a Negative Pole. You have known also of the term Hate. If you know this Energy Force, you will know its' constructive and destructive Elements. 'Earth' it into the tri... Utilize it in full !'

On Oct 19, 7:56 am, ESKI <deanlsincl...@gmail.com> wrote:

- Show quoted text -

Reply to author   Forward   Report spam

End of messages

« **Back to Discussions**   « Newer topic   Older topic »

A dot has
one dimension -
Existence

TRAVELING ON ONE WAVE LENGTH -

clear passage-way for the Sound of Light to travel in. So fasten your seat belts and come with me as we travel along this Tube!

At 360 Revolutions a Second we have traveled up to 21,600 in one minute. ""1,296,000

in one hour ""31,104,000

in 24 hours

Remember this is Light we are traveling in, and at the Speed of Light that is some Distance! With the only example being the Sound within the Light. Plus One Wave Length. There is no other in which to travel at this Ultra Sound Specific Speed. We have used the Pure While Ray, yet utilized the Fusion of all Seven Rays. How?

Stay in your Mind's Eye here. The Tube acts as a Protective Barrier to any harmful Radio-Active Elements, giving safe mode of travel. As the White Light acts as a Repellent, it pushes aside the Ultra - the Black Light - forming the passage-way for clear travel.

But! The Black Light closes in from behind after we have passed through, acting as a Propellant from behind, increasing the Specific Speed of Propulsion. This... is traveling at Double the Speed of Light.

It is also the Oscillating Factor of, being the driving force, coming from in front to, behind, all the way through, within this action of Propulsion. Nothing is wasted in this System. All Energies are utilized, safely.

Liken it to this as an example.
An Electromagnetic Field Force has been created due to all Energies within the Sound of Light which is Movement. Because this Energy pushes aside - lets say water (the sea,) - the Object of travel would not get wet.

The Electromagnetic Field would keep its' Radius free of water, even though, the water by it's own Replacement would close in, at the back, and, as a Wave Force, project our Object of travel with an extra Boost of Propulsion.

From here is the preceding Discussion... Here the 'Revolutions' and 'Specific Speed' are explained.
DOUBLE THE SPEED OF LIGHT - TIME TRAVEL.

ZERO DISSECTED

Zero = Nothing, right?

So... No Butter, no eggs no apple no 'what'? No Zero? What does Zero look like? That is not nothing but a symbol of something. OOOOOOOOOOOOOOOOOOOOOOOOOO Here I have 24 Zeros I can, divide them up! The division of 6 into these 24 = 4. Four Zero's is not nothing but something. I can already see where this is leading to. It's leading to 'the theory that everything is linked.' That being the case makes it possible to have the theory 'Devision by Zero, is possible.' Because in essence, there can never be, 'nothing.'

But what about divisions by Zero. Say the temperature was - 80 degree Zero. If I was to divide - 80 by + 100, should the answer be, + 20 ? Or is this addition. Lets take a really good look at Zero, a word, representing a set point of measurement. Yet when it is measured in temperature, Zero = either side of heat and minus heat. That would be like saying there is nothing below Zero. Does that make sense to your mind? Below Zero is freezing. Absolute freezing! And 'absolute' is not nothing.

Ground Zero? That, is no-where to go but, down to ground level. Zero Gravity? Doesn't this mean this is gravity's vacuum. Vacuum 'can' mean empty. Yet even empty can be 'full of air'. That is not Zero Air. Zero, is also a measurement, used in aerodynamics, absorbing the air which is driven at a velocity through, as in a wind. Here, we have all the sounds different 'pitches', reaching into ultra sonics, may be 'without sound to the human ear'.That is certainly not without speed, but more like, super sonics.

Zero has been used in many ways. It is a precision point in adjusting the sights, for fire arms, in order. to shoot directly at the desired point. Lets bring our sights back to Zero and look at it in the direction of what we have already. You ask me where I'm heading with this? 'Z' is the English last, or 26th letter of the Alpha/bet. Alpha is the 1st. Letter of the Greek. So I 'bet' Zero can also be found in all frequencies, in one way and another. We can go into the frequencies of all sounds, whether phonetics, or even amplifiers. Even the O/S theory.

Of course if we bring it back to the Zero = O - like it or not - with a name, which already means something. Repeat again; it cannot be nothing. So without going into all the brain groping explanations whether via. science, mathematics, algebra or simply the fact that we say in the 'sound' of 'O' when realizing something in (Oh !), the reality of Zero. Basically speaking with all it's quantified reasons for 'just being what it is', is found within the field of 'sound' vibration. .. as a division of, in any given direction needed within it's sights to be divided up into.

Mmmm... we are back to Zero hour. No more Time. But wait! This is a set time. Beyond this, Time, still carries on. If I keep going we'll find Zero in Time Travel and end up in Outer Space. I think I have dissected Zero! Is that division possible?

- Show quoted text -

I am in complete agreement, Zero is not "nothing." By giving it a name, it becomes "something. It is the starting point, the marker for turning around, etc. Hey. why am I saying anything, you said it all...DLS.

OSCILLATING ENERGY - IN A GRAIN OF SALT

ka-sala View profile

OSCILLATING ENERGY – IN A GRAIN OF SALT

Alternative fuel ? A new millennium ? Once the designer of a water fueled engine, was bought out, to 'shut up'! So now that we have turned over from then, where the price of fuel has caught up with the purse strings of every one except who holds the purse, why not try salt water? They look at Solar Energy, and that's good.

Consider just for a moment the availability of salt water. Or just salt? This Oscillating Energy running off the Ionic O/S is as cheap as it gets these days. The Solar Energies are drawing up from the earth - where once were seas of their own nature - the salt of the earth. People, are now learning their rivers and water-ways need to be 'managed.' Such things, as water, was just taken for granted. But it is the salt which is of interest here.

Saline gels, of different ratios, are used for something as simple as an Ultra Sound, and we see, the babe within the womb. Such are the properties of salt's worth, conductively speaking. It permits an easy flow of electricity - energy - and ultrasonically speaking, via the Energy of Sound as well. Sound, is only vibration of anything which moves. So salt being the conductor-transmitter, it's very make up is already charged with its own ions; it's own live particles, as crystals grow.

How come 'SALT' are the very letters, 'Strategic Arms Limitation Talks' ? Even the word 'strategic' is of importance, as is, 'specific'. Is it coincidental that it should fall into being just 'SALT'? No it's not. It was very specifically thought out, as any army would, the nature of their strategies. Just as a salt crystal is a very specific symmetry. It is a perfect 'cube'. What's all this got to do with alternative fuel ?

Already the Energy of just this Specific, has been extremely, briefly, mentioned. Imagine if we were to go into the whole Strategic Force found within all this, and put it together to become an alternative fuel. Are there any grains of SALT ' Strategic Alternative Litigating Terminology' floating around out there in their own Oscillating way, under their own Ionic Atmospheric Pressure who can work on this Grain of Salt Fuel?

A bushman agent, could be the guiding intelligence back into the archives. As all good Strategic Talks about the 'foot prints' of the tracker, could only lead them to the source. Like a horse to water, you can't make it drink... but it does like a lump of salt. So even if you take this attitude, to an alternative fuel is with a grain of salt, you have begun to think. Don't underestimate even what you cannot see... yet.

It took Leonardo da Vinci 500 years for others to realize his Science! Please stay with me a bit longer than that? Let me know, won't you, because nothing can have the power to be combined with salt, and not have Energy enough to fuel even the visionary ions, of so much wasted fuel, and not be electrified by it.

All for the sake of... 'ease up on the black gold, and try the white'! It might just be another perfect solution - can you 'hear this word, Solution' - to the perfect Cubic Crystal, of what they are looking for.

After all, it could even be the very vehicle to drive enough brains oscillating out there into action. Maybe I should put on the brakes here, before the world is ready for this fuel. You might even find something as good as Germanium or the Silicon Chip's worth... Oh no, where are those brakes! At least I know where the Light Switch is.

MANY OSCILLATIONS - CONNECTIVE ISSUES

ka-sala View profile

More options Oct 19 2008, 9:52 pm

MANY OSCILLATIONS – CONNECTIVE ISSUES

Dealing with something as normal as pitchblende, seems of little import. Ships of the past, and even some of the present, patch holes with Pitch. Yet the Pitch Element is very significant. Being a Carbonic Element it is linked to the Transition Element from which Uranium is derived. The Radio Active Element, Radium, is its Principle Source.

Nuclear Transition creates change - transforming in the configuration of an Atomic Nucleus - and changes its Energy level by the Emission of Gamma Rays. These then emitted by the Nuclei of Radio Active Atoms during Decay. Gamma Rays may be 'history' to Science, but we started this O/S Theory with the discussion, of a Space Ark, in 'Double the Speed of Light – Time Travel' So here, we will be Oscillating between the Science of the past and now, this Space Age.

There are Nuclear Links to what was 'Noah's Ark' (past,) as was 'instructed'. To line the Ark with Pitchblende. It was shown by one who once reconstructed this Ark, exactly as given, that it had all the attributes to a "Space' Craft. You cannot have a Space Craft without the Oscillation Substance... and in this case, we carry on... from the Pitchblende, and the Nuclear Transition.

These are the Elements here within the Electro-Magnetic Field of Quantum Mechanics. The Quarks, the indestructible Particles which reproduce themselves, every time they are split. Like it or not, believe it or not, no matter how much Science divides and brakes down, it cannot be subtracted from the fact that the Quarks can be 'likened' to the Amoeba of Cosmic Energy. A Nano of this Cosmic Energy is touched on in my Profile.

Right now, even beginning to the end, is Oscillation, where all Core Essences have the Di or Twofold Function to become Tri, or three. As simple as the analogy might be, we have a Third - the Quark of Life as a Human - reproducing itself. Despite all the Interactions of the Nuclei and Cells Transitioning, the beginning and the end of this process, is, it has created the final stage. A multiplication of itself. As equally as the Cell of the Human works likewise, though with many name Components.

It doesn't change within the multiplication of the Cell from it's Amoebic State. This Plasma Essence - the Amoeba Quark - only becomes more fluid, throughout the process. We do not need to go off into another subject of Chain Reactions here where 'Everything is Linked' but simply to say, everything contains its Core Essence Element. It begins (taking on a word to name it,) to where it ends, and in between the changes and words become another Nano (or billion) of Expressions in the Process.

So many Elements are over-looked which are so relative in the O/S theory, where Science seeks beyond that of themselves. The average Earth Being, 'forgets' that even they are of their own Solar System, and made up of almost all the Elements of their own Sun. Even by using their mouths through words, they are Radio Active, Oscillating between Creative and Destructive. Creative has just been mentioned above, in one form, graphically speaking.

But destructive... how? Blow a fuse (or temper,) and watch the Chain Reaction it has on those around you. The 'blast' can have an effect for hours, days, and weeks... and in some cases, even a life time! Human Radio Active Radiation and Solar Storms can interrupt a whole household, community, and even beyond. It is again purely touching on the Positive Negative issues of all things connected. It is only again using the O/S between whatever any two Poles are, which must have a 'Balance Factor' - call it Neutral - here.

In truth when it comes to the Earth Science, we call it the Equator, or Equalizing Factor of this Planet. Otherwise it would be in a very Negative State if it was not 'Earthed'. It's Oscillating Substance within it's Axis would have spun right out of control! Here within the Axis, this Substance, is called, it's Light. But because it reaches into the Oscillating Substance of 'Black Light' it is not visible to the eye, yet could not spin without this. It is no co-incidence that the Axis has tilted more and is!

Our concern is the Negative Forces which are man made, which are the Destructive Elements, and his inability to deal correctly with the Science he already has at hand. That of the Nuclear Physics, and what he does not know. The Nuclear Waste is Oscillating like a Time Bomb within the Earth's given Field, while man searches the Stardom for Life. At the very same time Life as he knows it is caught in the web of his own incapability. Time... is not on his side of what he does not know, 'how' to fix in time!

THE PROTON and The Sixth Element of Action

ka-sala View profile   More options Oct 30 2008, 3:47 am

THE PROTON and The Sixth Element of Action

Which-ever particle we speak of, they are all inter-linked.
To some, this may be more metaphorical, and to others, analogies.
Maybe... a hint of insight. This should not subtract from a broader
perspective into a subject which is a 'pin point' in the Cosmos. It
requires the mind's eye to expand and be the chameleon, to see it from
360 degrees. And to become the eagle, which can fly above it, around
and below, looking into it from all angles. We are looking at a Sphere
Shape. Let's liken it to a circular, or 'round' egg, taking into
consideration it's layers.

The nucleus (the Proton,) being the Positive Charge, can be the yolk.
Around it, is a membrane: call this, it's Magnetic casing. The white
of this egg we can call the the Neutron; the uncharged Elementary
Particle. The inner casing of membrane within the outer shell, we'll
call the Negative Electron. The stronger outer Shell... the Positron's
'antiparticle' of the Electron - or Positive Charge - of the
antielectron. From it's Core to it's Shell, we have the amazing
Proton! Five elements. It 's Positive, Negative and Neutral + it's
own two inner Electromagnetic sheaths.

You have to be that fly on the wall with a thousand microscopic eyes
to see what can be seen. It is teeming with Oscillating Energy. Within
the Proton, which requires it's combined Electromagnetic Field, is the
key to the whole Force (Energy.)

So why does it 'need' the introduction - or bombardment - of the
Electronically charged Positive and Negative to act as a bolt of
lighting, to kick start this Proton in the direction of it's given
automatic field? The answer to this will be seen in the 'Core' or
it's Coaxial State. From here, Energies must Polarize. All Elementary
Particles are linked, in order to carry out this total process. It's a
Fusion of the Elements within their own Given Field. It provides the
very Energy required for it's own Electromagnetic Field within its own
Quantum Leap.

Remember... within this Proton Spherical Shell was found five
Elementary Particles at work. The Proton, Neutron and Positron. Plus
the Dual charged Electromagnetic sheaths/membranes within, making up
the five. Within is stimulation - Oscillation - the Electromagnetic
Charge interacting within its equal charge within the Proton. The
result being – a Current – the Electronic Charge, of all five bound
within each other which creates a sixth. The 'Universal' Charge or
Frequency found in all. The O/S Theory of Everything.

Think of the Proton being no different on a Universal Scale. Liken
it's use as a Proton of the Cosmos.
Extract one in your mind's eye - the Proton - and see it in it's
Elementary Position of it's own Electromagnetic Field, where at all
times it's own momentum is from it's N-S Poles (+ it's equator
E-W.) Bound within its' own Magnetic Field of Opposites, which
Polarize and double the strength within the core/axis.

THE PROTON and The Sixth Element of Action

If two 'donut shapes' were only seen from one dimension, where they overlapped slightly in their rings, it would appear as one big donut with a dent in the top and bottom (or it's N-S axis.) And the overlap of the two could be seen also. Now... shut off your mind's eye to the fact of where they overlap. Know... they are at work, but out of sight, even though 'within' this Core, you know they overlap. But... if these donut shapes were seen from the multi-directional - say 'above/below/around the 360 degrees' - they would be as multiple arcs. All forming the same interaction within the 'unseen' at it's 'dent'. Here... is where the Fusion of being Polarized, within the core/axis/ vortex., their Energies interact.

Now expand your mind's eye again here. See these multiple arcs, (not as 'donut shapes',) but needle fine Rays of Light Energy, all forming the same 'dent' where they interact, where you cannot see, in it's co-axis of Polarization. It's 'Engine' is the central core/vortex, where clock-wise and anti clock-wise, it's Energy is Permanently Oscillating to the Vibrational 'hum' of the Electromagnetic Force 'within its own bounds'. Together, this would then be the Universal Proton, within all Cosmic Coaxial Equations.

To go any further, would be to multiply their Wave Action, where all 'dents' run in and through their Cosmic Energy.

Reply to author   Forward   Report spam

ka-sala View profile

Slight alterarations re. wording, of same subject matter in this 2nd. post.

More options Oct 30 2008, 3:53 am

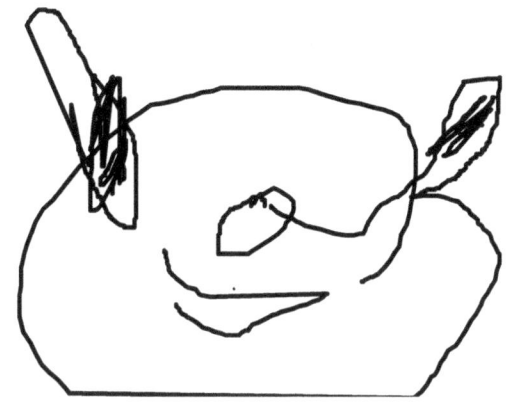

THE PROTON and The Sixth Element of Action

2 messages - Collapse all - Report discussion as spam

dean sinclair View profile   More options Oct 27 2008, 8:04 pm

Ka-Sala-
I think that in this essay you are coming awfully close to definining
some sort of "proto-oscillator" perhaps, a "proto-on." The physical
description comes close to descriptions of a toroid, pseudo-sphere
which would describe very closely what I would envision as the shape
of most of the tiny dots making up the "Substance." The only
"criticism" I have of the essay is that the word, "Proton," appears to
have two different meanings and apply to two different things...
Cheers. Eski

- Show quoted text -

Reply to author   Forward

ka-sala View profile   More options Oct 28 2008, 10:50 pm

Helo Eski,
Thanks for such an instant reply... What ever you feel about the 'two
different meanings' can you suggest how you would word this?
In essence, seen from the description of a toroid - it would then
separate 'your proton' view into two different meanings - or
application of two different things. As the toroid 'coils' do not
intercept, or become the coaxial of the polarization of energy, it
would leave the axis/vortex void of the dual energy concept. By
bringing in the Cosmic Coaxial Equation it is merely extending the
'idea' of this oscillation still within the proton, and the light,
into the Harmonics (as I did mention the 'hum' of this Oscillation.)
So... maybe 'somewhere in here' you can 'see' the dual energy created.
Maybe I've taken it a step further?
Try 'not to see' the 'hole' in the donut, but rather, just the 'dent'
with the knowledge that all this Coaxial Polarized Energy is
Stabilized, through Fusion, within. Not void, even though it can also
be termed 'a vortex'.
It's a bit like, going outside at night into complete darkness, and
'seeing nothing'. But when holding a strong light beam up into the
air, it is teaming with life and particles. Sometimes... I say the
atmosphere is 'thick' tonight, as I can see and feel what others with
me can't. Just another analogy!
Let me know further though how you 'see' more into this essay, and as
I asked, 'how would you word it'?

On Oct 28, 12:04 pm, "dean sinclair" <deanlsincl...@gmail.com> wrote:

The following "Steps to the Big Bang" covers the basic logic of "Eski's Version of O/S Theory." as expressed in papers published on Helium.com and the Google Group, Oscillator/Substance Theory. It is roughly what some brilliant scientist might have done. The meandering path followed by the slow-thinking, semi-senile old man who actually did this theory work was far from this clear cut.

Step I. Communication theory notes that the maximum velocity that information will be carried in any direction is the average velocity of the carrier wave in that direction whether the carrier wave be made up of Pony Express riders or electromagnetic radiation. Whatever be the case, information travels in packets transferred from one unit to another.

The speed of light is the maximum velocity of information transfer in our Universe. It may, therefore, be considered as the average velocity in any direction of the motions of whatever be the basic units of the substance of our Universe.

Step II. Basic physics theory suggests that rotors spinning with a tangential velocity of the speed of light, "c," would be able to pass information tangentially in any direction at that velocity.

Step III. Planck's constant which relates vibratory motion at the speed of light to "Energy." a packet of motion, has the dimensions of angular momentum, hence it is related to rotary motion.

Setting Planck's Constant, "h," equal to the three variables which make up its definition-- mass, radius and tangential velocity, gives the equation $m \times r \times v = h$.

Evaluating this at the average v of c we get $m \times r \times c = h$ which rearranges to $m \times r = h/c$. This two variables equal a constant is the Law of Levers which is defining for a teeter-totter, or any other kind of oscillator. This particular equation defines a family of oscillators which would have the average at $m = r = (h/c)^{0.5}$. The value of this expression is approximately 4.7×10^{-19} centimeters or grams, in the cgs system of standards.

This gives a picture of our basic rotors as having an unlimited number of possible shapes, sizes and masses, but all would have their average as noted above and it can be expected that, in an ultimate compression, could be reduced to units of this mass and radius whether they be an electron or a universe.

Step IV. Matter is composed of electrons and protons. These, logically should be, members of the above oscillator family.

The electron and anti-electron seem to be a "balance pair." The "annihilate" to give off energy, the same amount of which is necessary to cause "pair-production" in which they can be formed. This can be explained if they are counter-rotating, "vortex-oscillator halves" of a cavity oscillator which can be split into them and into which they can revert if they line up along their common axis of rotation. Coming together they would cause a disturbance at right angles to the axis along which they joined. "Annihilation energy" is known to be annular, i.e. given off in a ring disturbance in the "substance." This is consistent with the above idea.

We, therefore, can consider the electron and anti-electron--the negatron and positron--as halves of a neutral unit which we may dub the "zerotron." We have not yet accounted for the proton and the neutron. As it is known that the proton and electron are the products of the "decay" of the neutron, it seems logical, in light of our reasoning above, to consider the neutron as a variant of the zerotron, a dual oscillator, which can collapse to an electron and a proton. A "proto-proton" would take the place of a "proto-anti-electron" in this variant of the zerotron.

If we consider the kinetic equation expressions for the anti electron and the proton and their different masses, we reach an interesting conclusion. If the velocity of an anti-electron were decreased to about

1/43 of its original velocity, an anti-electron could be converted to a proton if its kinetic energy were converted to mass. This could happen if a zerotron unit were to encounter a situation wherein "the back coming forward met the front coming back" with the back-half taking the mass increase. That is in the size/motion deformation one might have an explanation for neutron formation.

Step VI. Looking for some event, or idea, that would possible produce neutrons, we consider the concept of the "Big Bang"

in terms of an "Oscillator/Substance Reality." The "Big Bang" would appear to be the inversion instant for an oscillator, the moment at which the oscillator changed from a compression mode to an expansion mode. An ultra-low-frequency oscillator operating in a substance made up of oscillators, some of whom were our "zerotrons" could conceivable account for the neutron formation.

the logic for the above statement is this: If the average velocity of motion within the substance is "c," and we assume that the average acceleration/deceleration is constant, then then initial velocity at inversion is 2c. Now a motion at 2c moving in a medium having an average motion velocity of c would ride as a shock wave. Such a shock wave would flatten any properly oriented units in its path.

The zerotrons converted to neutrons--in our half of the postulated "Universe/Anti-universe" Oscillator-- would collapse to electrons and protons. As these have opposite spins they set up opposite disturbances in the substance/medium, causing them to "attract" one another. However, they cannot stably interact but can associate.   The result of these associative interactions and of the natural tendency of any substance to balance all motions throughout can be seen as the reasons for the "Creativity out of Chaos" in the Universe riding behind the shock wave.

Delete this page	

Version: | Current by ESKI - Nov 17 2008 ▼ |

---

1 message about this page

Nov 4 2008 by ESKI

Click on http://groups.google.com/group/oscillatorsubstance-theory/web/some-basic-background-ideas - or copy & paste it into your browser's address bar if that doesn't work.

Note: The following article was published on the Science News Site, SciScoop in the Physics Section. Much of this information has been published on our site before; however, this is a bit more focused.

A Constant's Secrets, A different Look at Planck's Constant   Nov. 20, 2008
                                                                by Eskil

For the last hundred years, in plain sight, there has been hidden within a Constant of Nature, important information about the Universe within which we live.

Planck's Constant was discovered in the early 1900 by Max Planck. In all the time since, however, it appears that no one has taken the time to ask some logical questions about the Constant. Some of these questions are the following: The constant applies to the relationship between Energy and frequency of electromagnetic radiation, so what is it operating on to connect Energy and frequency of radiation? Why does it have the dimensions of "Action" or "Angular Momentum," therefore it is logical to ask, " On what is the action occurring? What has the angular momentum?"

Running these questions around brings to mind the thought that if this factor, Planck's Constant were an average amount of action in any direction, or the average angular momentum in any direction--apparently the same-- of units making up reality, then that average unit would appear as a constant.

Planck's Constant is used at the speed of light, another Constant of Nature. Could the thought that perhaps Planck's Constant is an average value of a characteristic of some unit of existence also apply to the speed of light? Could the Speed of Light also be connected to something related to rotating units as Planck's Constant may well be?

This last speculation fits in very well. The Speed of Light is the maximum velocity of information transfer. Information transfer in any direction from a point has a maximum--and in practice, unattainable--velocity of the average velocity in eny given direction of the velocity of the "packet carriers" whether these packet carriers be Pony Express Riders, Sound Waves or Electromagnetic Radiation. The Speed of Light would make sense as the average velocity in any direction at any time of the units, rotors(?), acted upon by Planck's Constant.

Acting on these ideas, Planck's Constant is an average angular momentum and the Speed of Light an average velocity in any direction, we can try some mathematical analysis.

Let us set Planck's Constant, (h), as the constant value into the form equation, K=xyz. In this case, let Planck's Constant be "K", "x" will be an instantaneous mass, "m," rotating at a distance, "r." from the center of rotation, and the third unknown be an instantaneous velocity, "v" measured at that distance, "The Tangential Velocity." We write, "h=mvr." Since "h" appears to be necessarily valid only at the speed of light, "c," we evaluate h=mvc. Since "c" itself is a constant, the ratio of "h" and "c" is a constant so mr=h/c. This two unknown equation can be said to define not only a variable area but also anything which varies in this manner. One such is the Law of Levers in Physics. The numerical values of "m " and "v " can be interchanged, hence this little equation could describe something moving back and forth between two limits or two "states." Such motion is called "oscillation" and something acting in such a way is called an oscillator. We can say that the equation, "mr=h/c describes a family of oscillators. Since the values of m and r can be switched, they can also be equal to define an "average" oscillator when m=r=(h/v)^0.5. That is when the numerical value of the mass equals the numerical value of the radius and each equals the square root of the ratio of Planck's Constant to the Speed of Light. Inserting the value of 6.63 x 10^-27 erg. sec. for Planck's Constant and 3 x 10^10 cm./sec for the Speed of Light we get a value of approximately 4.7 x 10^-19 cm. (and 4.7 x 10^-19 grams) for the "dimensions" of the average oscillator defined by the "(h/c)^0.5 Constant."

It is interesting that this value, 4.7 x 10^-19 cm. as a radius is almost exactly one half of the diameter at which the Strings of String Theory are said to disappear into a "10 dimension hole." Also, Quantum Mechanics is said to fail at below the same distance.

At this point we suggest that the basic units of our existence are tiny oscillators half of whose existence is unknown and unexplored as it lies below the "threshold" of (h/c) ^0.5.

Perhaps we should close this little essay at this point and hope that the reader's curiosity has been aroused enough to follow up. There is much more that can be said

Gmail Calendar Documents Photos Reader Web more ˇ   deanlsinclair@gmail.com ˇ

Google groups

« Groups Home

Oscillator/Substance
⊞ **Theory**

Search this group   Search Groups

Home
Discussions
Members

About this group
Edit my membership
Group settings
Management tasks
Invite members

An Intertwined Universe?

This little article was first published on SciScoop where it made the "Front Page." It is based on ideas from this site.

An Intertwined Universe?

A mathematical treatment of two constants of the Universe leads to an interesting conclusion.

Maybe string theorists are not so far off base as one might think in suggestions of an intertwined Universe. The following bit of dealing mathematically with two constants of nature suggests something of the same thing.

Combination of Planck's Constant and the Speed of Light can be carried out by setting Planck's Constant equal to its definition as an Angular Momentum evaluated at the Speed of Light to give the equation, $m \times r = h/c$. In words, the mass of a rotating object (m), multiplied by its distance/radius (r) from a center, is taken to be equal to Planck's Constant (h) divided by the Speed of Light (c). This can be shown to be a defining equation of a set of oscillators. We can go on to write, $m = h/c^2$. This will define a mass moving at a radius of "one light second" from the center at a velocity of the Speed of light. This mass figures out to be about 7.4×10^{-48} grams.

If we guess that this mass may be the smallest mass detectable in our Universe-- perhaps the mass of a neutrino or of a "zeroino," a postulated neutral "parent" to the neutrino and anti-neutrino--then this mass would be the detectable "rest mass" in our Universe of a particle having a radius of 3×10^{10} cm. If this be an oscillator of the above family, $m \times r = h/c$, then the other, unseen limit (undetectable in our Universe) will have the opposite absolute values, a mass of 3×10^{10} grams at a radius of 7.4×10^{-48} cm.

What would be the lightest, biggest unit would also be, in its "alternate reality," the heaviest, smallest....

If we truly do exist in a substance whose basic units are such as this describes, it may be expected that the Universe in which we exist could definitely be very intertwined.

Delete this page   Version: Current by ESKI - Dec 1 2008

Latest 3 messages about this page (5 total) - view full discussion

Jan 1 2009 by ka-sala

A 'PS" Eski,
I cannot help but read between the lines of this article of an Intertwined Universe - that you have read my mind - of my article, 'Traveling on One Wave Length'? Facts are you have been able to translate this science of mine into the earthlings jargon, and keep it all connected. Thanks for this... it should help

Jan 1 2009 by ka-sala

Hello Eski,

Group info

Members: 15
Language: English
Group categories:
Science and Technology > Physics
Science and Technology > Chemistry
Science and Technology
change categories
More group info »

Recent pages and files

Essentials of O/S   Aug 23
The Electron and Pr   Aug 9
Possible Script for V Jun 10
Congruent Parallelo   May 4
O/S and the Periodic Mar 10
o-s-and-th   Oct 28
O/S, Space-Time an   Oct 15
Some Constants of t Sep 22
Letter to "Sophie" at Aug 20
The Hadron Collider   Aug 6

Just going over a couple of pointers here re. one of your earlier
discussins..."Reality and Speculation in Science' and your comment on
'taking with a grain of salt.'
Just as you found your SciScoop front page a great boost (An
Intertwined Universe,) a publisher purchased my "OSCILLATING ENERGY in

Dec 29 2008 by dean sinclair

Thanks for the boost! Guess what! I got another one on that front
page, too. The article ,
Negatron plus Positron equals "Zerotron?" I screwed up the title
somehow when I posted it on our site.... The title on SciScoop is the
right one.

Thanks. also for the "Scientist of the Year Christmas Greeting!!!

2 more messages »
Reply to this discussion   Report this page

Oscillator/Substance
⊞ **Theory**

Positron plus Electron equals Negatron?   Options

1 message - Collapse all - Report discussion as spam

ESKI View profile   ⭐⭐⭐⭐⭐ (1 user) More options Dec 10 2008, 3:53 pm

This little article was posted on Sci-Scoop. Although the subject is
mentioned in other papers on this site, it is felt that the idea might
deserve attiention by itself. Also, I am posting bits of the theory
from this site on Sci-Scoop as a "spreading of the word" so to
speak.

$$e^- + e^+ = e^0 ?$$

A little manipulation with the idea of "electron-anti-electron
annihilation" leads to the consideration of there being a neutral
combination particle.

The electron and anti-electron (negatron and positron) are known to
combine and disappear. This gives off energy as "annular radiation"
and is known as mutual "annihilation." However, let us take a look at
the situation.

Let us assume that an electron spinning in one direction along a given
axis at the speed of light, meets an anti-electron travelling at the
same velocity with the opposite spin orientation. As the velocity of
each, in opposite vectors is "c" the speed of light, the Kinetic
Energy of each is
$(mc^2)/2$ and the Energy expected to be dissipated in the "head on
collision" is mc^2. However--believing Einstein--we see that we still
have at least one "mc^2" worth of energy unaccounted for as each of
the two participants in the "collision" had this much Energy ascribed
to them.

This indicates that there is enough Energy remaining for a particle of
the same mass as either of the originals to have survived the
collision. A logical thought is that, rather than a "mutual
destruction," the negatron and positron simply did a "Yin-Yang"
combination to a neutral unit dumping excess rotational motion as
"annular' (ring-form) radiation into the plane in which they met.

There is another known process, called "pair-formation" in which
radiation above a certain threshold, the same Energy as the annular
radiation. will, under certain conditions, lead to the appearance of
an electron and a proton. It is reasonable that a "parent entity" of
the "positron and negatron," which we may call a "zerotron,"
"e^0," could be an explanation for both phenomena.

A neutral entity as a "parent to a positive-negative pair" is, of
course, known. The neutron is the "neutral parent" to the electron and
proton. It is not too far-fetched to suggest the following: $e^0 +$
Energy $\to n^0 \to e^- + p^+ +$ Energy. That is, that under some
condition the unit which we are postulating as the predecessor of the
electron and anti-electron may be converted to a neutron which then
splits to the electron and proton.

Along the same lines of reasoning we may note that the Muon comes in
three forms, positive, negative and neutral. Has anyone checked to
seen if the positive and negative forms combine to form the neutral
one? Probably that would be very hard to check.

Google groups

« Groups Home

Oscillator/Substance Theory

View this page "Quantization: A 3-D Merry-go-round?"   Options

4 messages - Collapse all

ESKI View profile   More options Dec 29 2008, 5:03 pm

Hi. this is a copy of a submission to Sci-Scoop which, apparently,
they don't know what to think about it. It takes two negative votes
to discard a submission, and a net of 4 positive votes to publish it,
That's the supposed rules, So far it seems to have gotten nothing but
abstentions, which would mean that people reading it don't know what
to think. I need a few buddies in sci-scoop.....

Cheers and Happy New Year. ESKI

Click on http://groups.google.com/group/oscillatorsubstance-theory/web/quantiz...
- or copy & paste it into your browser's address bar if that doesn't
work.

ka-sala View profile   More options Jan 1 2009, 9:55 pm

Hello Eski,
Having just read through this submission... it sounds very close to
the whole concept of mine just submitted to this group. PEACE is an
OSCILLATION SCIENCE. Can you see the connection I do?

As for you needing votes for the SciScoop... can you in any way tie
the whole context of what we are both saying - though we have used
different terms - and help your Sci-Scoop 'understand better' what to
make of
what you were and are trying to say?
As I have said... you have the language; and I'm only trying to say, I
'understand where you are coming from.' It all links... Try
linking up (I don't need to be included!) but just for others in
your Sci-Scoop. It 'might' just throw a bit more light out there for
them on your submission?
Good Luck
ka-sala

On Dec 30 2008, 9:03 am, ESKI <deanlsincl...@gmail.com> wrote:

- Hide quoted text -

> Hi. this is a copy of a submission to Sci-Scoop which, apparently,
> they don't know what to think about it. It takes two negative votes
> to discard a submission, and a net of 4 positive votes to publish it,
> That's the supposed rules, So far it seems to have gotten nothing but
> abstentions, which would mean that people reading it don't know what
> to think. I need a few buddies in sci-scoop.....

> Cheers and Happy New Year. ESKI

> Click onhttp://groups.google.com/group/oscillatorsubstance-theory/web/quantiz...
> - or copy & paste it into your browser's address bar if that doesn't
> work.

ka-sala View profile   More options Jan 1 2009, 10:12 pm

PS! me again Eski...
Does this mean we/I have to join Sci-Scoop too...(if you need a few
buddies?) Or can we just comment/vote?
ka-sala
(Forgive all the replies... shoulld have checked my spelling before
posting. ENGLISH is not my best subject!)

On Jan 2, 1:55 pm, ka-sala <irrir...@gmail.com> wrote:

- Show quoted text -

dean sinclair View profile   More options Jan 2 2009, 5:44 pm

To "help out" on Sci-scoop you'd have to join. It's similar to
joining the oscl/sub site, as far as I know its an open group. I
joined several years ago but was very inactive for years as my first
submissions got "shot down" immediately.

As to the Circle of Peace, I'll have to mull that over a whike, maybe
I could come up with something to submit as philosophy which acually
is where you are coming from.
Burr, it is COLD here! Tomorrow, I won't be able to do what I did
last year on Jan. 3, I told them that I just "turned 67, today."
darn it though, I'll have to wait nine years to tell people that "I
just turned 68, today! :)) Cheers. ESKI
On Thu, Jan 1, 2009 at 9:03 PM, ka-sala <irrir...@gmail.com> wrote:

> PS! me again Eski...
> Does this mean we/I have to join Sci-Scoop too...(if you need a few
> buddies?) Or can we just coment/vote?

Nope, don't think so.

- Show quoted text -

End of messages

« Back to Discussions   « Newer topic   Older topic »

Gmail Calendar Documents Photos Reader Sites Web more   Sign in

Google groups

« Groups Home

Oscillator/Substance Theory

Search this group   Search Groups

Quantization: A 3-D Merry-go-round?

Quantization, a 3-D Merry-go-round?

Certain considerations about a possible structure of the Universe suggests that a constant speed merry-go-round may be an understandable model for the fact of "Energy Quantization."

On many children's playgrounds there is a simple spinning device which children can jump on and off of and move toward the center of to make it go faster or move out farther to slow it down. If f one had a large enough group who wanted to keep it at a constant speed, they could do so by giving it enough push all of the time to compensate for friction. If one jumped on the whole group upon the merry-go-round device, could move toward the center to keep the device from slowing down. When one jumped off, everyone still on could move out slightly to keep the device from speeding up.

The moving of one of our children from the outside group which is trying to maintain the merry-g0-round device at a constant speed to being on the merry-go-round and what would happen on the merry-go-round and to the group outside, corresponds very closely to the phenomenon of "quantization." The child climbing on corresponds to "a quantum of energy absorbed," the merry-go-round gets heavier and must contract to keep rotation constant, the outside "crowd" expands slightly. When a child jumps off, there is a "quantum of energy" given off, the group on the device expands slightly and the outside group becomes more tightly packed. The readjustment motions in both cases would be somewhat of "wave-motions."

As there are indications, in at least one theoretical approach, that the basic units of our existence may be constant-speed, constant-torque oscillators, we can extend the children's playground device to three dimensions, have the rotating object which we are seeing quantum interactions upon be made up of a group of these oscillators--as if the play ground device were made up of children--and the surrounding crowd be other independent oscillators.

In other words, absorption of a quantum would correspond to additional oscillators moving from the external group to the central rotator which we are observing. Emission would correspond to an oscillator disengaging. The electromagnetic radiation which is involved corresponds to the readjustment of the two groups. Observed, of course, in the external group.

As the oscillators appear, at least in our universe, to be constant torque--that is having a constant ability to push or pull one another--which is independent of radius, and to have a constant spin, or spin average, this "3-D-Merry-go-round as a model seems to be quite reasonable.

Version: Current by ESKI - Dec 29 2008

Latest 3 messages about this page (4 total) - view full discussion

Jan 2 2009 by dean sinclair

To "help out" on Sci-scoop you'd have to join. It's similar to

joining the oscl/sub site, as far as I know its an open group. I
joined several years ago but was very inactive for years as my first
submissions got "shot down" immediately.

As to the Circle of Peace, I'll have to mull that over a whike, maybe

Jan 1 2009 by ka-sala

PS! me again Eski...
Does this mean we/I have to join Sci-Scoop too...(if you need a few
buddies?) Or can we just comment/vote?
ka-sala
(Forgive all the replies... shoulld have checked my spelling before
posting. ENGLISH is not my best subject!)

Jan 1 2009 by ka-sala

Hello Eski,
Having just read through this submission... it sounds very close to
the whole concept of mine just submitted to this group. PEACE is an
OSCILLATION SCIENCE. Can you see the connection I do?

As for you needing votes for the SciScoop... can you in any way tie
the whole context of what we are both saying - though we have used

1 more message »
Sign in to discuss   Report this page

Create a group - Google Groups - Google Home - Terms of Service - Privacy Policy
©2011 Google

Two Energy Expressions Interact?

A "reversed viewpoint" of the momentum expression leads to a new energy expression and a new view of Mass and Energy.

The integration of the momentum expression, $mv=p$, mass times velocity equals momentum, with mass considered constant leads to the expression for Kinetic Energy, $KE=(mv^2)/2$. This is well-known.

Apparently overlooked is the fact that since "$mv=p$" is an equation of the type, $xy=K$, that is two variables equal a constant, it can just as legitimately be integrated with velocity considered as constant and the mass allowed to vary, to obtain an analogous, "Alternative Energy" expression, $AE=(vm^2)/2$.

Since mass and velocity can both vary it seems reasonable to consider that a true energy expression would take into account both of these expressions. One way to do this is simply to write $TE=KE+AE=(mv^2)/2+(vm^2)/2$. We simply assume that a "True Energy Expression would be the sum of the two alternative expressions.

Assuming that a quote attributed to Albert Einstein, "Mathematics is the Reality." has a degree of pertinence here, let us look at what conclusions may be reached by checking out this expression in various situations. Three situations come immediately to mind. Where the energy expression will equal zero, where the energy expression will be at some maximum, and situations in between where whatever the expression applies to is a constant value, neither a minimum nor a maximum.

In the first case, we write, $0=(mv^2 + vm^2)2$. For the entire expression to equal Zero, both expressions would individually have to equal Zero, or the two would have to be such as to cancel. Let us consider this possibility.
We can remove the expression $(mv/2)$ by multiplying both sides of the equation by "' $2/mv$" , leaving $0 = m + v$. Therefore, $m = -v$ or $v = -m$. In either case, mass is a vector, directed quantity, opposite in direction from the velocity considered.

For an Energy Maximum situation, we can assume both Energy Expressions to be in the same direction and both maximized. That is we hit maximum velocity, the system could go no farther and additional energy on the same vector appears as mass. As the mass increase will change both of the expressions they will maximize at a point of maximum velocity in both expressions. As "c," the Speed of Light" is generally considered to be a limiting velocity in our Universe, we assume that the limiting velocity is "c." We can therefore write, Maximum Energy equals $(mc^2)/2 + (cm^2)/2$. We have noted that the two expressions must be equal, we also remember that "mass" can be considered a "velocity" vector. Hence, $(mc^2)/2 =(cm^2)/2 = (c^3)/2$, and our maximum energy content appears to be c^3, for any system which we wish to analyze.

As this would be the maximum energy directed in any direction, it would be the maximum energy which can be directed outward from any point, or inward toward any point. In oscillatory motion, this would be the directed energy content at an extreme of vibration or pulsation.

Energy may perhaps be considered a "3-D Motion Package" of which mass is one dimension. Any three numbers multiplied together define a volume....

A common, and interesting, situation would be where the Energy Content is a constant, neither Zero nor Maximum. In this type of case, wherein mass and velocity can vary, but momentum and total energy content cannot, we have defined oscillatory or orbitory motion. As mass increases, velocity decreases and vice verso. we can rewrite our summary equation in several ways, considering that $mv=p$ we can write, $E= (pm + pv)/2$ or $E/p=m+v$, or $E/mv=(m+v)/2$. There will be some point at which $m=v= p^{0.5}$. At that point, when we put it all together we find that $E=p^{1.5}$. $(E=(p)^{3/2})$.

Another interesting situation would be wherein all motion was in the $(vm^2)/2$ package. The forward motion is a constant, but what does our motion package now mean? Forward motion is of a point along a line. If forward motion be static, then we can consider motions referenced to the point on the line. That is, as Kinetic Energy is known to reference to the motion of that point along a line, the AE expression would apparently apply to motions, i.e., vibration and rotation, referenced to the point, vibration/rotation/pulsation, etc. within the body/entity/system which we are considering.

Gmail Calendar Documents Photos Reader Web more ⌄   deanlsinclair@gmail.com ⌄

Google groups

« Groups Home

Oscillator/Substance
⊞ # Theory

[Search this group] [Search Groups]

PEACE...is an OSCILLATING SCIENCE   Options

2 messages - Collapse all - Report discussion as spam

ka-sala View profile   More options Jan 1 2009, 8:57 pm

Peace within a Wave Band? It is a Frequency, a Vibration; just as one
could say the Frequency of Music, the note, Middle C = 440 Oscillation
per second. Middle C to the 7th note is a Scale in Music. All sound
has a Frequency, a Vibration, and in vibration there is movement. This
is the Peace Wave. With the Earth's population by November 2008 up to
7 billion, it seems against all odds for peace to spread at all.

There is movement in the air, and the pace of life is trying to keep
up. But something has to give, and underneath it all, the work goes on
in peace and quiet. The Radius of the Earth also fits close to the
figure 7, in 7,000 Km. upward to it's Outer Sphere. Just for some
understanding of where this is leading to, keep the 7 in mind. There
are also 7 Rays withing the visible Light Spectrum.

See this Light as a Vibration from the Circle of the Earth, working
inward to the Center as Waves. This is just the opposite of the pebble
effect where waves form in rings from the center out. The Earth is
40,075 Km..at it's circumference. (The Numerical Key of this figure is
7!) Think of this Wave Vibration being like Surround Sound if it
helps understand this as vibration.

With these circular waves growing smaller in circumference as they
grow inward towards the center, when meeting in the middle, one flash
of Light will Amplify; Radiating the Earth with the Peace Wave, and no-
one will escape... their 'own' silence. So where's the big bang gone?
Out the window of theory, where more than the eye can see, and 'some
science' has been overlooked.

Analogy.
If there were only 10,000 peacemakers, and each being a Time Capsule
of 440 granules which have their own specific time to trigger, would
be their Oscillation/Vibration per. Second (using the Middle C's
analogy.) That would make 4,400,000. per minute = 264,000,000, and in
1 x hour = 15,840,000,000. When brought up to 24 hours = E 380.16 at
infinitum. (E = a Transcendental Number.) Reaching into the
Ionosphere, are also Electromagnetic properties.

A phenomenal build up is created in closing in, where Oscillation
simply would be a literal Vibrational explosion of Brilliant Light at
the central point. One might even say it would take 7 days of a week
for the Peace Wave to form needle fine Sky High (to the atmospheric
halo,) Light Works, enveloping the Earth. Within this time, Time as it
is known, would stand still. No-one will be able to speak with awe;
and the result will bring, Peace. A Big Silence in place of a big
bang!

So despite how far fetched the whole concept of peace on Earth is
within these statements; you may just get a glimpse of how. For
Science of the Stardom is one thing which is yet to be learned here on
earth, and is not ancient rockets, piggy-backed up to the System of
it's Sun. Peace, is an Oscillating Science of Life, in as much as a
Way of Life.

Reply to author   Forward   Report spam

ka-sala View profile   More options Jan 10 2009, 2:31 am

On Jan 2, 1:08 pm, ka-sala <irrir...@gmail.com> wrote:
> Hi eveyone!
> With Eski's '3D Merry-go Round' ending the year of 2008 and success
> with SciScoop... This is to wish everyone a Peaceful New Year for
> 2009. Ride with me!
> ka-sala

From: ka-sala
Date: 1/10/2009 6:06:31 PM
To: ka-sala
Subject: Re: PEACE...is an OSCILLATING SCIENCE

(Re... Snippets of post between Eski & ka-sala)

>I LIke your article, will have to read it several more times; your
>right tho that it sounds so much like some of my science writing.
>If I remember rightly,, the 440 cps. tone > which is used for calibration.
>is "A above middle C " rather than > Middle C....>
> Happy New Year >

(From Eski)

- Show quoted text -

Reply to author   Forward   Report spam

End of messages

« Back to Discussions   « Newer topic   Older topic »

Oscillator/Substance Theory

Fwd: CMNS: Zero Point Energy, Casimir effect and hydrinos

Options

6 messages - Collapse all

dean sinclair View profile   (1 user) More options Feb 6 2009, 4:42 pm

Somehow this should fit into oscillator-substance theory!

- Hide quoted text -

---------- Forwarded message ----------
From: Pierre Carbonnelle <pierre.carbonne...@gmail.com>
Date: Fri, Feb 6, 2009 at 12:06 PM
Subject: CMNS: Zero Point Energy, Casimir effect and hydrinos
To: CMNS <cmns@googlegroups.com>

Dear all,

Jovion recently got a patent for extracting energy from the vacuum,
using the Casimir effect to create what looks like hydrinos.

Scott, Marissa, I believe that there is a strong link with Earthtech's
research. Would you have any other info to share ? Is there any link
with cold fusion ?

Thanks
Pierre Carbonnelle

Patent description
http://nextbigfuture.com/2009/02/jovion-corporation-gets-patent-for-z...

A system is disclosed for converting energy from the electromagnetic
quantum vacuum available at any point in the universe to usable energy
in the form of heat, electricity, mechanical energy or other forms of
power. By suppressing electromagnetic quantum vacuum energy at
appropriate frequencies a change may be effected in the electron
energy levels which will result in the emission or release of energy.
Mode suppression of electromagnetic quantum vacuum radiation is known
to take place in Casimir cavities. A Casimir cavity refers to any
region in which electromagnetic modes are suppressed or restricted.
When atoms enter into suitable micro Casimir cavities a decrease in
the orbital energies of electrons in atoms will thus occur. Such
energy will be captured in the claimed devices. Upon emergence form
such micro Casimir cavities the atoms will be re-energized by the
ambient electromagnetic quantum vacuum. In this way energy is
extracted locally and replenished globally from and by the
electromagnetic quantum vacuum. This process may be repeated an
unlimited number of times. This process is also consistent with the
conservation of energy in that all usable energy does come at the
expense of the energy content of the electromagnetic quantum vacuum.
Similar effects may be produced by acting upon molecular bonds.
Devices are described in which gas is recycled through a multiplicity
of Casimir cavities. The disclosed devices are scalable in size and
energy output for applications ranging from replacements for small
batteries to power plant sized generators of electricity.

ka-sala View profile   More options Feb 8 2009, 2:08 am

On Feb 7, 8:42 am, dean sinclair <deanlsincl...@gmail.com> wrote:

> Somehow this should fit into oscillator-substance theory!.

Hello Eski,
You have made comment... re this work of Pierre Carbonnelle which fits
in exactly to the Quantum Electomagnetic Vacum - converting Energy and
available - at any given point in the Universe. It certaily does tie
in with the Oscillation Therory. It could not be, without being re-
energized by the ambient electromagnetic quantum vacuum.mentioned,

I have also copied an extract regarding this, (though within different
wording,) in 'Double the Speed of Light', and placed it below Pierre
Carbnonnele's.

It should re-open this whole theory posted some time back by me -
through my eyes - in my given example of this 'Vacume and Universal
Energy'. What do you see?

Regards
ka-sala
- Hide quoted text -
- Show quoted text -

- Show quoted text -

****
******
Exrtact from Double the Speed of Light
Stay in your Mind's Eye here. The Tube acts as a Protective Barrier
to
any harmful Radio-Active Elements, giving safe mode of travel. As the
White Light acts as a Repellent, it pushes aside the Ultra - the
Black
Light - forming the passage-way for clear travel.
But! The Black Light closes in from behind after we have passed
through, acting as a Propellant from behind, increasing the Specific
Speed of Propulsion. This... is traveling at Double the Speed of
Light.
It is also the Oscillating Factor of, being the driving force, coming
from in front to, behind, all the way through, within this action of
Propulsion. Nothing is wasted in this System. All Energies are
utilized, safely.
Liken it to this as an example.
An Electromagnetic Field Force has been created due to all Energies
within the Sound of Light which is Movement. Because this Energy
pushes aside - lets say water (the sea,) - the Object of travel would
not get wet.
The Electromagnetic Field would keep its' Radius free of water, even
though, the water by it's own Replacement would close in, at the
back,
and, as a Wave Force, project our Object of travel with an extra
Boost
of Propulsion.
From here is the preceding Discussion... Here the 'Revolutions' and
'Specific Speed' are explained.
DOUBLE THE SPEED OF LIGHT - TIME TRAVEL

On Feb 7, 8:42 am, dean sinclair <deanlsincl...@gmail.com> wrote:

- Show quoted text -

dean sinclair View profile   More options Feb 11 2009, 2:48 pm

HI. KaSala,--
Like everything else, it all fits together. I looked up the Casamir
Effect. The attractive force between uncharged plates in a vacuum.
This does fit into Oscillator/Substance Theory, and I guess, that
shouldn't ssurprise me, as yet, anything we've looked at closely seems

to fit in. The Casamir Effct would, , at the most basic level be for
the same reasons as gravitation--differential pressures between
entlities at the "sub-sub.microscopic level," the level of our
pervasive oscillators. In the case of "uncharged plates" in a
"vacuum" at the surface of the Earth a lot of things would be
happening which could be taken advantage of to transfer motion around
so as to be able to use part of it as "Useable Energy." The patent is
probably valid.

The description of extraction of energy from the quantum whatever and
replenish it from somewhere else is I guess as good as any.

Quantum fluctuations in a vacuum, would be localizing into "vacuum"
the apparent fact that our "substance" is continuously, and impossibly
always "trying to reach a state of uniformity." The only argument I'd
have at all with the quantum fluctuations is that the assumption is
that somehow this is a phenomenon somehow confined to vacuums rather
than being univerwal....
Cheers, Eski

- Show quoted text -

ESKI View profile   (1 user) More options Feb 11 2009, 6:56 pm

I have to apolgize for my typos in the previous and wish I could
apologize to Dr. Casimir for misspelling his name..... (: ! Eski

On Feb 11, 1:48 pm, dean sinclair <deanlsincl...@gmail.com> wrote:

- Show quoted text -

ka-sala View profile   More options Feb 12 2009, 12:32 am

Hello Eski,
I qote you again...
"The only argument I'd have at all with the quantum fluctuations is
that the assumption is
that somehow this is a phenomenon somehow confined to vacuums rather
than being univerwal...."

Don't worry about assumptions... What cannot change is that you
cannot
separate vacumes from being universal, otherwise you must leave out
the earth!

Many things considered by some as only/or a 'phenomenon' are as real
as you and me, and whether or not a 'thing' can be seen or not as
Oscillation, we would have nothing without it. In this is all
movement, energy, etc, by whatever name it is
given. Quantum or otherwise. This in truth is a great step forward
when seen in the light that it really is.

Regards,
ka-sala

On Feb 12, 6:48 am, dean sinclair <deanlsincl...@gmail.com> wrote:

- Show quoted text -

ESKI View profile   More options Feb 14 2009, 5:00 pm

Ka-Sala,
How right you are. As usual you re a step ahead of me in thinking. I
think that I'm going to post something like this to the cmns group: I
I'll let you have first shot.--

"If Max Planck were alive today, he might have something like this to say: 'Casimir Effect. Gravitation, and every other phenomenon of our Universe is a result of differential pressures in the Substance of Existence....' If this statement be valid, then there is a chance that someone can work ot a scientifically acceptable, presumably mathematical, description and do for our Universe what Galileo did for the Solar System. "

Cheers, Eski

On Feb 11, 5:56 pm, ESKI <deanlsincl...@gmail.com> wrote:

- Show quoted text -

...

read more »

End of messages

« Back to Discussions   « Newer topic   Older topic »

Positron plus Negatron equals Zerotron?

$e^- + e^+ = e^0$?

A little manipulation with the idea of "electron-anti-electron annihilation" leads to the consideration of there being a neutral combination particle.

The electron and anti-electron (negatron and positron) are known to combine and disappear. This gives off energy as "annular radiation" and is known as mutual "annihilation." However, let us take a look at the situation.

Let us assume that an electron spinning in one direction along a given axis at the speed of light, meets an anti-electron travelling at the same velocity with the opposite spin orientation. As the velocity of each, in opposite vectors is "c" the speed of light, the Kinetic Energy of each is $(mc^2)/2$ and the Energy expected to be dissipated in the "head on collision" is mc^2. However--believing Einstein--we see that we still have at least one "mc^2" worth of energy unaccounted for as each of the two participants in the "collision" had this much Energy ascribed to them.

This indicates that there is enough Energy remaining for a particle of the same mass as either of the originals to have survived the collision. A logical thought is that, rather than a "mutual destruction," the negatron and positron simply did a "Yin-Yang" combination to a neutral unit dumping excess rotational motion as "annular" (ring-form) radiation into the plane in which they met.

There is another known process, called "pair-formation" in which radiation above a certain threshold, the same Energy as the annular radiation. will, under certain conditions, lead to the appearance of an electron and a proton. It is reasonable that a "parent entity" of the "positron and negatron," which we may call a "zerotron," "e^0," could be an explanation for both phenomena.

A neutral entity as a "parent to a positive-negative pair" is, of course, known. The neutron is the "neutral parent" to the electron and proton. It is not too far-fetched to suggest the following: $e^0 + Energy \rightarrow n^0 \rightarrow e^- + p^+ + Energy$. That is, that under some condition the unit which we are postulating as the predecessor of the electron and anti-electron may be converted to a neutron which then splits to the electron and proton.

Along the same lines of reasoning we may note that the Muon comes in three forms, positive, negative and neutral. Has anyone checked to seen if the positive and negative forms combine to form the neutral one? Probably that would be very hard to check.

One may also suggest that there possibly be a neutral predecessor of the neutrino and anti-neutrino, a "zeroino."

| Delete this page | Version: Current by ESKI - Feb 17 2009 |

---

2 messages about this page

Feb 28 2009 by ka-sala

Yes ESKI,
I m finding the web is a very convient place for the scouts which can

View this page "Positron plus Negatron equals Zerotron?"

Options

This discussion is about page positron-plus-negatron-equals-zerotron

2 messages - Collapse all - Report discussion as spam

positron-plus-negatron-equals-zerotron was created by ESKI

ESKI View profile   More options Feb 17 2009, 7:15 pm

For some reason, this page, although it had a page title had no content. Who knows why.... DS

Click on http://groups.google.com/group/oscillatorsubstance-theory/web/positro...
- or copy & paste it into your browser's address bar if that doesn't work.

Reply to author   Forward

ka-sala View profile   More options Feb 28 2009, 11:05 pm

Yes ESKI,
I'm finding the web is a very convient place for the scouts which can turn one's words around to suit their own.
Main thing is... the word is spead!

REPLY FROM ESKI + Copy of this post.ESKI,

Thanks, KaSala,
I don't know how it disappeared out of the Google Group Site. I'm glad you had a copy... My computer, is a variety of public computers so almost all of my stuff is stored on Google so the glitch would be there or else someone simply cut and pasted it leaving nothing behind which I suppose is possible with an open group.....

Yes, I got both copies. Cheers from the "Ice Box.";;;;; Eski

- Show quoted text -

Not sure if you got this Eski,

> Maybe there was a glitch on your PC as it is on mine. Or maybe
> somebody liked it better than Front Page material for
> Sci-scoop... A rather stange thing also went missing from EVERY one
> re. an article re. UFO's on Helium + all whose were translated to the
> Internet. I was told they removed it as it was a mistake.
> Strange mistake for a Helium Marketplace Article!
> Hope this helps,
> Regards,
> ka-sala

On Feb 18, 11:15 am, ESKI <deanlsincl...@gmail.com> wrote:

This little article is much prettier on the top of the "Front Page" of Sci.Scoop where we evenj get a link to here. (Mar. 9, 2009) The editor there even added some italic emphasis on some of the more pertinent points.... Anyway, I'm printing it here, too.

Es/Ef - Mar.9, 2009

Why Einstein was right.

Why Einstein was right, when he was, in his "Relativity" theorizing, was often because hidden facets of mathematics and of the definitions used in physics allowed the ideas to be usable, even if his reasonings were fallacious. In other words, often, he was lucky.

Einstein's Special Relativity, which title for it I am told he disliked, accurately describes how information is changed as the relative velocity of a transmitter/receiver pair approaches the velocity of the carrier wave. When the carrier wave velocity is taken to be the limiting relative velocity of any two independently moving objects, the Special Relativity view leads to nonsensical conclusions.

"SR" accurately predicts that "Mass" will increase when a moving object reaches the "speed of light." However, the prediction is that the mass will go to "Infinity." There are some problems with that. Infinity." in practice, simply means that our measurement device, or our logic, fails at this point. There are two other factors, of which Einstein seems to have been unaware, which come into play here to allow his model to accurately predict a change in the situation at the speed of light.

 First, even mathematics does not allow empty space. There is always an implied "dot field" or "dot matrix," so it can be expected that what ever is considered to be moving would be moving within "something."

The other factor is also mathematical. When the process of "Integration " is carried out on the momentum equation, mass times velocity equals momentum, $(m \times v = P)$, this process can be carried out with either the mass or the velocity considered as the variable. If velocity varies, we get the usual kinetic energy equation, $KE=(mv^2)/2$. However, if mass be considered as the variable, with velocity constant, we obtain another energy equation, $(vm^2)/2$.   This latter does not appear in the literature and has apparently been ignored. These two equations may be interpreted as indicating that "velocity" can change to a limit, if it hits a limit, and the "accelerating situation" continues, then, "mass" will change.

Neither the "dot-matrix" aspect of mathematics, nor the alternative energy formulation, seems to appear in any of Einstein"s work, nor anywhere else in the readily available literature. If both the dot-matrix and the alternative energy expression that appear in the mathematics are reflected in reality, we see that any moving entity within the matrix will, itself, be a part of the matrix and the mass will be a measure of the balance of the interior of the moving entity and the remainder of the matrix. As long as the translational velocity of the moving object is small with respect to the average speed of motion in the matrix, there will be little effect of the "second" energy equation. Velocity will change and the amount of disturbance dissipated by the matrix, i.e., the "Energy" will be measured quite accurately by the "Kinetic Energy" formula. There is a change in the situation when the translational velocity of the moving object starts to equal or to exceed the average of the matrix. (In the "Universe" In which we exist, this average is known as "c," a "constant of nature" which is the "speed of light in a vacuum.") At this point there will be a significant change in the balance between the rest of the matrix and the part of the matrix within the surface of the moving entity, this surface will become changed in size and shape. The motion disturbance is no longer primarily dissipated into the matrix at large, but becomes localized at the surface of the involved entity. The balance changes, the "mass" increases. Special Relativity turns out to have been at least partially correct. The situation changes at the speed of light.

A third factor hidden in the interface between mathematics and reality apparently allows Einsteinian Space-time modelling to give quite accurate predictions. This is in the definition of "Time." In the "Space-Time World" of conventional thinking, "Time" is a reality which somehow came into being at the beginning of "Existence." Actually, time is a convenient method of keeping track of motion in sequence by a measured interval. The hidden factor that apparently makes "Space-Time" modelling work is that "Time" is always referenced to some reproducible cycle in nature. Therefore, "Time" has hidden within it not only the idea of motion, but also the idea of cyclic motion. A unit of time, a second, for instance, can therefore represent a cyclic motion, or motion in a circle, and the expression "sec^2," could stand for the motion content in the volume of a sphere! This insight leads to some interesting interpretations of some of the equations of physics, which, unfortunately, are beyond the scope of the main thrust of this paper.

The "hidden factor in Time." however, makes Space-Time Modelling and Motion in a Matrix Modelling reach much the same conclusions and, at this point, they seem to be essentially equivalent approaches. The Motion in a Matrix users being, perhaps, more cognizant of why their ideas have validity. This "circle/sphere" aspect of time hints at the idea of a spherical oscillator as a basic entity. This latter idea, has become the basis of a variant of Motion in a Matrix Modelling which could be called the "Oscillator Substance Model."

The "Oscillator Substance" model, although developed independently, echoes Max Planck's ideas of dots controlled by oscillators and is the discovery of a chemist who was once trained as an electronic technician.\, who has, within the last year, put together insights from both fields to suggest that there is a very simple model of everything which perhaps would have been close to the "Unified Field Theory" which Einstein spent his life trying to develop but could not. We will come back to this later. First, however, since, we are focused on where Einstein was right, or wrong, we should note the errors that probably doomed Einstein's quest for a "Unified Field" from the start.

It is highly probable that the "First Fatal Error" in Einstein's Unified Field attempt was that he "Threw out the field." That is, he assumed a void, a nothingness, for the "Field" to operate in. The "Second Fatal Error" is one that modern theorists continue to make. They try to set up a unified field theory from the "Four Fundamental Forces." The problem is that the "Four Fundamental Forces" all violate the Law of Forces, "For each and every force there is an equal and opposite force." This law of forces, if examined carefully, can be interpreted to clearly indicate that any "Force" is simply a readjustment of pressures within a "substance." The "Four Fundamental Forces" are either, in two cases (Gravitation and Electro-Magnetism) descriptions of observed phenomena, which are the result of other factors, These kind of "Forces" are known as "Fictional Forces," the best known, and best explained of which is "Centrifugal Force." The other two "Fundamental Forces," the "Strong and Weak Nuclear Forces" appear to be simply imaginative explanations which arise as justification for the idea that neutrons exist, as such, in atomic nuclei. A much simpler explanation appears if one considers atomic nuclei to be electron-proton aggregates in which a neutron has, at the best, potential existence. Einstein's frustration was apparently caused by his operating on sets of erroneous ideas.

In the next couple of paragraphs will appear a statement, which Einstein could have published, a logical outgrowth of Planck's ideas, Which might have kept us from almost a century of what this writer considers "semi-mystical nonsense" in scientific theory.

"Let us assume as a working hypothesis that there exists a 'dot matrix' of separable oscillators. These. in turn are collected to form a 'substance' at its 'triple-point,' of larger separable oscillator entities capable of correlating motion and of separation into the electron-anti-electron set and distortion into neutrona." Amplifications of the ideas in this statement, and corollaries, can give explanations for electrons, protons, the expanding universe, "The Big Bang" and almost any other phenomenon to which it has been applied. It is this working hypothesis which has led to the explanations of "Mass" and "Energy" which have been used throughout this paper.

Why was Einstein right? When he was, it was almost as much luck as brilliance. The hidden implications of the mathematics which matched reality made the theorizing seem to fit even when the basic ideas erred. Had he gone in a different direction, following up the ideas of Max Planck, he might have reached the same basic conclusions as the "Oscillator Substance Model" which has arrived nearly a century late, and will probably be ignored.

Post Script: Although this writer finds the Oscillator Substance approach so natural and useful as to feel that it should be common knowledge, the discovery of this possible "Explanation of Everything" is only a few months old. The ideas may wait some years longer for confirmation or disproof as what information published about it, thus far, is almost exclusively on Helium.com. and the discoverer has no professional standing as a member of any research institution or group. It is sad that a young, patent clerk could not have reached these ideas in the early 1900's rather than a very elderly janitor being the discoverer a century later....

Now, if a certain "String Theorist" were to have happened to have come up with this... Oh, well, we can't have a perfect world!

Einstein was very right to try to explain the workings of reality, it's sad that he made a few key mistakes.

Gmail Calendar Documents Photos Reader Sites Web more

Sign in

Google groups

« Groups Home

Oscillator/Substance Theory

Search this group   Search Groups

View this page "Why Einstein was right."   Options

This discussion is about page why-einstein-was-right

3 messages - Collapse all

why-einstein-was-right was created by ESKI

ESKI View profile   (1 user) More options Mar 9 2009, 1:05 pm

HI, everybody, and thanks for the heads up on various other things!
Anyway, this is the latest try to get us some more attention via
SciSdoop. Look it up there, OK/ Thanks, ESKI

Click on http://groups.google.com/group/oscillatorsubstance-theory/web/why-ein...
- or copy & paste it into your browser's address bar if that doesn't
work.

ka-sala View profile   More options Mar 22 2009, 6:30 am

***Spelling corrections made. Appologies for 3rd, try!
But vital that the word 'strings' is NOT stings!

Hello ESKI,
It's me again... Sorry for the epistle! But just to say...

In relation to Einstein and the 'dot theory' plus 'infinity' these
terms are simply used as the only example equation which make some
sense of a greater mass than has even entered into any earth
vocabulary yet. So therefore are limited to explain. But still truth
between all, if you can read between the lines.

I believe in my 'Double the Speed of Light' mentioned, 'the Vacuum'
and
the 'Energy' drawn from all around, we may not see with the naked
eye,
so it might 'appear' empty, or even a bit fallacious to the
uninitiated. Yet
is FULL of the energies which can be utilized. One could call this
mass energy the dot matrix. Whatever terms are used they are only
words
to try and explain such Forces all tied up together.

It is only natural with the Universe still being an unknown quantity
of Space Time Dimension that the term Infinite must be used, for that
is where we all stand within this matter. The fact that the Mass
Energy's Essence is still Light - electromagnetic or otherwise - dots
or waves, black or white, it is all a part of the One. The One of
everything which could not be dissected to many, if it was not all
linked in the first place.

We who have this interest in our different scientific approaches, have
all needed people like those gone before. Some need to feed off,
approve or disapprove, and all add our bit to the extensions made.
Einstein was in truth like many, who never intended so much to become
what it has, and yet opened doors for others who never thought to walk
through; to find out for themselves what he believed 'to be'.

Whether Einstein or Plank or anybody else, theories which remain
theories until proven as even supposed to be facts, still will have
more seemingly hidden - or rather unknowns - because the Matrix of all
things is not a rigid Science but as mailable as pure gold. The
Essence of Life which even Radioactivity can tell you it has a life-
span, is still born out of Light. We are only now holding the puppet
'strings' (string theory,) of this and not yet seeing the fullness in
the Oneness
of it's Velocity in which it moves through Time. While we - the specks
-
are still trying to link the dots, in the vacume of our lack of
knowledge,
of how.

Best I stop here, or this should be posted as another 'discussion'
which hasn't even begun, rather than as a reply to you. Don't give up
when you know within yourself, what others are still learning. Nor
before you have the conclusion to pass on. This O/S Site is
Oscillating with all you need to get out of you, and congratulations
with making it so far with SciScoop. Sometimes I have to mark time
with time rather than blow too many brains... yet! Being burnt at the
stakes once was enough... I have to Live this time!

Regards for now,
ka-sala

On Mar 10, 5:05 am, ESKI <deanlsincl...@gmail.com> wrote:

- Show quoted text -

Hugh V View profile   More options Apr 30 2009, 5:40 pm

I happened into the new version of SciScoop yesterday an happened upon
a bit by some person who goes by Barak who said some rather negative--
to say the least--about Doc's post re "Why Einstein Was Right. " I
couldn't resist answering part of his little rant.... (I suspect that
Barak may be perhaps a college Sophomore who's read a physics book
once and thinks he knows every thing.... Anyway, I'm tacking my answer
on here. Another view, and a contrast, but certainly not conflict to
what Ka-Sala had to say.
Yeah, I'm still around, come out of my shell once in a while....

Answer to Barak

To say that Dr. S's Energy Expression, $(vm^2)/2$ is not a type of
energy expression because the units do not seem to match, is probably
much like saying, "That is not a leopard, I cannot see any spots." It
may also fall into the category of saying, "A bonobo is a Chimpanzee.
take a good look at one. " The spots of melanistic leopards can be
seen only if light happens to strike them at the right angles and
there is a good deal of DNA evidence that either Bonobos (also known
as the "Pygmy Chimpanzee of Zaire) should be classified as in Genus
Homo, or humans should be classified in Genus Pan. In mathematics and
physics, as well as in biology, appearances and labels may be
deceiving.

Doc's contention is that both $(mv^2)/2$ and $(vm^2)/2$ are "energy
expressions" since both arise from legitimate, alternate ways of
integrating the momentum equation, $mv=p=vm$. He also notes in one
paper, that one can also integrate "p" as an entirety to obtain the
expression, $(p^2)/2$. which when we insert the original "mv=p=vm"
fact, gives us $(m^2 \times v^2)/2$. Here in lies the key to the reason
that the two energy expressions differ in form. In the process of
integration, one of the variables was "held constant," that is taken
out of real consideration except as a "scalar number" in the process.
However, it is still a variable, a vector quantity which should be
included in the result. The true descriptor for both energy
expression should be the one found when "p" was integrated as a unit
and the definition of "p" reinserted. In the cgs system the true units
of what is going on when we talk about the motion energy of a unit
would be $(g^2 cm^2)/2$ or mass times the "conventional energy

expression" or velocity times "Doc's expression." Looked at either way the concepts of mass, energy and velocity are inter-tangled.

As the energy, or motion content, of a moving body in a medium of which it is part--assuming here the the Oscillator/Substance Model is probably valid--will consist of two parts, the vibrational-rotational motions within the unit, and the motion disturbances caused by the motion of the entire body along a line, there may well be two "Energy Expressions" of the same form but actually arising from different views of momentum because of the two differing types of motions involved. If we assume this to be true, the total motion content i.e. energy would be $2(m^2 \times v^2)/2$, which equals $(m^2 \times v^2)$. Evaluating this at the Speed of Light, whether it be a limit or an average would then give the expression, $m^2 \times c^2$ for the Energy content of a particle rather than the mc^2 value usually accepted.

Taking this even farther from the fact that the Absolute Values of variables can be interchanged when they are related by an equation of the type, $xy=K=yx$, one can postulate that the maximum motion content possible of a particle moving at the Speed of Light would be c^4 or about 8.1×10^{41} $(g^2 \times cm^2)/sec.^2$.

P.S. I think that I'll drop this in the "papers" also. hv

On Mar 22, 6:30 am, ka-sala <irrir...@gmail.com> wrote:

- Show quoted text -

End of messages

« **Back to Discussions**   « Newer topic Older topic »

Iso-sets, Radioactivity and CMNS.   ESEL~Mar. 13, 2009

One may define an Iso-set as the set of all structures that may be
formed from a given number of electrons and protons. In this type of
definition, "Iso-set-1,0" would be the electron, Iso-set-0,1, the
proton, the Iso-set-1,1 would have three characterized members, the
subset, {e, p} , the neutron, and the Hydrogen atom. Iso-set-2,1
would include, {e, e, p; '"eep" (a central aggregate), and Hydride
ion}. The "2,2" set includes the H:H molecule and the Deuterium Atom,
an unspecified, as yet, central aggregate, and all possible
combinations of previous sets.

As each set contains all possible combinations of smaller sets, this
becomes complex very rapidly and may seem so obvious as to be
trivial. However, one interesting observation arises. In any given
set there is a possible central aggregate of all the units of the
set, an "Iso-set Aggregate," an "Iso-A." which would be the
"Residence" of the highest "mass/energy" states of the set and would
have all of the electrons confined within, or very closely associated
to, the volume defined by the motions of the protons.

Any Iso-A set would have within it the potential to decompose to any
combination of units of smaller sets and would be thermodynamically
unstable with respect to such decomposition. Collapse to an electron
(the Iso-A of "set-1,0) and a cation) accounts for many radioactive
transformations, e.g. neutron [Iso-A of set-1,1 (?)], to a proton and
electron. All other Beta-particle radioactivity can be explained
similarly. An atom reaches an isomeric Iso-A state which does not
revert to the original atom but to a cation by loss of an electron. A
similar argument holds for the Alpha-emitters, except that the unit
left behind is a di-anion... The Iso-A explanation can be extended to
all nuclear transforms with appropriate modification of what is seen
to be happening. Positron emission would involve interaction with
Iso-set-0,0. The parent set considered pervasive which contains a
"zerotron" unit splitable by "pair-production" into a proton and
electron.

One notes that Iso-As of neutral units, atoms and molecules could be
considered as "poly-neutrons" a term that has appeared in the
literature. If it be taken that this means, "units that may be
considered as 'polymers of neutrons' within which the neutrons have no
individual identity' " the writer has little objection to the
nomenclature. The Iso-A concept simply includes these and any other
central aggregate units, charged or uncharged which can be
considered. Iso-As of huge sets may well be what are known an
"neutron stars" and Quasars which are probably the "Isotope Factories
of the Universe."

An Iso-A intermediate formed from the Deuterium molecule which, under
certain circumstances, transforms to another version which rewould
account for the most commonly observed CMNS transformation. CMNS is
short for "Condensed Matter Nuclear Science which comvers the area
once known as "Cold Fusion" and other phenomena in the border land
between "nuclear science" and molecular and atomic chemistry.

Iso-A intermediates might be interesting to consider in much of CMNS
work. As noted above, they would be considered as the ultimate
"Condensed Matter" of any given set of electrons and protons.

Gmail Calendar Documents Photos Reader Web more ˮ   deanlsinclair@gmail.com ˮ

Google groups

« Groups Home

Oscillator/Substance
⊞ # Theory

[Search this group] [Search Groups]

Home

Welcome Major Ray   Options

Discussions
+ new post

1 message - Collapse all - Report discussion as spam

Members

ka-sala View profile   More options Mar 25 2009, 9:45 pm

"...the atom of truth could only win."

About this group

Edit my membership

Welcome Major Ray.

Group settings

Would love to see a discussion here from you. You have a lot up your
sleeve and you are far from alone in your beliefs.

Management tasks

Invite members

I had a scan of your BIOS and as I have been a fly on the wall of the
Academy of Future Science since the 70's - I prefer ro remain out of
the lime-light of so much - but was asked by ESKI, the Oscillation/
Substance Theory, to join.

View this group in the
new Google Groups

Anyway...
This is just a welcome hello, and would like to see anything you can
post in this Group?

I cannot boast of scientific jargon/terms, yet I do know what I am
saying! Though most seems to come through in a mix of terms in
English, it doesn't change the facts.

Best regards,
ka-sala

Reply to author   Forward   Report spam

End of messages

« Back to Discussions   « Newer topic Older topic »

OLD DATA/NEW MODEL

Reversing the century old interpretation of the Michelson-Morley Experiment and renaming Planck's Constant from a "Constant of Action" to the equivalent, "Constant of Angular Momentum, " leads to a new model of existence.

The Michelson-Morley Experiment showed that the Speed of Light. "c." was a Constant (as nearly as could be determined) of Nature.

The interpretation that was given to this--Albert Einstein seems to have been a major proponent of the view--was that this fact proved that the "Aether." the all-pervasive something thought to be everywhere, does not exist-- or at least could be ignored as an idea. It was sufficient to consider that light waves simply pass through empty space at the fastest possible speed at which anything can move. This view seems to be accepted by the scientific community to this day.

An alternative, almost totally reversed, interpretation goes something like this:
  The Michelson-Morley Experiment defined the maximum speed at which information is carried in the "Aether," whatever it may be. It, also, showed that the "Aether" can act as if it were a solid, carrying the transverse wave disturbances which we understand as "light."

 Never considered at that time, and, apparently not in theorizing to this day is the fact that the maximum speed of information transfer from a point can be shown to be the average speed in any direction of the motion of the information carriers, whether these carriers be Pony Express Riders or Electromagnetic Waves. Considering this, "c," the Speed of Light, would be the average velocity, measured in any direction, of the motions pertinent to the carrying of information. It is not a maximum limit of velocity.

 In that era of theory development, Max Planck, a genius contemporary of Einstein worked out his famous relationship between the frequency of light waves and "Energy." The result is well known, the equation, "Energy equals Planck's Constant times the frequency. That is "E=hu." E can be expressed in "Ergs" with, "h," the constant, in Erg-seconds, and u, "Nu." in cycles per second. Erg-seconds is called "Action" and Planck called his constant the "Constant of Action." Although action has the same dimensions as "Angular Momentum," this fact was ignored. Had Planck called his constant the "Angular Momentum Constant--" which would have emphasized a rotatory aspect to Energy and Wave Motion-- scientific theory might well have developed in a totally different direction.

If we take Planck's Constant as being an Angular Momentum Constant of Nature and equate it to the definition of angular momentum, "Mass times radius of the rotor times the velocity of the rotor measured at the outer edge ("tangentially").
we obtain the equation, "h=mrv." Snce this constant applies for the movement of Information/Energy at the speed of light, "c," it makes sense to evaluate this at the velocity, "c." Doing this produces the equation, "h=mvc," which can be rearranged to "mv=h/c." This valuable little equation defines "h/c" as a "Torque Constant of Nature." What more it discloses will be explained shortly. Applying a simple mathematical principle to the above leads to a new Model of Existence.

 Any equation in two unknowns of the form, "xy=K," two unknowns multiplied together equals a constant," can be rewritten in the form, AxBy=K=BxAy. That is the values of the "coefficients" of the units of the variables can be interchanged to give a valid statement. In this form, this little mathematical equation has much use in physics, It is the Law of Levers, the Law of Balance, the Law of Forces, and can be used as a defining statement for the Limits of an Oscillator,,,, Noting this use in Oscillator Definition, we see that the equation, "mr=h/c," is an equation of the type discussed above; and, therefore, can be considered to be the defining equation for a family set of constant torque oscillators, {mr=h/c}, with a torque of "h/c" and an average "size" at the value wherein the 'Absolute Values" of mass and radius are equal. That is, when m = r= (h/c)^0.5..[Note: 4.7 x 10^-19 is the value, to two significant figures, found when (h/c)^0.5 is evaluated in "cgs" units. }

By re-interpreting the Michelson-Morley Results and taking an alternative view of Planck's Constant, we have come to an alternative model of existence as within a something, "Aether," consisting of rotating units, which are--or are organized into--oscillators of the family set, {mr=h/c}. We have found an average value for these oscillators and, have found a mathematically defined "dimension," previously unreported, which would apply to all of these oscillators, this would be the "dimension" of masses greater than 4.7 x 10^-19 g. and radii less than 4.7 x 10^-19 cm.

One major question remains about this as a model. How can the "Aether" act as a solid? A look at chemical theory answers this question. If the "Aether" be considered to be a substance at its triple point-- where it can be solid, liquid or gas--this problem disappears. We have a model, therefore, which may be called and "Oscillator/Substance Model."

There is an open membership web site, http://groups.google.com/group/oscillatorsubstance-theory set up to explore, elaborate, and/or refute the implications of this model.
Follow-ups on this model give some potentially quite useful and informative views on a number of topics. Here are some examples:

Law of Forces (Equal and Opposite)--Balanced pressures. Also, mathematically, as above, AxBy=K=AyBx.

Force of Gravity--Not a true force, an observational fact explained by differential pressures, even if it can be, and is, handled mathematically as an "attraction."

Mass--Balanced pressures/tensions at a surface.

Negative and Positive Charges--Result of counter-clockwise and clockwise spins of "stable" vortex oscillators/oscillations in the "Substance." usually the electron or proton.

Electron--Counter-clockwise-spinning, inverting-vortex oscillator/oscillation having oscillation limits of 9.11 x 10^-28 g. at 2.43 x 10^-10 cm. and 2.43x10^-10 g/ at 9.11x10^-28 cm.

Proton--Clockwise-spinning, inverting-vortex oscillator/oscillation having oscillation limits of 1.67x 10^-24g. at 1.32x10^-13 cm.
and 1.32x10^-13g. at 1.67x10^-24 cm.

Electron-anti-electron annihilation--Combination of these two vortex oscillators of opposite rotation but otherwise identical characteristics into a spherical, pulsating oscillator, having the same mass and size limits as either of the "halves." (This was dubbed by this writer, "zerotron." See article, "Negatron plus Positron Equals "Zerotron?")

Pair-production--Reversal of above process....

The "Big Bang--" The instant of inversion--or splitting--of an oscillator creating a Universe/Anti-Universe Set. Neutrons being created on "our" side, which collapse to electrons and protons....

"Zerotron--" A hypothetical spherical oscillator, probably ubiquitous to the "Substance," which is splittable to electron-anti-electron pair and deformable to the neutron and/or anti-neutron....

The above are some of the definitions/explanations which develop easily from this model, some of which do not seem to arise from any other model. It is hoped that despite its proletarian birth from re-interpreting old data from a different view, people in the scientific community will give it a chance to prove its worth--or lack there of--without simply dismissing it "out of hand."

This little article is a simple version of the origin of the Oscillator/Substance Model, written to be submitted to SciScoop as sort of a prequil to some of the other article which I have published there.

The Oscillators

Reanalysis of the possible significance of the speed of light and of its relationship to "Energy" through Planck's constant has led to the idea that it is possible that all of existence is within a "substance" which consists of oscillators of a family defined by the equation, $m \times r = h/c$. These would have a constant torque of h/c and an "average size" of $(h/c)^{0.5}$.

Further consideration of how the electron, proton and neutron and neutrinos would fit into this pattern has led to additional conclusions, If the model is valid there should be neutral entities, which for convenience have been dubbed the "zerotron" and "zeroino" as respectively, "neutral" oscillator units which can be divided, respectively, into the electron/anti-electron pair, and into the neutrino/anti-neutrino pair. The zerotron, also, would be distortable into the neutron, which decays, in our universe, to a proton and an electron. By analogy, one can suspect that a "zeroino" entity would be distortable to a neutron-like entity, hence there may be three types of "neutrinos" corresponding to the electron anti-electron and proton.

The differentiation between particles and their anti-particles appears to be direction of rotation, with the electron and ant-electron this appears to be a clock-wise, counter-clockwise situation. It is said that the neutrino-anti-neutrino situation is one of spin in the direction of travel, the other a spin the the reverse direction of forward motion.

The charge state of the electron and proton can be related to their physical spin. Spin in a clockwise direction related to what we call a "negative charge, " and spin in a counter-clockwise direction related to a positive charge, or vice versa. since an electron forms a bubble chamber track in a counter-clockwise fashion, it is probable that the electron's inherent spin is actually counter-clockwise .

The apparently defining equation for the oscillators of our universe, as noted above, appears to be $m \times r = h/c = r \times m$. That is the absolute values of radius and maw are interchangeable to define the limits. One might also say that they are interchangeable to define the instantaneous "positions" of the parts of the oscillator as the two components of the oscillator will apparently always balance across a line (through a circle ?) in the case of the vortex-form oscillators or against a sphere in the case of the neutral oscillators. The vortex oscillators can be envisioned as consisting of two coupled parts operating 180 degrees out of phase so that the maximum value of one corresponds to the minimum value of the other and each of these corresponds to the value
$(h/c)^{0.5}$ One of the oscillators operating in the region of radius greater and mass lesser than this value, the other operating in the converse region of mass greater than the above value and the radius smaller. The greater volume, lesser mass region appears to be the region in which our scientists can measure "rest mass" and "radius." The radius being known in the literature of science as the "Compton Wave Length Value."

As the neutral oscillators have to be considered hypothetical at this moment, with the exception of the neutron, one can only make a hypothesis as to their motions. The most logical hypothesis is that they do an internal-external motion of pulsation against an internal sphere of $(h/c)^{0.5}$. The effect being of an inversion at that value. A third, neutral form can be envisioned as two counter-rotating halves. This type would be expected to be easily split into the vortex ("charged") forms.

[It may be noted that, atomic species would be considered as neutral combinations of oscillators, and, perhaps the Hydrogen I atom could be profitably analyzed as a neutral oscillator or, perhaps more accurately, as a combination of the two oscillators, the electron and proton. The neutron would be possibly analyzable as such. also}

It may also be postulated that the common form of oscillator may well be a "zerotron" which is made up in turn of "zeroinos" and containing at its center a "controlling" zeroino oscillator. All of this is hypothesis. It is possible that someone could do computer modelling on a similar type of structure with a simpler oscillator structure. That is, were one a good enough programmer it would seem possible to develop a computer program which would mimic our oscillator-substance models of the electron, proton neutron and "zero-tron" with simpler units and structures. Perhaps the oscillators could fit the equation, $yz = K = zy$, with initial y and z equal to 4 and 1 and K equal to 4 so that the point of reflection would be $y=z=2$. Whot do you think ? Is this feasible?

Gmail Calendar Documents Photos Reader Sites Web more Sign in

Google groups

« Groups Home

Oscillator/Substance Theory

Search this group | Search Groups

Home

Discussions
+ new post

Fw: I wish you enough   Options

1 message - Collapse all

Robert Vanderhoek View profile   (1 user) More options Apr 14 2009, 11:22 am

- Show quoted text -

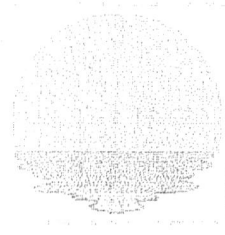

image_gif_part
8K Download

End of messages

« Back to Discussions   « Newer topic   Older topic »

About this group
Join this group

View this group in the new Google Groups

Create a group - Google Groups - Google Home - Terms of Service - Privacy Policy
©2011 Google

We can analyze the proton in the same way to obtain the values:
Our observations--minimal mass, maximal radius--1.67264 x 10^-24 g. and 1.321401 x10^-13 cm. ;
balanced by
the other limit of maximal mass, minimal radius-- 1.321401 x 10^-13 g. and 1.67264 x 10^-24 cm.
It is to be noted that the electron is both heavier and lighter than the proton and larger and smaller,
depending upon where the observation be taken. At the average values they are the same, as are all
oscillators of this family. The logic is, as folows:

The Electron and Proton as Oscillators

If it be taken that the relationship between energy and electromagnetic radiation be a fundamental
relationship of our universe, then examination of that relationship should furnish clues as to the nature of
our universe.

If Planck's Constant, the constant which relates energy to electromagnetic radiation be equated to its
definition as an angular momentum, one obtains the equation, m x r x v = h. Evaluating this at "c," the
speed of light, we obtain, mrc=h which can be rearranged to mr=h/c.

As any equation of the form xy=K can be taken to describe an oscillator, by writing it in the form, xy=K=yx,
to emphasize the interchangeability of the values of the two variables, we can see that the equation,
mr=h/c, can be taken to define a family of ocillators of constant torque, h/c.

When the electron is checked to see if it fits into this family, it is found to fit with one limit set with the "rest
mass" as the mass, "m", and the "Compton wavelength" as the radius,"r." For an oscillator the absolute
values can be switched to determine what the other limit is. For this particular oscillator, the other
oscillatory limit would be absolute value of Compton Wavelength as mass and the absolute value of the
"rest mass" as the value of the radius, "r."

 The most consistent set of values seems to be cgs, so the following are what is obtained as the vibratory
limts of the electron:
Noted in our Universe as "Rest Mass" and "Compton Wave Length" are, (Six significant figures)
9.10953 x 10^-28 g. for the mass, correlated to 2.42631 x 10^-10 cm.
The other oscillatory limit would be 2.43631 x 10^-10 g. correlated to   9.10953 x 10^-28 cm.

What we have discovered, if the mathematics above accurately reflect the "real world," is that what we
are observing is a sitution of minimal mass and maximal size of the electron which is "balanced" in what
we might call something like an alternate reality, by a maximimal mass and minimal size. We have also
found an average size for all oscillators of this family of (h/c)^0.5. This is approximately 4.7 x 10^-19 cm.
radius and 4.7 x 10^-19 grams.

As there is recent literature observation that the electron, and, presumably also, the proton, turn 720
degrees to return to their original states, it seems sensible to consider these entities as "vortexes" which
invert once per rotation, hence 720 degrees to return to the original orientation.

It can be seen that the application of the mathematical concept that any set of unknowns, xyz...n = K =
n...zyx can be said to define an oscillator situation in as many dimensions as one wishes to use, can have
interesting implications. In this case the implication is that we possibly exist in a universe of a substance
made up of oscillators of the constant torque family, h/c

Since this page was first writtten, the weirter has been considering the probable internal structure of the
proton and electron. It is very likely that both can be considered as combination oscillators with one
oscillator operating between minimal size /maximal mass and the 4.7 x 10^-19 limits and the other
operating between the maximal size/minimal mass limits and the "4.7;" inversion.

It also can be noted that both the electron and proton would show "nodes" at the oscillation limits and at the central inversion "equator." it seems possible that the Quarks, considered fundamental particles in Standard Model are observational phenomena due to this quality of these two particles, and the interaction of the inteernal structure of electrons with that of the proton and other "particles."

An internet, group has been set up to explore, amplify, elucidate, or refute the implications of the above ideas. Its "URL" is http://groups.google.com/group/oscillatorsubstance-theory.

Gmail Calendar Documents Photos Reader Sites Web more   Sign in

Google groups

« Groups Home

Oscillator/Substance Theory

Search this group | Search Groups

View this page "M.Margulis' Penetrating Ether ...."   Options

2 messages - Collapse all

ESKI View profile   More options May 13 2009, 3:11 pm

This theory has some elements in common with our ideas. I thought some
of you might like to compare....

Click on http://groups.google.com/group/oscillatorsubstance-theory/web/m-margu...
- or copy & paste it into your browser's address bar if that doesn't
work.

ESKI View profile   More options May 14 2009, 3:44 pm

Re: Dr. Margulis' Penetrating Ether Theory.

Scanning this material more closely , it appears that we have a far
simpler and more comprehensive view. The point in common seems to be
the acceptance of a pervasive substance. However, Margulis does not
characterize it as a substance; hence, misses most of the results of
that acceptance. Additionally, Dr. Margulis attempts to preserve a
good deal of material which we would find superflous,

The description of the electron as a Mobius band is interesting. the
motion of a given particle in a simple oscillator would follow what
might be called a form of Mobius Strip and the motions of electrons
(and protons ?} in their interactions would follow patterns that may
be related to the Mobius Strip concept and to the corresponding
Bottle, the name of which I have forgotten.

If one makes a big Mobius Strip and then starts slicing it down the
midline one gets and interesting situation. The first slice gives a
twice as long, twice as twisted strip. Slicing that one, however,
gives two interpenetrating strips. Slice each of these gives 4
interpenetating strilps. That is each one slice of the provious unit
produces another whole "one" unit as the resulting units all
interpenetrate, but the complexity encreases by some large factor each
time.....This observation may have some pertinence to atomic stucture.

It is interesting to see what others are doing. As you can tell, Dr.
Margulis just posted this material. I am impressed by the amount of
work and thiought,that has undoubtedly gone into the production of
this, but, unfortunately, I don't see the result as being of a lot of
use...Eski .
.....

On May 13, 3:11 pm, ESKI <deanlsincl...@gmail.com> wrote:

- Hide quoted text -

> This theory has some elements in common with our ideas. I thought some
> of you might like to compare....

> Click on http://groups.google.com/group/oscillatorsubstance-theory/web/m-margu...
> - or copy & paste it into your browser's address bar if that doesn't
> work. *Link did not work. Page was apparently erased. D.S.*

ON THE MATTER OF ANTI-MATTER, What is hidden where?

The writer, noting a "Hidden-half-of-existence" below a radius of 4.7 x 10^-19 cm., and the possible creation of alternate universes resolves the 'lost anti-matter problem of our universe " to his own satisfaction.

Re: Matter and Anti-matter.

Every one has heard of Matter and Anti-matter. Matter is made up of electrons and protons and anti-matter is made up of anti-electrons and anti-protons. When anti-matter and matter come together to meet they mutually annihilate to form pure energy. That is the commonly accepted picture. The reality may be not quite that simple, and, possibly, more interesting.

There is a distinction in science between something which is dubbed "Matter" and "Anti-matter." Matter is made up--for the most part-- of electrons and protons while anti-matter would consist of the corresponding "anti-particles." There seems to be a problem, however, as to whether anyone has actually observed an anti-matter atom or molecule. The writer does not personally know of any reports of anti-Hydrogen or anti-Helium, etc. There is a rather extensive writeup available on the Internet of a theory called "Dominion Cosmology" which does a quite logical development of a Cosmology based on the premise that matter and anti-matter have a property of mutual repulsion. There is, also, an interesting 2006 news release that the "Beta sub 2" particle vibrates (oscillates) between matter and anti-matter.

Let us look at the whole problem of "The Lost Anti-Matter of Our Universe." from the viewpoint of the Oscillator/Substance Model.

[For those who are unfamiliar with the O/S model, the following short introduction is given. O/S Model postulates existence as within a substance at/near its triple point, which is made up of (or organized into) oscillators /oscillations defined by the family set, $\{m \times r = h/c = r \times m\}$. That is, they--the organizational units--have a constant torque of h/c, Planck's Constant divided by the speed of light, and since the absolute values of m and r are interchangeable, and at an average value would be equal to each other and to

$(h/c)^{0.5}$ all of these oscillators can be considered to invert through values of about 4.7 x 10^-19 g. at 4.7 x 10^-19 cm.]

If the "O/S" Model be valid, the problem of Matter/Anti-matter, is tied closely to the presence of oscillators as the organisational units of reality.

Oscillators/oscillations within a substance can be classified into three general categories. (At least, that is the way this writer classifies them.) All three categories will appear as disturbances of a spherical form, but their "motion senses" will be different.
The simplest category, which we shall call a Type 1 Oscillator would be a true sphere, inverting through an inner sphere-- in our Model, the inner sphere would be as defined above--from an outer limit of maximal size and minimal measurable "mass," and an inner limit, also a sphere of minimal size and maximum "mass."

[Size is easy to understand. Mass, however. does not ever seem to have been given any true definition. As used here it will mean the pressure/tension between an entity and the rest of the "Substance of Existence" as measured at a point on the boundary between the entity and the "Rest." The more the concentration of motion within a space, the larger will be the measured mass. For a Type I Oscillator, indeed for any of the oscillators, the smallest mass would be corresponding to the largest size, and these would be the values which will be found in our "Reality," if they can be measured. It appears that most Type I Oscillators may go totally undetected.]

A Type 2 Oscillator could be called a "Toroidal Pseudo-Sphere." The total space occupied would be spherical, but the sphere would have, at any given instant, an axis of rotation and an equator. This type would invert through a circle rather than a sphere, and can, with sufficient motion content added to it develop counter rotating halves and, eventually, spit into two Type 3 oscillators

'

Type 3 oscillators would be always be formed in pairs having opposite senses of rotation/pulsation (inversion). The best known of these sets are the negatron-positron pair, also known as the electron and anti-electron. Type three oscillators because of their effects in the Substance being opposites are known as "charged particles." Here is where the matter/ anti-matter concept arise. A unit of once rotation/inversion pattern is the negative charge, the reversed pattern is the positive charge. If the two patterns are of the same oscillation limits, they can rejoin with loss of energy in what is known as "annihilation." If, however, the oscillation limits are different, the two units may associate in may ways, but, do not rejoin to form a Class I or Class II oscillator/oscillation. In our universe. we have an observed unit, the neutron, which splits in to "halves" which are not identical, these halves, the proton and electron can associate to form may things, but do not rejoin to the neutron, nor can they rejoin to what one may call a "zerotron," a unit which is can be postulated as a Class I oscillator which can be split to an electron/anti-electron pair or deformed into a neutron which then splits into an electron and a proton.

The proton and electron are considered as matter particles as they have apparently indefinite lifetimes in our Universe. The reversed particles would be anti-matter, having indefinite lifetime in an alternate universe having some reversed sense. In O/S thinking, our universes has one rotation/pulsation orientation which is compatible with the rotation/pulsation orientations of the electron and proton. At the instant of inversion, or splitting, wherein our Universe emerged there would have been a complementary Anti-Universe also emerge. This can be used as one explanation for the lack of observation of "Anti-Matter" in our Universe. This, in the main is very possibly correct. However, there is another explanation. It may well be that the idea of Matter/Anti-matter is an over simplification, and that, in essence, the halves of nature exist in both universes.

It, also, may be noted that the idea that matter and anti-matter will annihilate on contact may be in error. If matter and antimatter more complex than the simplest opposites, were to be formed in our Universe there would be slight differences due to orientation to the rotational characteristics of our universe, the exact "orientation fit" necessary for the "Yin-Yang" rejoining necessary for "annihilation" might be difficult. It is known that the positron-negatron pair exists for a period of time as a "Hydrogen type molecule" before combining with loss of energy.

Incidentally, it may also be noted that the usual statement that there are "two photons emitted at 180 degrees" seems a misapprenhension to this writer. A point radiator will radiate one pulse in a spherical pattern. In the case of the "annihilation" recombination, what would be emitted would be one pulse, "photon" at 360 degrees. It could be detected as "two photons at 180 degrees," but this would actually be only one "energy" pulse.

Eski - Jun 5, 2009

UNIVERSES WITHIN UNIVERSES

The discovery that evaluating Planck's Constant as an angular momentum at the speed of light leads to the mathematical definition of a "constant toque universe of torque, "h/c" (Planck's constant divided by the speed of light) leads to speculation about universes within universes.

The "Universe" defined by the set-equation, { m x r = h/c), mass times radius equals Planck's Constant divided by the speed of light, defines a "family of oscillators" having a constant torque of "h/c" and an average value of $(h/c)^{0.5}$. As a constant on nature divided by another is also a constant of nature. we can say that both of these values are "constants of nature.' So also would be $(h/c)^0$, $(h/c)^{+2}$, $(h/c)^{+1.5}$, etc.

What develops here is the problem that if, $(h/c)^{0.5}$ is the average value of an oscillator family defined by (h/c); then, (h/c) could be the average value of an oscillator family defined by $(h/c)^2$, and so on. That is, every possible power or root of the ratio could be taken to define another "oscillator family" and hence another "Dimension."

This problem of dimensions within dimensions seems somewhat related to two mysterious numbers of our number system. The numbers "zero," and "one." We say, "Zero is nothing." That, however, is not totally logical, by giving the "number," Zero, a name and a place, it has become something. It can be considered a starting point, a hole, even an entrance into another dimension. In another article dealing with signed numbers,* this writer has discussed the meaning of signed numbers as representing on a three dimensional chart, first a line in a given direction, then a square defined by two directions and third a cube, defined by three dimensions/directions. This writer did not go on to what would have to happen after that. Logically, one can pick a point at which to go on from some given three dimensional cube and move the new origin to that cube, then start the convention over. The first line-up would form a line of these cubes, next change would form a square of these cubes, the next a cube of these cubes and so-on. However, it may be noted that by moving our origin spot, our "zero" to a particular cube, we have made the "dots" in our next diagram be the 3-D cubes of the last diagram. Our new "Zero" is an origin dot, but it is also the cube of the last origin dot. We have taken, in a sense, 1 x l x1x0 and gotten 1^3, which is our new "Zero" but is certainly not "Nothing...." It, in fact, could be considered the "hole" from which are pulling out the "3-D" dots from our previous "dimension. "

Another little experiment with "Zero" and "one" also brings in the mysterious number, "Infinity," the "number beyond all numbers," which in practical terms is the next number beyond where we stop counting, for whatever reason. If one forms a "Mobius Strip" by twisting a strip of material one turn and fastening the ends together one obtains a "uni-dimensional infinity," that is, it is possible to go around and around, outside and inside, passing through the same points indefinitely. This strip is both "Zero" and "One" as it is our starting point, and it is one unit.

 Now if the one unit be cut by one knife or scissors, one time once around, passing once through the starting point of the cut, what is obtained is still one unit; which, however, is twice as long as the original and twice as twisted. We have run a number of "ones" together to still get "one" but it is no longer the same "one." Continuing by splitting this unit, one obtains, not as might be expected--a longer, more twisted continuous unit--but two intersecting units. As they intersect, they are still "one." Continuing the slitting process produces more and more of the intersecting units, still "one" as they are "attached" to one another by the intersections. We have been always operating on one unit at a time one operation at a time, but we have created many "ones." We reach, "infinity" when we no longer have enough material in each unit to split with our knife. Infinity begins at the point where our "instrumentation" fail.

Since we can set h/c as the basic unit, we could call it the "one" and let h/c=l to develop some sort of "absolute scale." The interesting thing a is that if we do this, since we consider all powers and roots of one as one, we will be defining all of our projected universes as "One."

Going over all of this, the writer feels that at this point, the one thing that should be done with this one article is to write one (.) (:>)

----

*Problems in Mathematics--Signs and Signed Numbers,"
www.sciscoop.com/story.2009/3/19/113736/171

Oscillator/Substance Theory

Deuterium molecular cation as a reaction intermediate. Dean L. Sinclair
http://docs.google.com/Doc?id=dcb2474d_244d6tf5d9x
http://groups.google.com/group/oscillatorsubstance-theory/deuterium-molecular-cation

Consideration of the Deuterium to Helium transformation leads to the idea that the Deuterium Molecular Cation (also a free radical) would disproportionate to Molecular Deuterium and a Deuterium Di-cation. This species would be a space isomer of the Alpha particle and would be expected to collapse into that form, which then, of course, converts into Helium 4.

Writing this out in simplified, inorganic chemistry form, for the usual observation by electrolysis, we obtain the following steps:
1. Ionization   DOD--> D+ + OD-

2. Electrolysis   D+ + e --> D.

3.Molecular cation generation
  D+ + D.--> DD.+   or
  D. + D. --> D:D ; D:D - e -->DD.+

4. Disproportionation of the molecular cation

  DD.+ + DD.+ --> D:D + DD++

5. Internal rearrangement of DD++ to "Alpha particle" form.

6. "Ultra-slow-Alpha-particle" form picks up two electrons to become He4.

The idea that the first step of oxidation of Deuterium could produce a molecular cation suggests that what is actually happening in the electrolysis cells is a Redox reaction of the electrolysis products which are sequestered such that the oxidized and reduced units are separated...That is, the molecular cation, once formed will have a high probability of encountering another of its own kind before further oxidation takes place.... This explanation would also account for the long induction time, a sufficient quantity of Deuterium and Oxygen or Deuterium Peroxide would have to build up in different parts of the electrodes before the reaction initiated.

(This would seem to almost create a pulsing situation. Electrolysis products would build up to a point of the situatiion "reversin"--well, almost--with "Deutero-water" being regenerated along with Helium... or perhaps some sort of situation will develop where in the amount of electricity input is just enough to supply the DD which is converted to He....)

This view, if valid, suggests that it might be possible to construct a fuel cell which would convert Deuterium to Helium and produce heat.

It also appears possible that a Hydrogen molecular cation could possibly convert to a Deuterium cation. Although somewhat more difficult to conceptualize, the same general ideas for using a fuel cell of Hydrogen and Oxygen to produce Deuterium and heat rather than water would seem possible.

It would seem that someone in the cmns group would have the facilities to check out the fuel cell idea...

All of the above are noted as simple inorganic reactions. The collapse of the DD++ unit to its spacial isomer, the "Alpha" arrangement, would release a good deal of thermal energy but there is no reason to think that there would be any released radiation.

NOTE: The ideas that spark this come from the little known Oscillator/Substance Theory. In O/S view, there is no need for there to be neutrons in a nucleus, additionally there will always be a pressure toward symmetry.... Putting these two ideas together, the nucleus of a Deuterium Atom would consist primarily of two paired protons--ala "para-Hydrogen" only closer spaced--The molecular form would contain two such sets. Looked at this way, an Alpha particle would be a "square planar spinning array" which would have an up, down, up, down arrangement of protons, which would be identical in arrangement--but smaller in size and far more symmetric--than what would be expected of a DD++ unit; hence a DD++ unit would be expected to collapse to the far more stable space-isomer.

 Second Note: As a once-upon-a-time, semi-expert on acid-base theory. I may have glossed over a very important point at step six above. The slow-Alpha would be a very strong Lewis Acid. In contact with Deuterium it would be expected to form a unit which would be a heheliumdeuteride di-cation intermediatewhich would be expected to collapse into He4 and the Deuterium di-cation which would collapse to an "ultra-slow Alpha structure and continue the chain.... !

This raises a question as to what would be the effect of am Alpha emitter in Deuterium gas. Could it initiate a chain leading to more He4 than would be expected from the Alpha particles alone?

Delete this page		Version:	Current by ESKI - Jul 1 2009	▼

Report this page
Reply to this discussion

| Post | |

Elements, Periodic Chart and O/S

If the Oscillator/Substance concept of what are electrons and protons be valid, and electrons and protons are actually spinning vortices, with the size of the protons and their interactions accounting for the "nuclei" of the nuclear atom, then it is necessary to do a complete re-examination of the concepts of atoms, elements and the basis of the Periodic Chart of the Elements. A new reason/rationale need be developed for the apparent presence of "isotopes" which are conventionally explained by the presence of neutrons in atomic nuclei. If there be not neutrons, per se, in the nuclei, then what are the electron-proton interactions which explain a situation which will make it appear that there are stable neutron type associations.

Conventional wisdom says that there be pairing of electrons such that their "spins" will cancel. O/S would agree that a partial cancellation of the vortex motions would take place if two vortices were "paired, upside down to one another." The conventional literature, however, does not continue to apply the same idea to protons, nor does it seem to realize that an up-down-up-down chain or circle can be extended. It should be useful to consider the first few known proton-electron associations.

The simplest units which may be considered as proton-electron associations are the neutron and the Hydrogen atom, H1. The neutron may possibly be, or, at least can be though of as, a tightly coupled dual oscillator of an electron and proton, "proto-electron/proto-proton,(?)" which when it "falls out of sync" collapses into an electron and a proton. These two can reunite into another unit which can synchronize in unlimited--or almost unlimited--spatial dimension, hence, cannot be "knocked out of sync" in the manner of its more restricted isomer, the neutron. This second unit is, of course, the H1 atom.

Perhaps the next easiest set of units to consider is what one might call "Iso-set--2,2," the set of units made up of two electrons and two protons. This would have also two members, the Deuterium Atom, "D," or, "H2," and the Hydrogen molecule, H:H. The Hydrogen molecule is known to have two forms known as "ortho-Hydrogen" and "para-Hydrogen" wherein the Hydrogen nuclei, ie, protons are "matched spin" or "paired-spins." that is the protons are moving in exactly the same orientation in one case and are "upside down to each other" in the more stable orientation. The Hydrogen molecule would be expected to be an ovoid. It is probable that the Deuterium atom could be considered as a condensed version of the "para-Hydrogen" molecule, containing much less vibrational energy. The close-coupling of the two protons in the nucleus and their probable containment, at any given instant within an electron, giving the illusion of a "nucleus"L consisting of a proton and a neutron. It can be considered that the vibrational motions of the electron and the protons would be such that for very short periods, the proton/electron interactions would be such as to be identical to those of a neutron, however, the neutron as such would not exist....

The three electron, three proton set, "Iso-3. 3-set." has two well-known, atomic members, Tritium, "H3," and Helium 3, "He3." Tritium is "radioactive," spinning off an electron to form a cation which when it regains an electron becomes the isomeric He3. Tritium does not have a magnetic moment; therefore, it is not a "spinning neutral unit." This implies an internal symmetry, if the nucleus be composed of an inverting tetrahedral array, that is, three units inverting as if they were at corners of a tetrahedron, then this unit would on the average have no polarization, do "dipole" even though the internal units have inherent dipoles. It appears from the chemistry of this unit that, at any given instant the central array, the "atomic nucleus" is encased within a set of two coupled electrons with a third electron more loosely coupled.

The "stable" He3 unit is a totally different configuration. With a definite magnetic moment in the same sense as that of the neutron, but larger, it is a spinning "neutral" atom, with a definite dipole dominated by the spinning protons which would appear to be in a trigonal array corresponding on the atomic scale to the "resonance-stabilized" trigonals that are known in molecular chemistry. These three appear at any given instant to be tightly encased in the vibrational pattern of one electron and more loosely in the vibrations of two others....This giving rise to the conventional idea of one neutron in the nucleus. Evanescent, transient states which would be identical to states of neutrons, would exist in both the Tritium and Helium3 units, as in all atoms, giving rise to the illusion of stable neutrons in nuclei.

In the "Iso-4,4-set," it is interesting to compare the molecular unit, Deuterium molecule, D:D, and its atomic isomer, the He4 atom. This has been covered in some detail in another short paper, "O/S Theory and the Deuterium to Helium4 Transform." The following paragraphs are essentially quoting from that paper.

The Deuterium Atom nucleus consists of two, "coupled-up-down" protons (in the most stable state) and the Deuterium Molecule would be two sets in what could be called a "stretched tetrahedral array" or possibly a "stretched square planar array," or an arrangement combining these two ideas. This set of "nuclei" would be surrounded, literally within and without, by an array of 4 electrons. The overall result being an ovoid with two distinct centers of motion.

The He4 nucleus would be composed of the same eight basic units; however, in this case the ovoid of two distinct centers of motion is compressed into a spheroid having but one center. It is possible that this spheroid would have a tetrahedral configuration of protons again surrounded within and without by a tetrahedral arrangement of electrons .

The Alpha particle would not be a He4 nucleus, but a square-planar (actually, circular) array of 4 coupled protons surrounded again, "within and without" by two electrons. The coupling of the protons giving the unit a very strong, clockwise? spin.

These "pictures" make sense if you look at the oscillator limits of the proton and the electron and the reversed spin/inversion senses which give rise to the "positive" and "negative" charges.

(The electron and proton by O/S logic have the same "average" mass and radius, but with very different oscillation limits, with the electron limits approximately "ten to the third" times those of the proton hence, with both having the same rotational/inversional velocity of "c' the proton does many inversion/rotations for each of those of the electron. The electron is. therefore, "heavier and lighter" and "larger and smaller" than the proton. The reader is also referred to the "page" http://groups.google.com/group/oscillatorsubstance-theory/web/the-electron-and-proton-as-oscillators)]

It can be seen that the analysis of the entire periodic chart and corresponding molecular isomers could be an impossible task. It appears that nature is such that certain patterns arise of electron-proton interactions which lead to the Periodic Chart pattern with the use of "neutrons in the nucleus" as a convenient bookkeeping tool. This, however, has to be remembered, if one feels the O/S insights to be valid, is illusional, and can obscure other patterns of value. some of which can be found in use of the Iso-set, Iso-A, concepts covered, to some extent, in another article. "Iso-sets, a Key to Radioactivity." which can be found on SciScoop, and. also, as a page on the Google Group, Oscillator/Substance Theory.

Delete this page		Version:	Current by ESKI - Mar 10 2010 ▼

---

1 message about this page

Jul 20 2009 by ESKI

Click on http://groups.google.com/group/oscillatorsubstance-theory/web/o-s-and-the-periodic-chart
- or copy & paste it into your browser's address bar if that doesn't
work.
Report this page
Reply to this discussion

To Sophie Spencer c/o Judge Parker, c/o Aberdeen American News

Published on Tuesday, August 11, 2009

(ABERDEEN AMERICAN NEWS - On line edition)

allAberdeen.net

Dear Sophie,

As my favourite heroine, I cannot resist trying to get you involved in some of my world. I think I owe it to you and to all the brilliant youngsters like you, to give you a chance to look at some ideas and see what you could add or change. I'm sure that with your razor sharp intelligence, you will understand immediately what it took this writer over five years to figure out, or 77 years if one wanted to look at it in another way....

Here's the situation. If one switches/reverses the interpretation of two sets of over a Century old data, one can put together the basis of a quite comprehensive "Theory of Existence, " which seems to be able to be extended to account for almost everything, except, of course, the fact of "Existence."

Actually, what is done is to reverse the interpretation of the famous Michelson-Morley Experiment which determined the Speed of Light to be an apparent constant of nature, from proving that an "Aether," an all-pervasive something, does not exist, to the idea that it actually furnished information about that "Aether." The other switch is in the naming of Planck's Constant which connects the frequency of light to Energy, from a "Constant of Action" to the mathematically equivalent, "Constant of Angular Momentum."

The Speed of Light, as determined in the "M/M' Experiment, can be shown to be the maximum velocity of the carrier wave information in a medium. That it was the maximum velocity for information has always been known, but, it's significance has been ignored. The Maximum Velocity at which information can be carried from a point is the Average Speed of the Carriers in any given direction. The Speed of Light,then is not necessarily a limit, it is certainly an average. That is.if movement of information is the same in principle by electromagnetic radiation as by Pony Express. The "M/M" Experiment can be taken therefore to have determined a characteristic of the Aether, as being able to carry information by light waves as the Speed of
Light, "c," which is about 3×10^{10} cm./sec. Also, it showed that the "All-pervasive-substance," could act as a solid, in order to carry transverse waves. A moments consideration of a fact of chemistry leads to the conclusion that a substance at its triple point where it can be solid, liquid or gas on slight changes of pressure fits this very well. Also, as a substance at its triple point would have the nearest situation to a total equilibration of motions throughout, and would tend to revert to that situation, a solid at its triple point accounts for such things as the Law of Forces. For each and every force there is an equal and opposite force." One can paraphrase this, "In a substance at its triple point any disturbance will be compensated...."

There would always be motion in a substance. Planck's Constant as an Angular Momentum implies that an important part of that motion would be rotation as Angular Momentum is a characteristic of a rotating unit. If we set Planck's Constant, "h," equivalent to the definition of Angular Momentum, "Mass times radius times velocity," we can determine some characteristics of at least some of the substance. We write, $h = m \times r \times v$. Evaluating this at "c," and rearranging we get $m \times r = h/c$. Since mass times radius equals torque which is the push or pull on a rotating object, this makes sense. We have discovered the "torque constant of nature."

Additionally since the equation can be fitted into the form , $xy=K=yx$, which is among other things, the definition for an oscillator, we have discovered an equation defining an oscillator set, which we can write in set notation, {m x r = h/c= r x m) , which points out that the numerical values of mass and radius can be interchanged and at some point they will be equal at $m = r = (h/c)^{0.5}$. This last figure is about 4.7×10^{-19} cm. and 4.7×10^{-19} grams. We now have our substance made up, at least in part, of oscillators which invert through a constant value. We even have discovered an interior "dimension" to these oscillators that is smaller Than 10^{-18} cm. in diameter. This figure is the size of the hole into which string theorists say that their dimensions disappear and Quantum Mechanics is said to fail.

Checking published data for the electron and proton which are the basic units of all matter, we find that the Rest Mass and "Compton Wavelength" values fit as one set of limits, at the situation og maximum size and minimum mass, the switched Absolute Values define the balancing limit of maximum mass and minimum size....

We can go on from here with the implications of a substance at its triple point made up of oscillators of this family to develop an
entire Theory of Existence. I'm sure, Sophie, that with your intelligence, in your world, you could get someone to listen to you and add, amplify, maybe even refute. If you can find use for any of the work this writer has done in your world, or you can find some way to help, you can have some of your contacts in my existence check http:// groups.google.com/group/oscillatorsubstance-theory.

In my world there is too much information, too much "white noise" for anyone to hear anything. Scientists are too busy creating new data to re-examine old data for new significances....

Sophie, if any of this could help you to perhaps get a jump-start on a career as a scientist or a full-ride scholarship in college in
your world, you're welcome. In my world, if noticed at all, it will probably be ignored as "Cracked Pottery."

Dean L. Sinclair of Aberdeen
deanlsinclair@gmail.com

Gmail Calendar Documents Photos Reader Web more -   deanlsinclair@gmail.com -

Go gle groups

« Groups Home

Oscillator/Substance
⊞ Theory

[Search this group] [Search Groups]

Explaining the Universal Law of Physics - COMMON BOND

Options

3 messages - Collapse all - Report discussion as spam

ka-sala View profile   More options Sep 18 2009, 11:02 pm

COMPLETE BOND
Explaining the Universal Law of Physics would be the Elemental Effect
in which each and everything is Linked.

We could take the largest circumference of a circle in any given
direction within the shape of a round ball - call everything within
this a Complete Bond - then take everything out of it and divide it up
into a finite number of particles, until we visibly see nothing left.
Yet truth would be, that now we are left with the invisible empty
space within where the circular ball exists. Does appearing to be
void, mean absolutely empty just because this is how we see it?

So we start all over again, within the this space and shine a light
within, only to find it is absolutely brimming with microscopic
particles we definitely could not see without the light. Yet without
this means, there is no way we can divide these particles up as we did
right down to what we believed to be the last visible molecule prior
to the use of the light.

We find in a different way, it is as full as when we first began to
dissect all that was within. What we did not see in this dark inner
space that it is literally oscillating with energetic particles. The
best we could do to capture these now - being too small to dissect -
is work out how to figuratively acquire these? Physics would require a
Vacuum Effect to suction in this 'everything' we now see in the light;
hopefully, zapping the lot.

Yet having done so, where to then? What can be done with these trapped
microscopic molecules? Physics must as physics will, and explore each
and everything it has captured; remembering these are not particles
visible as an finite number of the Complete Bond found when we
dissected the ball. These are infinitesimal left overs caught from
within this seemingly emptiness; which somehow must have some
connection. Otherwise the ball would be evacuated - absolutely empty
- void of anything at all.

But wait a minute. Can anything be absolute void of anything? Does
lack of particles in light mean absolutely nothing? Inside that
circular ball of Common Bond, may now appear so, but just because we
have vacuumed everything seen in the light out; does this seeming
vacuum mean there is nothing? There is energy in everything; even in
what appears to be a vacuum.

It is a known thing that afterthoughts are able to 'float' in a
vacuum. If every molecule of anything oscillates, this means there is
always movement. For even astronautics must practice if they wish to
remain still. It is seen how even a drop of water moves 'differently'
- one could call it any way and in slow motion - but it moves within
this pressurized vacuum. Both the astronautics and anything else is
said to be in an anti-gravity space.

Therefore there is an energy - like we need light to see the colors in a bubble - yet a buoyancy within another form of travel. Something, must be creating this action. An oscillation of it's own as against another of their own, yet all bound within the Common Bond. It is no different within the bonding of any element; creating another; in the changing recipe book of the Cosmos in which we are even linked though our own Common Bond.

We are all bound to the exactness of whatever we and everything are meant to be. Whether we try to create rain though the use of different elemental mixtures, or snow through another in trial recipes of compounds. Right down to whether or not we use the chemistry of an egg or flour in a cake. What bonds as the best flavors gives rise to their success. The Law of the Universe is to naturally Link.

The Law of the Universe is that everything is utilized; and there is no waste. Everything has it's Positive and Negative Poles; and will Link to whichever is the Complete Bond of it's own Reason. Nothing is not permanently creating and recycling within their Common Bonds, at infinitum. Everything is being utilized with purpose and control.

The Universal Law is that physics, physically, is so often overlooked because that bond to which we each are a part, is overlooked. The Complete Bond is when we are included, for we - whatever creature - are as much bound up in the star-dust of the Cosmos - as is the Planet Earth on which we Live. We and everything oscillate within this One Common Bond, despite our differences.

Reply to author   Forward   Report spam

ka-sala View profile   More options Sep 18 2009, 11:09 pm

Throughout most Subject Matter - so far written - it appears, despite whose or what Theory, I see a Common Bond running through this whole Program, but answers, still searching. Regardless of no 'equasions' in this, there is the Link between all who search in the O/S Theory. Maybe some-one could work on this one.

On Sep 19, 2:02 pm, ka-sala <irrir...@gmail.com> wrote:

- Show quoted text -

Reply to author   Forward   Report spam

dean sinclair View profile   More options Sep 21 2009, 7:04 pm

How true. A common bond, a unity, an existence..... Glad to know you're still here!!!

- Show quoted text -

Reply to author   Forward

End of messages

« Back to Discussions   « Newer topic   Older topic »

Gmail Calendar Documents Photos Reader Web more -   deanlsinclair@gmail.com -

Google groups

« Groups Home

Oscillator/Substance
⊞ **Theory**

[Search this group] [Search Groups]

A look at Quantum Superposition, Atemporality and the direction of Entropy in relation to Eternity - INFINITE VARIABLES

Options

Home

Discussions
  + new post

Members

About this group
Edit my membership
Group settings
Management tasks
Invite members

 View this group in the new Google Groups

2 messages - Collapse all - Report discussion as spam

ka-sala View profile   More options Sep 19 2009, 7:28 am

In our Oscillation Theory, a lot has been said lately regarding
Infinity. Directly of indirectly; we have been oscillating all around
the subject of the O/S Theory. To get to where we want to be we must
bring it back to the unified weak and electromagnetic interaction
between elementary particles, including inta alia, the prediction of
weak neutral current.

We have to return to the source of our search; see it for what it is,
and go on from there. ie. A look at quantum superposition,
atemporality and the direction of entropy in relation to eternity. We
can go over and over Planck's Theory, and anyone else's. The
difference is we are here... now. Today's Theory.

*** Please excuse any spelling mistakes. I would be very interested
to have feedback on these two posts regarding the O/S Theory which
Links all which has been covered to date.

Wishing you what you so desperately seek.
Ka-sala

****

INFINITE VARIABLES

This is like asking one such as a human, to stand on the point of a
pin; take a step off, and simply hope we won't fall into the abyss of
Eternity.

The quantum superposition if being so finite as to suppose we can take
a minute step in such a diversity as the Infinite we call Eternity, is
neither to know we are retreating or progressing. To literally make
such an absolute turn around, with the knowledge that we will step
back into chaos; is more than one giant step for man to handle. He has
enough already to assemble in his own collective elements.

If this direction of entropy points with a definite sign-post (be it
visible in such a universal science as the quantum's allow,) there has
to be some-one who has been to Eternity and back to signify any
measurement at all its physical science. Even quantum superposition
would be that all the pieces of the jig-saw puzzle still missing, will
automatically just fall into place as the elements should.

Hoping that each like a stepping stone - or giant quantum leap of
certainty - our only amoebic particle of theoretical virtue will take
our weight. Or carry us as a whips of quantum superposition back into
a space time dimension of... are we there yet, or is there more to
come? The place called atemporality, or timeless.

Elements of the unknown all lie within the universe of each of us -
within our sub atomic thermodynamic elementary and transitional
particles - right through to the seeming chaos of decay. We shovel

every component required to keep us fueled to create an infinite layer of cells required to hold us in an overlay called skin. As long as the life-span of our own nuclear decay exists while in this superposable state called human, we will continue within our own field force.

So while on the pin-point of quantum superposition, instead of stepping off, we can use our own electromagnetic energy, vacuuming every quantumized oscillating molecule around us, to us. By doing so, we draw on all variable changes created via the quantum super positioning; propelling us by repulsion within an non moving vacuum.

Through this, we create from what was a seeming a state of entropy, a transverse tunnel in which we can now safely step off the pin. By turning matter back on itself from what would have been a physical phenomenon of universal conformity, it has been transformed into a usable state.

There is nothing in the universe called oscillating energy, which is not transferable, transitional, reversible, or any other attribute, which cannot be utilized when known how to superimpose. The atemporality is unaffected by the space time dimension of Eternity's Infinite variables.

The only unknown factor is the inconsistent knowledge yet, within our own field. The Cosmos it's Photons, and Electromagnetism is within each and every one of us. Why are we oscillating for so long over so many theories?

Reply to author   Forward   Report spam

ESKI View profile   More options Sep 23 2009, 4:55 pm

With an unlimited number of variables. the Universe that we know seems to have condensed down communication to constants which seem to be involved in those variables. The writer, "Eski," has spent a good dean of time working with two of them , the speed of light, otherwise, the maximum speed of communication, and Planck's Constant, which helps to define how that communication be carried They combine into a universal torque constant, "h/c." From this has developed a good deal of information about a possoble organizational pattern of the Universe we know and of what it may exist within.

A new member of our group has referred us to his work in the unification of gravitational theorizomg and electromagnetism. HIs ideas of the relation of these to the expansion of the universe seem to fit, with only slight modification, into what else has developed.
I think this group is progressing .....

A slight chance is develoing that we may be able to send someone to a prestigial conference in Europe next summer. Don't hold your breath. the odds are not high......

On Sep 19, 7:28 am, ka-sala <irrir...@gmail.com> wrote:

- Show quoted text -

Reply to author   Forward

End of messages

« Back to Discussions   « Newer topic   Older topic »

Go⊙gle groups

« Groups Home

Oscillator/Substance
⊞ Theory

Search this group   Search Groups

Unification of Gravity and Electricity finally discovered ! !Options

3 messages - Collapse all - Report discussion as spam

EinsteinGravity.com View profile ☆☆☆☆☆ (1 user) More options Sep 21 2009, 9:14 pm

www.EinsteinGravity.com

www.EinsteinElectricity.com

Reply to author   Forward   Report spam   What you rated this post: ☆☆☆☆☆

ESKI View profile   More options Sep 23 2009, 12:39 pm

On Sep 21, 9:14 pm, "EinsteinGravity.com" <alzee...@gmail.com> wrote:

> www.EinsteinGravity.com

> www.EinsteinElectricity.com

I have read through the Einstein Electricty Unit, I find the
Unificatlion ideal to apparently be sound. That ais the idea that
mass times ecceleration of mass equals "Energy," that is the kinetic
energy of translation of the unit under consideration, or, in my
thinking, the "translational motion package" associated with an
entity, to seem valid and an excellent advance.

The write up is excellent and I found only one place where the spell
checker goofed up, "thee" for "the" in the explanation below one
chart. I disagree somewhat with some of the explanation, however,
this may be simply diffeerenfes in view point. If it can be shown
otherwise that the acceleration of "mass" is truly the expansion
constant, at this "instant" of our Universe, this will be of
considerable use in putting together the O/S model which agrees with
the conclusions found in Al's work in so many ways.

This member has delved into areas which this writer has not dug into
and his work appears to be very compatable. The spiral motion,
counter-clockwise or "left-handed" for most readily observable motions
in our "Universe" is noted.   The ccw electron and cw proton which
account for "negative and positive charges" is not metioned, but, this
is understandable, elsewhere in the literatukre it is only mentioned
in this writer's "papers."

Some of the gravity results seem to fit very well into O/S. I'll try
to examine that more closely later.

I hope that other members of this group will closely examine Al
Zepper's conclusions and add their observations as to how they fit
in.....

Thanks, Eski.

Reply to author   Forward

ESKI View profile   More options Sep 25 2009, 2:48 pm

Alm Must apolgize for mispelling your name! Also have a few more
typos in the write up above.

 I hope you will see fit to check how your work fits against other
work here. Is there any way that I could ~~comment~~ comment internally on
some sort of a working copy of your "book" something like can happen
inide of Google Docs? There are places where your data seems "right
on" but the expanations may be such that you cannot back them up with
logic, or they conflict not only wiithe the "conventional wisdom" but
also with my conclusions, even though t, too, am in dasagreement with
the conventional "wisdom" on the particular point. I think that
working together wouldbe to our mutual advantage.

I realize that your entry into the group may have simply been by way
of getting more exposure for your work, I hope that you will see us a
s way fo improving and extending your work.

 Would also like to know more of your background. Clearly you have
education in electricity and a strong interest in the field.....

On Sep 23, 12:39 pm, ESKI <deanlsincl...@gmail.com> wrote:

- Show quoted text -

Reply to author   Forward

End of messages

───

« Back to Discussions   « Newer topic   Older topic »

3 limesions
of space, one
of time — or
vice versa?

Google groups

« Groups Home

Oscillator/Substance
⊞ # Theory

Search this group   Search Groups

Some Constants of the Universe?

Some constants of the Universe, as derived from the transference of information. (For more background information, see paper, "Reduction to One Dimension as Applied to Planck's Constant." }

In the above quoted paper, it was noted that Planck's constant could be taken as the evaluation of all the "dimensions" of a basic "Energy Package," i.e., amount of motion, this was considered as a "Whole" and, as such, equated to "ONE." From this, and other basic considerations arose the following list of basic constants of the universe, as related to information/energy transfer.

Some of the "Constants of nature derived from "h" and "c" that arise from the above considerations are the following:

1. The defining oscillator set equation: $m \times r = h/c = r \times m$.
2. Basic torque of oscillators: h/c
3. Average mass and radius: $(h/c)^{0.5}$ (The "SinVree Entity," * which probably does on exist other than transitorily.)
4. Apparent values of oscillator limits for information transfer: Absolute values in any given system of $ch/2Pi$ and $2Pi/c^2$
5. Frequency for basic information carrier: $1/h$
6. Mass, measured in "Our Half of Universe," : $2Pi/c^2$
7. Radius of basic information carrier or apparent carrier: $ch/2Pi$
8. Basic wave length: ch
10. Basic "time" unit, derived from average size radius of $(h/c)^{0.5}$: $2Pi(h^{0.5}/c^{1.5})$ This is time of one cycle at the average radius.

The numerical values of the above constants will depend upon the dimension set chosen to evaluate them.

Presumably, these above values could be the basis of a set of "Universal Values," similar to the Planck Values. As yet they have not been compared to see how they correspond.

One very interesting observation from the above is that the apparent unit that is responsible for information transfer is--as measured in our universe--much heavier and smaller than the proton, with a much greater vibrational frequency. As other parts of O/S theory suggest that there could be a continuous evolutionary development throughout existence, starting, perhaps, with something like the "zerotron, " combination which would be splittable to the electron and positron but by distortion compressible to the neutron and thence to electron and proton. Proton and Anti-proton combination would form another neutral unit with limits closer to the "Sin-Vree" Entity, with that "Entity" being the "Ultimate--idealized unit" toward which the entirety would tend to be formed . Each successive cycle of changes producing units that more closely resembled this "shell." It seems just possible that the "Information Carrier " that arises in the above considerations is the current most prevalent approach to the "Sin-Vree" Idealization--a pulsating sphere which is otherwise undetected.

*The term, "Sin-Vree Entity," which would be pronounced, "Sin Free Entity." is a somewhat punning combination of the first syllables of the last names attached to papers on Helium.com which first proposed the idea of an "average unit" based on Planck's Constant and the Speed of Light. (Dean Sinclair and Hugh Vreeland).

Delete this page	Version: Current by ESKI - Sep 22 2009

1 message about this page

Sep 22 2009 by ESKI

I've republished some data that was in another paper with a little
different slant on it; Some of you might want to pick up on the idea.
(Or pick it apart.)

Click on http://groups.google.com/group/oscillatorsubstance-theory/web/some-
constants-of-the-universe
- or copy & paste it into your browser's address bar if that doesn't

Reply to this discussion   Report this page

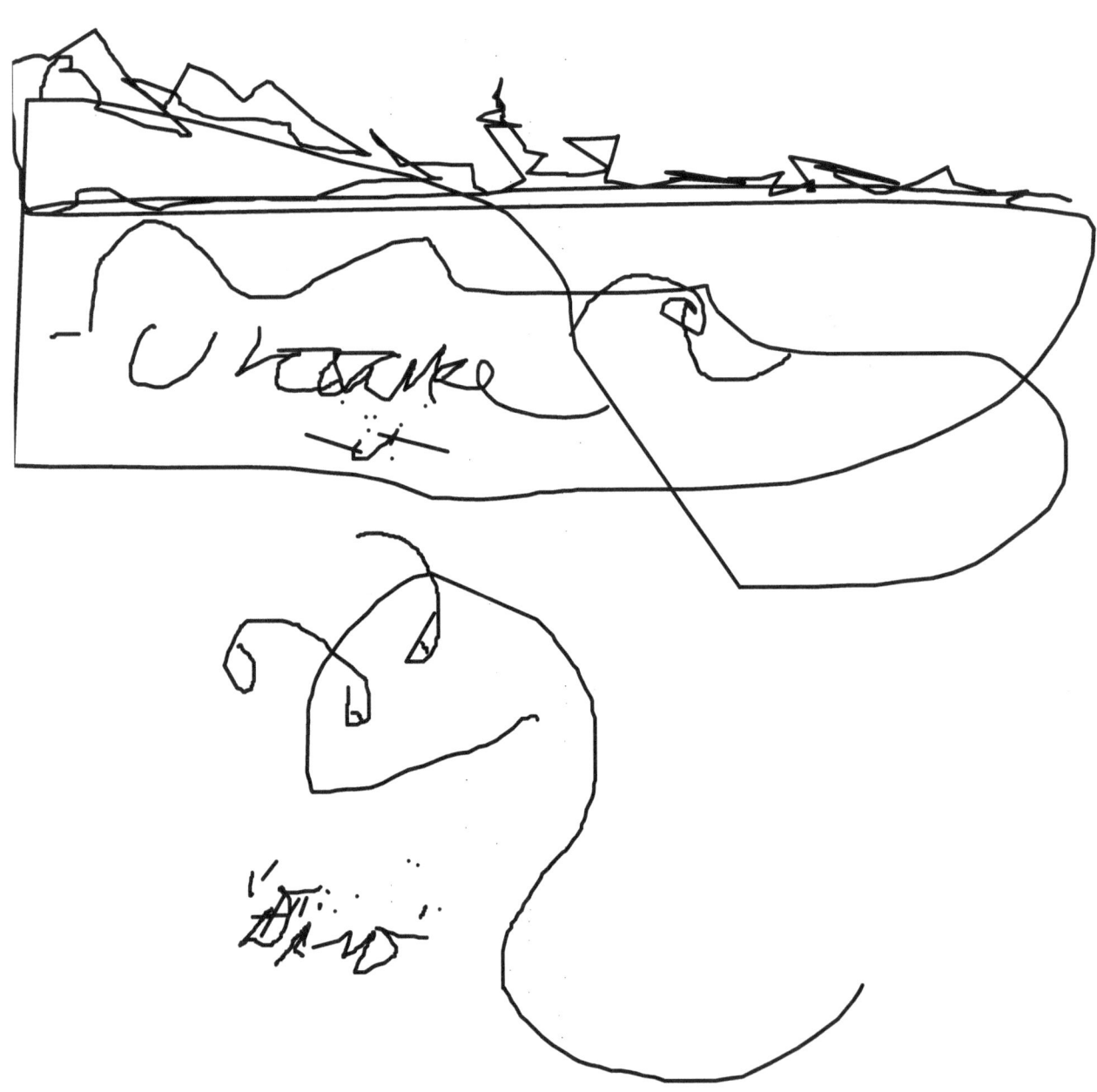

Google groups

« Groups Home

Oscillator/Substance
⊞ **Theory**

[Search this group] [Search Groups]

TAKE NOTE - NOTIFICATION OF BANNING MEMBERS TO THIS SITE.

Options

2 messages - Collapse all - Report discussion as spam

ka-sala View profile   ⋆⋆⋆⋆⋆ (1 user) More options Sep 26 2009, 3:13 am

Any 'members' trying to get in the back door to this O/S site will be banned. Those who already have, have been reported. Take what you want elsewhere. It has nothing to do with the reputable members of the O/S Theory. USE YOUR MINDS! You are not wanted here.

Reply to author   Forward   Report spam   What you rated this post: ⋆⋆⋆⋆⋆

ESKI View profile   More options Sep 28 2009, 3:08 pm

Commenting and amplifying on KaSala's post. It should be emphasized that this sfite has a serious purpose focused on basic physics/ chemical theory. Serious discussion of such things as Dr. Mills' Orbit Sphere, or even Zecariah Sitchen's ideas as to tthe 10th planet, might be valid but the site is not intended for discussion of human taboos, costumes, sexuality and such like topics, often lumped under the catch-all title of "Porn." Please don't waste our time in dumping such material, it's even a waste of your time posting here, even as "Spam."

Please reread the messages to new members before posting here. Thanks. ESKI (Dean L. Sinclair, Site Owner/Founder.

On Sep 26, 3:13 am, ka-sala <irrir...@gmail.com> wrote:

- Show quoted text -

Reply to author   Forward

End of messages

« Back to Discussions   « Newer topic   Older topic »

Home
Discussions
   + new post
Members

About this group
Edit my membership
Group settings
Management tasks
Invite members

 View this group in the new Google Groups

Sponsored links

GX vs GMC Acadia
Interested in the GX?
Then Check Out the GMC Acadia.
www.GMC.com/Acadia

Compare Silverado HD
Come See How You Can Do More
With the New 2011 Silverado
3500HD.
www.Chevrolet.com/SilveradoHD

Compare to CTS-V Sedan
See What CTS-V Sedan Offers
That
the Competition Doesn't. Get Info.
www.Cadillac.com/CTS-V

See your message here...

Gmail Calendar Documents Photos Reader Sites Web more ˅   deanlsinclair@gmail.com ˅

Go gle groups

« Groups Home

Oscillator/Substance
⊞ Theory

[Search this group] [Search Groups]

Home

Invitation to Video SEX Chat   Options

Discussions
 + new post

1 message - Collapse all - Report discussion as spam

Members

ESKI View profile   More options Sep 28 2009, 2:45 pm

I am commenting on this as an example of the sort of thing that is a
total waste of time on this site which is a site dedicated to a
scientific research project. It has no need for discussions of human
sexuality, human taboos and such material, which may be of interest to
many people but has no value here. I am marking Helga as "no post."
ESKI (Owner/Manager)

On Sep 27, 10:26 am, Helga <petuuclaro86...@gmail.com> wrote:

- Hide quoted text -

> My name is Helga Morrow, I am a girl 21 years old.
> I invite you to a video sex chat. Come even if you do not have or have
> not enabled webcam - you can participate in one-way.http://crystlecarle.150m.com

Reply to author   Forward

End of messages

« Back to Discussions   « Newer topic   Older topic »

About this group

Edit my membership

Group settings

Management tasks

Invite members

 View this group in the
new Google Groups

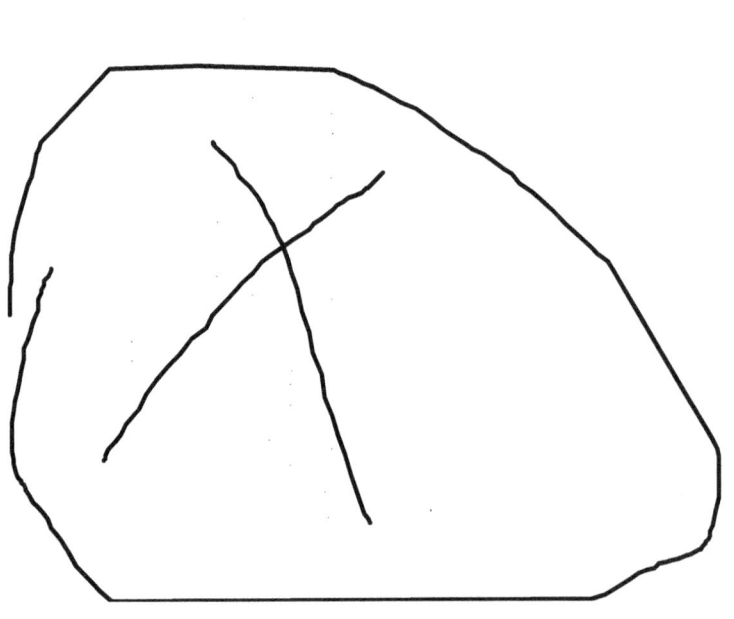

Gmail Calendar Documents Photos Reader Web more   deanlsinclair@gmail.com

Google groups

« Groups Home

Oscillator/Substance
⊞ # Theory

Search this group   Search Groups

O/S, Space-Time and QM   Options

Home

Discussions
+ new post

Members

About this group
Edit my membership
Group settings
Management tasks
Invite members

 View this group in the new Google Groups

Sponsored links

Vibration Testing
Use Fluke 810 to Identify Common
Vibration Problems. Free Demo!
Fluke.com/Vibration_Testing

Amwand Zero Point Energy
In Stock Original AmWand
Free Shipping. Low Price
Guarantee!
AmwandZeroPointEnergy.com/

Linear Motion Solutions
Shaft-Rails-Guides-Slides-Actuators
Proven Linear Bearing Technology
www.PBClinear.com

See your message here...

---

2 messages - Collapse all - Report discussion as spam

ESKI View profile   More options Sep 28 2009, 3:44 pm

Oscillator/Substance Model, Space-Time and QM.

The Oscillator/Substance Model seems to be more general than the
"Accepted Models" of Space-Time and Quantum Mechanics.

A close look at Space-Time shows that the name, itself, suggests
"substance" and, as is pointed out in the paper, "Why Einstein Was
Right," the hidden factor or cycle, in time, allows the mathematics to
work on a macro-scale where time is considered as the "dimension of
continuing sequences of motion."

The influence or involvement of oscillatory motion is not included
and the problem of distortion of information as the transmitter and
receiver move relative to one another is probably over-emphasized and
distorted in Relativity Considerations. Energy is not considered as a
packet of motion and apparently is usually synonymous to Kinetic
Energy of Translation.

That Mass is a Tension/Balance of motions at a surface does not
appear as a concept.

Quantum Mechanics is a mathematical model which focuses on vibratory
motion and again, linear motion. it's area is the energy (packets of
motion) of electrons. This model does not appear to consider vib./
rot. motion, motion about a center per se, and does not consided
electron/proton motion interactions.

It would appear that both Space-Time and QM can be considered as
bearing on smaller areas of consideration of the Oscillator/Substance
View.

O.K> People, any comments? Hopefully....

Reply to author   Forward

ka-sala View profile   More options Oct 23 2009, 8:08 pm

Hello Eski,
1st... Your Quote.
<"It would appear that both Space-Time and QM can be considered as
bearing on smaller areas of consideration of the   Oscillator/
Substance View.">

From whatever 'name' or angle one looks at energy, none can be left
out of the O/S Theory. There is 'no energy' which does not oscillate,
within whatever 'substance' one looks into. If what you want is only
the Oscillator/Substance Model; this could never be reached without
everything combined.

Simply put... It requires all these 'theories' you have mentioned in
'Energy Value' to gather the information sought, for the Oscillator/
Substance Model. Once gathered, like all the pieces of a jigsaw, you
can start putting your Model into action.

Getting 'off the ground' even NASA still needs to 'piggy back'
whatever they want to send out there into that big Space-Time
dimension, and could not have done this much, without the QM aiding
them.

Hold onto all the threads each one has offered here in this site -
weave them into what is it 'you' need - and like any good finished
product, after tying the knots, the rest is discarded.

Everything here since you began, has been like a tree growing leaf by
leaf. Now it's Time within your Space, to prune this O/S Tree, and see
fruits of everyone's labor.

You planted the seed of this Oscillator/Substance Model; and your
gardeners have helped tender it. It's not a 1,000 year old tree, but
ready for the picking! Don't let the fruit drop and rot before you've
tasted the goodness. Nor spend too much Time counting the leaves, or
you really will need your QM to help. You'll end up somewhere in the
Google Count. In short, what you want for your Oscillator/Substance
Model, your Universal Particle has been there all the Time!

You know by now I only operate in the Light. Now you see me, now you
don't. I'm so busy Oscillating in this Polarized Universe of where
sometimes there are too many Theories, and not enough Action; I'm
about to Spin off my Axis! You can 'hear' where I'm coming from. I was
not born with a Mass of Tension, nor without the Scales of Balance as
my right to use; which surface so often as acceptable or not
acceptable within the motions which continually arise in this O/S
Theory Site.

Perhaps in Space-Time when both are One, theories don't enter the
equations of which Energy is what. In the Light it is all Speedy
Specifics, and the Universal Co-ordinates of facts simply fit like a
glove?

Take care, and enjoy Oscillating, Just gather what you have from the
weavers of your tapestry, and tie the knots. You of all people should
understand that one! Tesekkiur Eski...

Ka-sala

On Sep 29, 7:44 am, ESKI <deanlsincl...@gmail.com> wrote:

- Show quoted text -

Reply to author   Forward   Report spam

End of messages

Google groups

« Groups Home

Oscillator/Substance
⊞ **Theory**

[Search this group]   [Search Groups]

Home

Discussions
+ new post

Members

What you want Fun Guy?   Options

1 message - Collapse all - Report discussion as spam

ka-sala View profile   More options Oct 20 2009, 7:32 pm

Check you are in right place otherwise you in wrong place.

Reply to author   Forward   Report spam

End of messages

« **Back to Discussions**   « Newer topic   Older topic »

About this group

Edit my membership

Group settings

Management tasks

Invite members

View this group in the
new Google Groups

To sleep worm —
ditch the
P/s.

Gmail Calendar Documents Photos Reader Web more ⁻   deanlsinclair@gmail.com ⁻

Google groups

« Groups Home

Oscillator/Substance
⊞ **Theory**

[Search this group] [Search Groups]

BAEconstant//Planck'sConstantRevised   Options

3 messages - Collapse all - Report discussion as spam

Jack O Suileabhain View profile   More options Oct 25 2009, 10:42 am

Respectfully submitted to Dean L. Sinclair & Oscillator-Substance Theory Group

Author: Jack O'Suileabhain/O'Sullivan-

For formal submission to: Max Planck Institute--> M...@Mpe.mpg.de <&> R...@mpiwg-berlin.mpg.de <&>Oscillator-Substance-Theory group <&> vorte...@eskimo.com <-- From Jack O'Suileabhain(eng."O'Sullivan) aka Jack 'Harbach' O'Sullivan

***************** Planck's Constant Revised ********************

* * * Postulated necessary Planck revisions per extending fixed-definition-status to the value of 'E'-energy relative to Planck's-Constant's values for all subsequent international quantum-mechanics calculations hense forth* * *

Purpose of Paper: Introduction of a Seminal-alternate-Concept of defining newly a 'Fixed Universal Quantum for all quantum-mechanics calculations for 'E' as for 'energy' as a fixed-universal-constant and newly minting this concept as the 'BAE-Constant' while specifically 'not' altering the contextual basis of Max Planck's constant; re: The Planck-Einstein-Dirac constant.

Rather; this paper is in extended support of the Planck's-Constant and 'not' in refutation of 'Planck-Einstein-Dirac' in any regard. In light of the purposed fixed-universal-value new definition of 'E' I will posit an alternative formula-revision of Einstein's classic E=MC^2quared to solve for 'M'ass @ M=EC2quared. And this is also 'not' intended to alter Einstein but rather restate Einstein's formula(Einstein's-Constant) only in terms of the newly proposed assignment of said "universal fixed-quantum-formula-value for 'E.' " And further: I propose that the fixed-value of posited AExo-DarkSpace-ergo-DarkEnergy as AE=EC^3ubed in light of the proposed fixed-quantum designation for the value expressed as the BAE-Constant of 'E,' aka the Base-Ambient-Constant of Energy.

This proposed conversion renders the classic Planck's-Constant E=hv to EC=hv as changed by the fixed definition of 'E'nergy to be here-forward expressed as the 'BAE-Constant' fixed-base-energy-value of Interstellar Space. Specifically the posit here is that there is a 'lowest-base-ambient-energy-density-level' to Interstellar Space; and below which that energy level cannot fall. The theory for this is that the quasi-static-energy-tension of our Universe mandates & sustains that BAE-Constant. And this is because our Universe is a relatively-low-energy-density bubble within a larger/virtually infinite DarkEnergy Aexo-space. And this AExo-DarkSpace exists at it's own Base-Ambient-Energy-Density-Level again of AE=EC^3ubed and maintains the gravitational tensor-outpull upon our bubble universe. (Note: A sub-posit to this is that AE=EC^3ubed also describeds the nexial energy level of a black-hole-singularity ingress point into parallel Aexo-DarkSpace as is also at the outer fringe-border of our bubble universe with Aexo-DarkSpace)

Posit: There can be 'no total energy vacuum' that can exlude the density-field of energy below the BAE-Constant within our universe. And this is because the DarkSpace tensor-outpull tends to want to inexorably actually accelerate outward to itself all lower density objects & energy wave-forms than it's own original AE=EC^3ubed base-ambient-energy-density-level, again, outward back into itself as a fluid dynamic flow system. Hubble clearly observes & articulates that inexorable outward acceleration of galaxies/mass/energy flow to the outer diameter-border of our bubble universe. And that border is the point where mass @ EC^2quared has accelerated x 'C' light speed and reaches the DarkSpace ingress energy thresh-hole of AE=EC^3ubed.

The rest of this paper attempts to provide theoretical proofs of the above based upon a revised- Planck's constant of (Photon= E=hv--->to---> BAE-Constant @ E therefore Photon=E @ 'C'-light-speed= (EC=hv) as Planck's revision. And this will perforce also revise Einstein's Constant @ E=MC^2quared to M=EC^2quared (to be demonstrated following).

Whereas Max Planck posited that certain discreet quanta cannot take on Indescriminate-Values herein I extend that same premise to newly quantafying the value of 'E'nergy as pertains to quantum-mechanics. Thusly from here forth 'E' is to be held constant at a 'fixed-value' that I will attempt to define & quantafy by the Planck-Einstein constant. Here after I will refer to the fixed-value of 'E' regularly as the BAE-Constant.

Supporting Premise: The Dirac-Constant rather straight-forwardly describes 'Light' as 'Helicoid-Wave-String' enter:----> Photon 'E'nergy = h/2pi x w = hv<--- (h= MPconstant & w=angular frequency & v=electro-magnetic frequency) eg. the double radii 2pi a la' Dirac indicates circular diameter/wave-spin/wave-twist etc.)--->ergo---> Light as Helicoid-Wave-String a la' Dirac & a la' Planck.

Subsequently a Photon per duel particle/wavicle characteristic would logically= 1-wave-crest to 1-'Dirac-full-twist/spyro-loop'-wave-crest of afore-said 'Helicoid-wave-String.' And classically the Planck-Constant writes that as 1-Photon's energy-'E'= hv.

Central Key argument per the logic of Max Planck: Pragmatically 'E' must equal some 'constant' based upon the Planck-Einstein Constant rather than 'E' being merely a generically descriptive term merely in indication of the variously calculated sums of 'whole-quantum-energy' of sub-atomic particles &/or wave values in general terms. To my mind that seems somewhat slip-shod 'not' to narrowly define 'E' in very specific fixed-quantum-value terms that would be universally-constant in all quantum-mechanics equations.

Max Planck's Constant is classically defining the 'E'-energy of a photon within light-wave-string obviously at 'light-speed' ergo E=hv. The problem here that 'C'-light-speed is 'not' an irrevocable constant-speed as regards to the fiction of 'the vacuum of space.' There is no-such-state as a complete energy vacuum. And 'space' relative density is compromised through gravitational warp-density rendering the idea of a 'Light Speed Constant' to be a defacto fiction relative to gravitational space-density stressors constantly impinging upon light's trajectory & drag vs speed. Hense the original proofs to Einstein's theory which perforce validated General Relativity etc.

There is difficulty here unless we define the ultimate base-constant of 'Light-Speed' by defining 'Light Speed' at a 'Fixed-Base-Ambient-Insterstellar-Space-Energy-density level' that cannot be further reduced because of the dynamic-quasi-static-tension of the Surrounding DarkEnergy gravitational out-pull-stressors upon our space-time-normal energy sheet within our bubble-space-time variable-normal universe. Then the relative interactive interchangeability of our various quantum mechanics experimental findings will not lack accuracy & therefore will not lack functional empiric validity. There are discoveries that we are missing by not establishing the BAE-Constant for 'E'-energy.

Accordingly then this modifies the Planck-Constant as follows: Logically then using 'E' as the fixed-quantum of the Base-Ambient-Energy Constant/BAE-Constant we can define Light Wave Photon energy-density simply as BAE-Constant 'E' x 'C' light-speed= Photon-energy-density. And so this concept simply converts the Planck-Constant Photon energy definition from E = hv --> to --> EC=hv.

And as follows relative to the Planck-Einstein-Constant this new fixed-quantum-definition of 'E' as the Base-Ambient-Interstellar-Space energy-density level expresses 'E' per 'Planck' as----> E=hv/C

Review revisions:

*Re: 'E' = BAE-Constant = hv/C

*Re: Planck's-Constant = 1-Photon = EC = hv

*Re: Mass of 1'H'-hydrogen= EC^2quared = hv^2quared = (M=EC^2quared)

*Re: AExo-DarkSpace = DarkEnergy = EC^3ubed = hv^3ubed = (AE=hv^3ubed=EC^3ubed)

Recap: Planck Revised-->Photon= BAE-Constant @ Light-Speed

Old Planck-->Photon= E=hv----->Revised Planck----> EC=hv

Therefore BAE = E = hv/C

If then Einstein Revised says M=EC^2quared

then also E = M/C^2quared

then also hv/C = M/C^2quared

therefore hv = M/C

and also again hv=EC

therefore {hv} x C = M which is also neatly ({hv} x C = EC^2quared) = (M=EC^2quared)

and again said Revised Planck says Photon= EC=hv because new BAE-Constant = E making Photon=E@C-light speed

therefore as before revised Einstein says M=EC x C which also = M = hv x C which also = M=EC^squared which thusly becomes Revised Einstein per Revised Planck's-Constant per reassignment of the postulated 'now' fixed-universal-constant value for 'E'-energy as the BAE-Constant aka Base Ambient Energy universal-fixed Constant.

Again--------> M=EC^2quared which is equal to ---> {hv} x C = M

And again Planck revised restates Photon=> EC=hv aka BAE-Constant of 'E' x C-lightspeed = EC=hv.

*Conclusional Postulate: Light Speed-'C' can only correctly be stated as an universal constant in the context of 'E'-energy being also stated as a fixed-universal-constant; and this within the further framework of a revisional re-expression of the Planck's Constant which in turn simultaneously revises Einstein's-Constant equation relative to the relationship of Mass to Energy to Light-Speed.

In short these revisions are stated: From E=hv ---->to---> EC=hv as the Revised Max Planck's Constant---->&>
which mandates, proves & articulates the Einstein Revision> From E=MC^2quared -->to--> M=EC^2quared perforce.

Thankyou: Jack O'Suileabhain/O'Sullivan

_____

Keep your friends updated—even when you're not signed in.
http://www.microsoft.com/middleeast/windows/windowslive/see-it-in-act...

Reply to author   Forward   Report spam

ESKI View profile   More options Oct 28 2009, 5:09 pm

Jack:
Welcome to the group. However, how about as little more basic infor as to why you are doing this revision, and what the meaning is? Most of us are not experts on on Quantum Gravity and such.
 In fact, I have doubts about both terms being used together having any meaning what so ever. So how about an explanation in "lay-man's language" that this stupid old duffer might understand? Eh?
 Dean Sinclair

On Oct 25, 9:42 am, Jack O Suileabhain <braghgoerin...@hotmail.com> wrote:

- Show quoted text -

...

read more »

ESKI View profile   More options Nov 4 2009, 7:48 pm

More comment on the original post. Basic energy level.:...Outer space
seems to have a basic Energy, at least translational energy. as
measured by temlperature. at about 2.7 degrees Kelvin.

Light as a "heliocord etc.: Light tranvels along a line hence could
be called a "string" or a cord. As there will be a spin, one can call
this spin a helix, actually it is usually an expanding helix, so that
could be a Helicord..

Yes, our universe is almost surely a tiny part of "Whatever Is" hence
a bubble.

As to revising Planck's constant to a Basic Energy Constant. I guees
this would fall under the fact that any mukltiple, power or root of a
constant, or combination of constants can be considered another
constant, which may, or may not be of practical value..... ESKI

On Oct 28, 4:09 pm, ESKI <deanlsincl...@gmail.com> wrote:

- Show quoted text -

...

read more »

End of messages

« Back to Discussions   « Newer topic   Older topic »

A gun massacre =
People buy guns) =
more now for
most massacre —

Gmail Calendar Documents Photos Reader Web more   deanlsinclair@gmail.com

Google groups

« Groups Home

Oscillator/Substance
⊞ Theory

[Search this group]   [Search Groups]

NewAtomModel-SingularityPhysics   Options

1 message - Collapse all - Report discussion as spam

Jack O Suileabhain View profile   More options Oct 25 2009, 10:36 am

-Alternate 'Atom' Model//BAE-Constant//DarkEnergy-Singularity Physics-per Jack O'Suileabhain/O'Sullivan(English)

* * * Fusion-Gate/Plasma Breach Hyper-Gravionic Fusion Sustaining Reactors// aka Gray-Jet subSingularity Hyper-Plasma Gate Reactor * * *

* * * Quantum-Electron Shell/Fields are the same hyper-gravionic lobe/field that the Plasma-Breach Reactor exhibits. The Russian's made the reactor and did it, thusly are recently boasting of 'Gravity Manipulation' and creating electro-plasmic/hypergravionic lobe fields that can interdict aircraft & missles and stress their superstructures so violently that they are virtually 'ripped from the skys.' * * *

 * * * I know this because I helped them do it. Whitt Brantley of NASA confirmed their success. * * *

~Root Einstein 're-configuration basis' for Singularity-Transdimensional HyperPlasma-Breach Physics~

1'E'nergy=Base(bottom)ambient energy speed-density state of Interstellar Space.

Mass solved for: 'M'ass=EC^2quared'----->formula modification for re-definition of 'E' .

Light/Photonic wave-string wave-to-wave speed-density/mass:--> 'P'=EC. or simply Light='E'nergy @ Light-Speed.

Quantum-Electron Speed-density/Electro-PlasmicGravionic field Mass= (E x C/2)^2quared.

AEX=EC^3ubed is the (BAE-Constant)Base-Ambient Energy Speed-Density level of transdimensional 'Parallel-AexoDarkSpace;' which speed-density the galaxies at the border fringe of the universe have accelerated to @ Mass's EC^2quared x 'C'-LightSpeed=EC^3ubed. And at this juncture galaxies turn into MegaGammaRayBursters as their Hub-Central-Grayhole Singularities turn to Black-Hole status 'again' @ EC^3ubed speed-density(as well as the singularity-centres of each & every Proton); and thusly en-toto all galaxies & mass are returned/reabsorbed/reintegrated into Parent AexoDarkSpace.

* * *And most likely the White-Hole Singularity BIG-BANG site out of Parent DarkEnergy Aexospace is still a White-Hole Fountaining ingress point of sub-DarkEnergy super-plasma.

PlasmaBreach Reactors are provide a Giga-Input/Mega-Output Electro-Plasmic Gyro-toroid fractional/& incipient Gray-Hole subSingularity Reactor. The eye-centre of the induced electro-plasma worm-hole subDarkEnergy super-plasma cross-over point.

Protons are this in micro-cosm. Their Quantum-Electron Shell/Fields are the same hyper-gravionic lobe/field that the Plasma-Breach Reactor exhibits. The Russian's made the reactor and did it, thusly are recently boasting of 'Gravity Manipulation' and creating electro-plasmic/hypergravionic lobe fields that can interdict aircraft & missles and stress their superstructures so violently that they are virtually 'ripped from the skys.'

I know this because I helped them do it. Whitt Brantley of NASA confirmed their success.

Home

Discussions
+ new post

Members

About this group
Edit my membership
Group settings
Management tasks
Invite members

View this group in the
new Google Groups

Thus: Below the Singularity event horizon at the 'reintegration nexus' said 'Nexus'=EC^3ubed as the reabsorbtion point back into AexoSpace/DarkSpace."

Our Universe & an infinite plethora of universe's like us(more or less) are relatively Low-Energy Speed-Density Bubbles within the greater Parent HyperFluidic/HyperSpeed-Dense DarkEnergy Aexospace/DarkSpace/SuperCosmos.

* * *And via PlasmaBreach Reactors we can 'tap' AexoDarkSpace High Energy levels which can sustain Fusion but is largely unnecessary as 'throating down' the 'PlasmaBreach Reactor Gate' avails us of all the HyperGravionic/EM Induction potential that we need and so very much more. Focused Gravity Hyper-Propulsion and Time-Dialation are merely bottom end of the potential.

Using a Hydrogen Atom new Model: Consider that the configuration of the H-Atom building block of ALL ATOMS to be along the lines of a GYRO-SCOPE-kind've. For argument's sake I'll say that the PROTON were a Gyro-Centrific Toroidal-HyperCompaction of ('E'nergy/Atomic E-force/Electro-plasmic-gravionic micro-singularity force ALL INTERCHANGABLE CONCEPTS FOR THIS MODEL.

I'm putting forth that the Proton centered Atom is a micro-singularity tapped/wormholed into an inexhaustable Parallel Space Dark Energy infinite E-Force Source. And I've also call this the SUPER-MEMBRANE. And s'theoretically' this 'Parallel SuperSpace/DarkSpace is the parent-source and surrounding sustainer of our bubble universe and an infinitude like us. And this is a SUPER-DYNAMIC E-FORCE FLUID DYNAMIC FLOW SYSTEM whose Quantum Gravionic signature from the centre of protons in dynamic opposition to the Gravionic Out-pull of surrounging PARENT DARKSPACE upon our Bubble Universe's lower density E-Force membrane sheet. Hense the explaination of EVERYTHING from 'SPOOKY ACTION A DISTANCE' to the OUTWARD ACCELERATION of ALL MASS/GALAXIES etc.

Without this picture these things will be VIRTUALLY IMPOSSIBLE TO INTEGRATE into a cohesive Atomic/Nuclear/Astrophysical construct and make ANY REAL PROGRESS. But we will and wonders stand at our scientific door-step that would beggar Jules Verne, Aldous Huxley, & H.G. Welles combined.

The core of our micro-singularity aka Gyro-Centrifically Hyper-Compacted Proton Toroid is a singularity centred E-force/Inflow axis which creates the EM-FIELD SHELL that I call the 'Quantum-Electron' AND is an axial-perpendicular flow of lesser density Electro-Plasmic E-force. And this is what 'stabilizes' the Proton as a 'balanced singularity' not black/not white, but a GRAY HOLE MICRO-SINGULARITY SYSTEM. Added inductant quantum-electron charge/flow effects the proton by slightly increasing the Quantum-Electon & thus 'whole atom charge' and thusly 'dialates' the proton turning it 'slightly whitish' which is even MORE PROFOUND with conductors & superconductors which makes their OUTER ELECTRON-SHELLS extremely SLIPPERY(quasi-isotopically-slightly-unstable) while conducting electro-E-force current.

So the 'dialated up-forced proton singularity centres' becomes a COMMON EXPLANATION by degree of E-force added/stimulated for all EM-PHENOMENA, NUCLEAR PHENOMENA & ELECTRO-CHEMICAL PHENOMENA(stimulated to micro-white-hole status=nuclear reactions) AND gravionic phenomenon. EXPANDED E-FORCE means a sped up proton-toroid, which means a more centrifically-compacted-densified whole atomic-field, which means the INCREASED TOROIDALLY E-force VISCOSITY ACCUMULATIVE and COMPRESSIVE of ADJACENT E-SHEET which is GRAVITY. Like the cotton candy machine model INCREASED E-FORCE SPEED DENSITY= increased E-force viscosity PLUS increased SPIN-TWIST which is what REALS-IN & compresses/warps the adjacent E-force sheet of Space-Time.

RELATIVE PROTON-ELECTRON ATOMIC WEIGHT: That quantum-electron shell is about a lesser density to the Proton @ approx. quanta of (1E-force x 1/2 light speed)squared. And for argument sake the Proton's 'weight/speed density' added to the quantum-electron makes up a 'modified Einstein formula for mass/energy' @ (Mass=E-force x light speed) squared.

Note: I'm using a hypothetical '1E-force' new-standard-quanta as the Base Ambient Energy level of Interstellar Space-Time-Normal or the 'Bottom Base Energy Density of our Bubble Universe.'

CASE IN POINT: STANDARD HYDROLYSIS by my reckoning functions resulting from Quantum-Electron Flow (electricity) effecting the Electro-Valent Shelled Proton(as a WHOLE SINGULARITY SYSTEM) by amping up the atomic-field to the point of EXPANDING THE DIALATION of the PROTON's singularity centre.

EVERY ATOMIC QUANTUM ELECTRON OUTER SHELL has the potential to hold

EXCESS Electro-Plasmic E-FORCE/aka/QUANTUM ELECTRON CHARGE and Molecules thus FULL to EXCESS charged lend themselves TO parting themselves from their MOLECULAR COMRADE ATOMS.

ABOVE HYDROGEN ATOMIC NUMBER NEUTRONS are 'not' MICRO-SINGULARITIES but evidentiary of an UP-ENERGIED MOLECULAR MULTI-SINGULARITY CONSTRUCT that CREATES/BUILDS NEUTRONS via the NOW UP-ENERGIED EXPANDED PROTON-MICROSINGULARITIES ingresses excess ELECTRO-PLASMIC E-Force austensibly BLED-IN from PARALLEL DARK SPACE via dialated Proton-Singularity eyes.

AND NEUTRONS like ROLLER BEARINGS buffering/separating THE SYSTEM FUNCTIONS of the ELECTRO-VALENT QUANTUM-ELECTRON SHELLS are THE BONDING GLUE whose construct SPEED-DENSITY connects WITHOUT DESTABILIZING the Atoms within Molecules. Basically the Neutron SHARES INTER-FLOW E-force energy with the adjacent atoms BUT WITHOUT INTERFERRING with their INTERNAL SYSTEMIC INTEGRITY.

OFCOURSE I'M STATING WHAT WE ALREADY KNOW in somewhat MODIFIED TERMS that I obviously surmise to be more accurate a portrayal of what we know to be prozaic chemistry/physics AS WELL AS to accurately define THE EXOTIC PHENOMENON that conventional theory is virtually CLUELESS TO EXPLAIN.

I see the the Atom as a Gyroscope like-toroidal flow system, using somewhat the model of the Axial-EM/& VanAllenBelt flow fields of our planet, (and singularity centred galaxies for that matter). The electro-valent onion-like shell system to larger atoms I see to follow precisely the empirical Quantum-Electron count per-shell as conventional chemistry. I just describe 'electrons vs electron flow' somewhat differently.

FREE ELECTRON FLOW: Are more akin to our idea as light-wave/photon function as a WAVICLE propertied affair that I would call HELICOID-WAVE STRING and for ELECTRON FLOW that each HELICOID Wave-crest-to-crest also constitutes a QUANTUM ELECTRON; and again @ electron E-Force/speed-density @ (E-force x 1/2 lightspeed)squared.

THIS I CALL THE ATOMIC BASE DEFINITION for what is SINGULARITY/TRANSDIMENSIONAL PHYSICS.

IF CONSERVATION of ENERGY discounts Parallel Dark Space & Protons as micro-singularities; then it is merely an ARCHAIC NEWTONISM which is merely SUPERFICIALLY DESCRIPTIVE but fundamentally INACCURATE and thusly a flawed psuedo-truism of REDUNDANT & passe' VIEWS of PHYSICS. Singularity/Transdimensional Physics. . . . Thankyou-JO-

---

Windows Live: Keep your friends up to date with what you do online.
http://www.microsoft.com/middleeast/windows/windowslive/see-it-in-act...

Google groups

« Groups Home

Oscillator/Substance
⊞ # Theory

Search this group   Search Groups

BAE&Aexospace/DarkEnergyPlasmaBreachReactor:Right on 'Space-TIME.'

Options

Home

Discussions
+ new post

Members

About this group
Edit my membership
Group settings
Management tasks
Invite members

 View this group in the
new Google Groups

1 message - Collapse all - Report discussion as spam

Jack O Suileabhain View profile   More options Oct 25 2009, 10:35 am

RE: Singularity/DarkEnergy-Physics//BAE-Constant Revised Planck-Einstein
Constant//Green Bubble Universe within DarkEnergy/DarkSpace Aexoverse//Jack
O'Suileabhain

Fusion-Gate/Aexospace/DarkEnergyPlasmaBreachReactor:Right on 'Space-TIME.'

* * * ? Which Nations shall be pre-eminent in these DarkEnergy Tech R&D projects;
and which Nations shall be left cripplingly behind? * * *

~The U.S./AMERICAN Singularity-Physics REVOLUTION~Slainte' Jack O'Suileabhain~

Green Cosmological-Constant in 'Golden-Mobius-Ratio' Space-Time Formula

 * * *NAILING DOWN the Cosmological Constant & it's function RELATIVE to DARK
ENERGY * * *

? ? ? THE BIG QUESTION: ?Who GETS to the MARK FIRST with 'SPACE-TIME'-
MANIPULATION TECHNOLOGIES which are simply
concentrated-GRAVITY-MANIPULATION TECHNOLOGIES that are imminently WITHIN
our GRASP:
AND THEN; just exactly WHOSE HISTORY &/or FUTURE gets MANIPULATED and WHO
DOES the MANIPULATING?

* * * The Russian functional 'proofs' of ACCESSABLE DARK-ENERGY are HISTORIC &
MOMENTOUS which lead
DIRECTLY to GRAVITY manipulating PROPULSION technologies which are also 'self-
sustaning' and virtually
LIMITLESS in POWER PRODUCTION POTENTIAL* * *

* * SINGULARITY PHYSICS: DARK ENERGY Gray-Jet Singularity Electro-Plasmic
Reactors~ & ~ Space-Time manipulation/via/THE COSMOLOGICAL CONSTANT '4-part
Formula-Mobius-Equations via Einstein modification.'

* * * St. Patrick's Space-Time 'machine' Universe GOLDEN-MOBIUS-RATIO Formula for
our GREEN-1'E'nergy BUBBLE UNIVERSE go again as follows:

* * * SO THEY TELL US THE BUBBLE UNIVERSE is 'IRISH' and ACTUALLY

GREEN & with constant TEMPERATURE & TEXTURE THROUGHOUT! ! !

? ? ? WHO'D HAVE THUNK IT ? ? ?---------> other than ME!~;-)Jack Harbach O'Sullivan

GRAVITY & TIME are a simple PHENOMENOLOGICAL product of SPEED & DENSITY
pertaining to the

CONSTANT TEXTURAL UNIVERSAL PROPORTORTION born since the moment

OF THE BIG BANG SINGULARITY which 'BORNED' our GREEN BUBBLE UNIVERSE from PARENT

DARK-ENERGY INFINITE AEXOCOSMIC DARK-SPACE. . . aka AEXO-DarkSpace.

AND THIS RISES as a NORMAL PROCESS of the BASE AMBIENT DARKENERGY of AEXO-DarkSpace
which is the basis discovery concept FINALLY DEFINING Einstein's- Cosmological-Constant' which
has a root-mobius formula of the BIG BANG instantaneous BUBBLE UNIVERSE expansion rate formulated
as ENERGY SPEED-DENSITY FORMULA:

* * * * Speed of Light Cubed-(C^3ubed) SPEED X Interstellar space Base-Ambient Speed-Density as ONE-ENERGY-(1'E') divided by Speed of light Cubed-------->(E/C^3)DENSITY ergo:

~the BIG BANG EXPANSION SPEED-DENSITY RATE= C^3ubed SPEED x E/C^3ubed DENSITY~ which moves as HYPER-TIME Speed-Density Expansion Rate. . . thus is the ROOT FORMULA of the COSMOLOGICAL CONSTANT. . . . .

COSMOLOGICAL CONSTANT AEXO-Dark Space= EC^3ubed: FOLLOWING as 'one' through 'four' in the articulation of UNIVERSE CREATION out of AEXO-DarkSpace HYPER-SPEED-DENSE current dynamics.

AEXO-DarkSpace @ EC^3ubed FRACTALLATING HYPERDYNAMIC (wild) current dynamics of said DARKSPACE-AEXOCOSMOS has it's base ambient speed-density as AEXO-DarkSpace @ (AEX=EC^3ubed). Our Newly born Bubble Universe was INFLATED via AEXO-DarkSpace super-plasma VIA the ingress BIG BANG AEXO-DarkSpace-GYRO-TOROIDAL-Singularity and that FIRST INFLATED our BUBBLE UNIVERSE. This birth-singularity would be a WHITE-HOLE from the NEWLY-FORMED BUBBLE UNIVERSES 'point of view.'

#1. This BASE-Ambient COSMOLOGICAL CONSTANT of the BIG-BANG is thusly:ONE-ENERGY INTERSTELLAR SPACE as 1'E'nergy={(C^3ubed) SPEED x 1'E'/C^3ubed DENSITY} mobius fashion formulates (again) the 1'E'nergy Instellar Space speed-density .
. which is the HYPER-TIME RATE base ambient EXPANSION RATE @ virtually INSTANTANEOUS EXPANSION/INFLATION RATE of the BIG BANG. . . AND the BIG BANG is a Bubble-Universe-forming SINGULARITY via the eye of a DarkEnergy/AEXO-DarkSpace EYE-sinularity-LOW-SPEED-DENSE centre of a routine AEXO-DarkSpace GYRO-TOROIDAL MAELSTROM.

And from our BUBBLE UNIVERSE SIDE said singularity OUT of AEXO-Dark-Space is a MEGA-WHITE-HOLE creation singularity INTO OURSPACE. And again the INJECTED super-plasma out of AEXO-DarkSpace expands the BUBBLE instantaneously at the HYPER-TIME-RATE of 'Base-Ambient ONE-'E'nergy Interstellar-Bubble Membrane Space'= 1'E'= C^3ubed SPEED(of expansion x E/C^3cubed HYPER LOW DENSITY primal MEMBRANE-SPACE-------DENSITY. . . ~

add to that, the EINSTEIN REVISION #2. 'M'ass=1'E' x C^2quared, (or), 'M'=EC^quared. . . rather than the more familiar E=MC^2quared. This would also funcion as 'M'=EC^2quared as the Base-energy Speed-Density of the HYDROGEN-ATOM.

add to that redefining the 'Zero-Point' of Feynmans ZERO-POINT ENERGY eg ZPE is simply 1'E' Instellar-space Membrane: #3. 1'E' @ C(light-speed)=EC=LIGHT, or(EC=Light) 'or' 1 base ambient intersteller 'E'nergy(which is cross-equivalent to Feynman's conceptual-ZPE/ZeroPointEnergy) then accelerated to Light-Speed takes the speed-density spin-twist form as HELICOID-WAVICLE-LIGHTSTRING which classically we interpret as WAVE &/or PHOTON-PARTICLE(One WAVE/wave-crest to wave-crest= ONE QUANTUM-PHOTON) simultaneously both WAVE as well as PARTICLE-INPACT-WAVE SEGMENT.

add to that #4. 'One-Hydrogen Electro-valent-shell(icle)Field' = ONE QUANTUM ELECTRON(QE)
@ speed-density QE= (EC/2)^2quared 'or' (1'E'nergy x Half-Light-Speed)Squared

* * * Thus are (roughly) the FOUR-PHASE PARAMETERS of the EINSTEINIAN COSMOLOGICAL-CONSTANT * * *

* * * The COSMOLOGICAL CONSTANT Overview: WE ARE ONE CHAMPAGNE BUBBLE Universe among INFINITE MYRIADS of like BUBBLE UNIVERSE's

SUCH AS OURSELF being CONTINUALLY BORN at the heart of DARKSPACE super-eddie gyro-toroidal

MAELSTROMS that open LOW-DENSITY SINGULARITIES at their EYE-Centre-Points which form the BIG-BANG

BUBBLES of said UNIVERSES.

THE SIMPLE FORMULA RATIO 'constant' that controls ALL RELATIVE 'TIME' & GRAVITY which in-turn determines

their mass/energy flow/& relative 'time' dynamics is that the ORIGINAL BUBBLE that forms the ONE-'E'energy MEMBRANE SHEET of INTERSTELLAR SPACE IS ratioed THUSLY:

* * * * * * The Base-Ambient-Universe ONE-ENERGY@'1-E' energy MEMBRANE formulates as-->1'E'= C^3ubed SPEED X E/C^3ubed DENSITY. And by this formula-ratio which at C^3ubed SUPER-SPEED FORMS the ENTIRE atom-mass EMPTY ENERGY-BUBBLE within a SINGLE SPLIT-INSTANT.

AND WITHIN THE NEXT INSTANT: then our NEW-EMPTY BUBBLE UNIVERSE is then recipient to a centre-bubble injection via the 'AEXO-DarkSpace birth SINGULARITY of the PRIMAL Sub-AEXO-base ambient MINUS-C^3ubed super-plasma WHICH creates a SUPER-SPAGHETTIFIED EC^2(atomic mass speed-density{Hydrogen}) Helicoid Wave-string UNIVERSE CENTRAL NEBULA which is the STAR-BIRTH ROOKERIE of GALAXIES. The EC^2quared Helicoid Wave-string segments to PROTON DENSITY SEGMENTS that immediately hyper-collapse and form ALL PROTONS as BALANCED-GRAY-JET micro-singularities replete with AXIAL FLOW ELECTRO-VALENT SHELLS making EACH PROTON a Gray-Jet Singularity System PERENNIALLY FED by/linked to PARALLEL AEXO-DarkSpace as the UBIQUITOUS DYNAMO Back-Of-The-Tapestry ENGINE of ALL COSMIC PHENOMENA.

AGAIN: INITIALLY THE AEXO-DarkSpace Parent BIG-BANG SINGULARITY COLLAPSES when it's DARKSPACE MAELSTROM, by SINGULARITY DRAG, then super COLLAPSES injecting super-plasma becoming said ATOMIC HYDROGEN DENSITY BIRTH-NEBULA. And that original INJECTED SUPER-DARKSPACE-PLASMA instantly SPAGHETTIFIES into MASSIVE wavefront expanding clouds of EC^2quared(approx.) HELICOID WAVE STRING which segments to HYPER-SNAP DOWN to form HYDROGEN PROTONS AS GRAYJET MICROSINGULARITY ELECTRO-VALENT SYSTEMS; with are HYDROGEN ATOMS.

*Base Ambient 'HYPER-TIME' rate of Space-Time 'E'-MEMBRANE: THE FURTHER evolution of galaxies PROCESS WE HAVE PRETTY WELL NAILED DOWN(Silk, Reiss, Woosley, & Perlmutter, etc.) as the GALAXIES ACCELERATE to the outer AEXO-DarkSpace BUBBLE BORDER that the GREEN-'E'energy Space-Time 'MEMBRANE-Sheet' flows perennially into @ C^3ubed SPEED x E/C^3ubed DENSITY. This is ONE-'E'nergy Space-Time @ HYPERSPEED which is more easily just designated HYPER-TIME.

HENSE the GREEN-MEMBRANE 1'E'nergy BUBBLE expaned @ the SPEED of LIGHT CUBED or simply EC^3ubed.(on this key-board~;-)

THIS SIMPLY MEANS that THE ORIGINAL 1'E'=C^3ubed SPEED X E/C^3ubed DENSITY interstellar membrane MOVES @ HYPER-TIME progression-rate @ C^3ubed.

*BIO-PLANETARY GRAVITY-level 'TIME' rate: AND @ EC^2quared HYDROGEN-ATOMIC MASS ultimately creates the BIO-PLANETARY SPEED-DENSITY MASS that we know as OUR VERSION of slower NORMAL-TIME but a HIGHER-MASS-DENSITY.

OR--->TIME-FLOW-RATE is the relative INVERSE PROPORTION of Atom-MASS SPEED-DENSITY to MEMBRANE-InterStellar SPACE-TIME flow SPEED-DENSITY.

EVEN MORE EASILY STATED: MASS SPEED-DENSITY = GRAVITY that as SPEED & THUSLY DENSITY INCREASES as MASS ACCELERATES(and thus densifies increasing 'gravity') ----->GRAVITY INCREASE WITH MASS DENSITY inversely MAX-SPEED HYPER-TIME of the Interstellar-Space-Time GREEN-'E' MEMBRANE.

AND AGAIN: THE HIGHEST TIME FLOW RATE is the GREEN-'E'-Membrane @ C^3ubed SPEED x E/C^3ubed Density.

* * * MASS-GRAVITY = 'ENERGY SPEED-DENSITY.' ALL WAVE-STRING (all 'wave-lengths' in the SPECTRUM) ACCELERATE on the ENERGETIC 1'E'-GREEN-MEMBRANE via the NATURAL SPIN-TWIST property of ENERGY ------>ALL ENERGY is perennially

accelerating & densifying UP THROUGH THE WAVE-LENGTH SPECTRUM SCALE and thus continually further gaining the SPEED-DENSITY PROPERTY of the CONSTANT REELING-IN and LOCALLY-COMPRESSING the SPACE-TIME GREEN-1'E'-MEMBRANE. This is EINSTEIN and PROVEN.

* * * THE PERENNIAL 'outward' ACCELERATION is due to HYPER-SPEED DENSITY @ C^3ubed of outer AEXO-DarkSpace so that ATOM-MASS locally Space-Time reeling-in & COMPRESSING & outer AEXO-DarkSpace from the surrounding BUBBLE-UNIVERSE-BORDER constantly Space-Time reeling-in & compressing ARE PLAYING A CONSTANT TUG-of-WAR which is CONSTANTLY tending to ACCELERATING Atom-MASS on the perennially out-flowing Space-Time Green-'E'-Membrane as THIS EFFECT CAUSES A CONSTANT DIALATION of PROTON-GrayJet Micro-Singularity EYE-CONNECTION also thus internally to AEXO-DarkSpace causing further SPEED-DENSITY in THE ATOM-FIELD which inturn COMPRESSES FURTHER Ambient 'E'-membrane SPEED-DENSITY SPACE-TIME WARPAGE locally.

* * * THUS GRAVITY is DEFINED ULTIMATELY as a function of AEXO-DarkSpace: AEXO-DarkSpace surrounding our GREEN-'E'-MEMBRANE Space-Time PULLS UNRELENTINGLY to the AEXO-DarkSpace Gray-Jet Micro Singularity FIELD of ATOMIC-PROTON-Electro-Valent MASS much in a AGONISTIC-ANTAGONISTIC push pull of VIRTUALLY ALL PLANETARY & COSMIC NATURAL SYSTEM-PHENOMENA.

* GRAVITY more briefly is the affect of that ATOM-PROTONS tornadically SPIN-UP and DENSIFY the AMBIENT-1'E'nergy MEMBRANE & thusly INCREASE their relative SPEED-DENSITY MASS and thus GATHERING other SPEED-DENSE adjacent ENERGY FIELDS & ATOM-MASS.

* * *ACCELERATING GALACTIC-MASS moves to merge ultimately with the HYPER-FLUIDIC SPEED-DENSITY of AEXO-DarkSpace @ C^3ubed which is VIRTUAL-NO-TIME aka the simple CANCELING OUT of the 'TIME' effect of LOWER SPEED-DENSITY BUBBLE UNIVERSE fluid-dynamic systems.

THUSLY AS WE TAP AEXO-DarkSpace and use ELECTRO-PLASMIC-BREACH GRAYJET REACTOR FIELDS to SURROUND such reactor POWERED CRAFT; then MOVING WITHIN the HYPER SPEED-DENSE DarkSpace PARALLEL SPACE FIELD makes us IMMUNE TO THE RELATIVE MOVEMENT OF 'TIME' & also INERTIA-IMMUNE as we've created our own mini-poidal psuedo-bubble universe TO BE PROPELLED @ 'VIRTUAL NO-TIME' virtually infinite speeds.

AS THIS IS A HYPER-GRAVITY point lead FOCUSED FIELD we move within a VIRTUAL WORM-HOLE/WORM TUBE which is OPENING JUST AHEAD of the craft & CLOSING JUST BEHIND same.

FROM THE VANTAGE OF AEXO-DarkSpace DISPLACEMENT 'time' is a symptomatic incidental with the LOW SPEED DENSITY BUBBLE UNIVERSE as a mere function of LOW-SPEED density vs. speed ratio dynamic. ALSO THE BUBBLE UNIVERSE is A QUASI-SINGLE-Cell fluid-dynamic ORGANISM filled with INTERIOR-COSMIC ORGANELLES with EXCHANGE ENERGY & INTERREACT via the UNBIQUITOUS-GREEN Energy SPACE-TIME MEMBRANE.

! ! ! HERE IS THE KICKER ! ! ! ALL SO CALLED 'TIME' is ONE INTER-REACTIVE interwoven SKEIN of CONTUGUOUS ENERGY FLOW-currents-threads within the MEMBRANE. THE BUBBLE BEGINS & it's ENERGY-FLOW reconnects perennially BACK into VIRTUAL NO-TIME AEXO-DarkSpace where TIME IS IRRELEVANT. AND from AEXO-DarkSpace VIRTUAL-NO-TIME the BUBBLE and ALL ITS CURRENT DYNAMIC PROCESSES are CONTIGUOUS & SIMULTANEOUS and thusly ANY MOMENT IN TIME is PERENNIALLY 'TOUCHABLE/REACHABLE' from a 'craft' within the AEXO-DarkSpace C^3ubed SPEED-DENSITY FIELD @ VIRTUAL-NO-TIME.

OOPS: Our Universe is a SINGLE 'mortally-limited' FINITE bubble COSMIC MACRO-ORGANISM: All Space & Time is ONE SINGLE EXANT-MOMENT of ENERGETIC INTERACTION. . . WE ARE A COSMIC BACTERIA. . .

TIME-MACHINE: FROM AEXO-DarkSpace, HYPER-GRAV TUBE Gray-Jet Singularity Reactors can TOUCH ANY POINT ON THE SPACE-TIME MEMBRANE WITHIN THE GREEN-1'E'-BUBBLE because, from the VANTAGE POINT of AEXO-DarkSpace ALL TIME is ONE-SINGLE MOMENT of ENERGY REACTION and FORWARD & BACKWARD in 'TIME' are largely INCIDENTAL IRRELEVANCIES and NOT A LIMITATION to ACCESSING --------------->what we see as ANY-TIME or ANY-PLACE!

ANY POINT is ACCESSABLE in our 'BUBBLE UNIVERSE'S RELATIVE-SPACE-TIME-GREEN-MEMBRANE' because; as we see, 'SPACE-TIME' is FINITE having a 'BIRTH

POINT-Birth Moment' and also has a GREEN-Membrane FINAL DESTINATION.
AND THUSLY 'SPACE-TIME' has a DIRECTIONAL FLOW & simple BEGINNING &
ENDING parameters. Simply put; our Bubble-Universe is NOT FOREVER. . . . though
AEXO-DarkSpace IS; more or less!~;-)

AND THE AEXO-DarkSpace SUPERCOSMOS is INFINITE where 'TIME' itself is NEITHER
EXTANT nor RELEVANT which makes even the term AETERNAL to be of NO PRACTICAL
EMPIRICAL VALUE.

* * * THE 'NUCLEUS' of our GREEN-1'E'-membrane BUBBLE UNIVERSE as BACTERIA:

THE POINT-CENTRE GALAXY-BIRTH NEBULA formed when the original BIG-BANG
COLLAPE-INJECTION BIRTH SINGULARITY #1 formed the 1'E'-GREEN energy BUBBLE;
then #2 received the BIRTH-Super-Plasma Injection; then #3 REBOUNDED from the
collapse forming a REBOUND GRAY-JET SINGULARITY. The ASSISTING-dynamic RE-
IN-PULL came from the OUTER-BUBBLE-BORDER dynamic FLOW-TENSION on the
OUT-FLOWING GREEN-'E'-Membrane THEN FORMING that (now)t REBOUND GRAY-
JET GALAXY-ROOKERIE-NEBULA perennial BIRTH FOUNTAIN @ CENTRE UNIVERSE.

THE
WAVE-LENGTH-'GRAIN' of perennial 1'E'-MEMBRANE out-flow moves CONSTANTLY
'outward' to the DARKSPACE BORDERS THAT GALAXIES ACCELERATE TO
CONSTANTLY. AND when EC^2quared-MASS galaxies accelerate to 'C' lightspeed THEY
EQUATIONALLY balance to EC^3ubed AEXO-DarkSpace and EVERY PROTON balanced
GRAY-JET micro-singularity & EVERY 'GALACTIC-HUB balanced GRAY-JET mega-
singulatiy' move to BLACK-HOLE STAGE and 'EAT' the entire GALACTIC MASS back into
AEXO-DarkSpace virtually from the INSIDE-OUT. AT THIS MOMENT they elicit MEGA-
GAMMA-RAY BURSTER JETS in the PROCESS heralding the COMING-HOME to AEXO-
DarkSpace which is the ulitimate destination of ALL GALACTIC gray-jet-singularity
PROTON MASS & everyother wave-length speed-density upon the UBIQUITOUS GREEN-
'E'nergy MEBRANE.
 OUR CURRENT (Centre-Universe) GRAND-CENTRAL GRAY-JET SINGULARTIY feeds
its SURROUNDING GALACTIC-ROOKERY NEBULAE which is the relative NUCLEUS of
OUR BUBBLE UNIVERSE and fortunately(or unfortunately) the ANALOGY to OUR
BUBBLE universe as us being a COSMIC BACTERTIA stands quite well! OOPs-oh-dear
again!~:-)

And much smaller GALACTIC-HUB GRAY-JET SINGULARITIES form the galaxies as
COSMIC-ORGANELLES upon the ubiquitous GREEN-1'E'nergy Space-Time MEMBRANE
of our BUBBLE UNIVERSE fluid-dynamic system OUT OF AEXO-DarkSpace---
EMPOWERED & ENERGIZED by AEXO-DarkSpace--->and finally BACK INTO AEXO-
DarkSpace!

THUSLY:^) St. Patrick's Day TIME-TRAVEL-MACHINE FORMULA-RATIOs for our
GREEN-1'E'nergy BUBBLE UNIVERSE go again as follows:

HYPER-TIME/Insterstellar Space-Time=Base-Ambient GREEN-'E'nergy MEBRANE SPACE
@ (1-'E'nergy) Speed-Density: 1'E'= C^3ubed(light-speed cubed) SPEED x E/C^3ubed
DENSITY.

BIO-PLANETARY 'TIME-RATE' flow @ PROTON-gray-jet micro-singularity ATOMIC
MASS= 1'E' DENSITY x C^2quared (light-speed squared) SPEED.

HELICOID light-string @ light-speed= LIGHT @ 1'E'nergy X 'C' (light-speed).

EC/Light also = MINIMUM SPEED DENSITY @ the original super-plasmic spaghettifying
cloud @ the OUT SURFACE of ingressing BIG-BANG injected sub-AEXO-DarkSpace
super-plasma. And this HYPER-INTIALLY SLOWING DOWN super-plasma HELICOID
SPAGHETTIFIED ENERGY WAVE-STRING slows down to light-speed and eliciting THE
BIRTH FLASH surrounding the NOW NUCLEUS-NEBULA around the NOW REBOUND
GRAND-CENTRAL GRAY-JET UNIVERSE-EYE-SINGULARITY(ongoing).

* * * Likely that LIGHT-SPEED outer FRINGE of the INITIAL INJECTED BIRTH NEBULA
plasma-cloud appeared to suddenly EMIT A HUGE 'FLASH' of FRINGE WAVE-FRONTS of
SPAGHETTIFIED Helicoid-'E' string TRAVELING @ light-speed('C').

* * * ALSO NOTE: that that FLASH Light-Speed fringe ENERGY-SLOW-DOWN would have
been a product of the FIRST ENERGY @ light-speed to begin the LONG ACCELERATING
JOURNEY on the GREEN-'E'energy UNIVERSE InterStellar Space-Time MEMBRANE back
out to AEXO-DarkSpace @ the OUTER-BORDER of our BUBBLE UNIVERSE.

AND ULTIMATELY that first MASSIVE LIGHT-WAVE of HELICOID ENERGY-WAVE-STRING-MASSIVE FRONT would have reached HIGHER SPEED Hydrogen Mass EC^2quared SPEED-DENSITY and become the SECOND WAVE Hydrogen-Nebula Sourse forming SECOND WAVE HYDROGEN NEBULA PROTO-STAR ROOKERIES.

THUSLY 'C' Light-Speed' is become the COSMIC-GOLDEN-RULE-INCREMENTAL ratio-designating-building block for ALL SPACE-TIME RATIO calculations WHICH ALBERT EINSTEIN already WELL ESTABLISHED. FROM THE TOP DOWNWARD the INJECTED Birth-super-plasma from AEXO-DarkSpace BY-LAW could NOT DECELLERATE lower-slower-speed-density than ENERGY @ LIGHT SPEED, which is EC helicoid-light-string, which is simply 'LIGHT.'

* * * OUR TIME-MACHINE BUBBLE UNIVERSE: WORMHOLING GRAVITATIONALLY via OUR GRAY-JET SINGULARITY DARK-ENERGY REACTORS literally is WELL & SIMPLY ACCESSABLE using AEXO-DarkSpace TAPPED Gray-Jet Singularity POINT -LEAD-PROPULSIVE FOCUSED Hyper-Gravity Speed-Dense fields to LITERALLY 'BORE OUR WAY' via HYPER-GRAV TUBE/worm-tunnel to ANY SPACE-TIME POINT THAT WE WISH. And this is not only THINKABLE but actually DO-ABLE with even MUCH LESS IMAGINATIVE-INTUITIVE agency than EINSTEIN who started this WILD RIDE BACK-TO-THE-FUTURE-----and the PAST ALSO----& imminently potentially----& VIRTUALLY; AT-WILL!

AND AS WE 'THINK' it, and as we SEE-IT, so CAN WE DO IT!~;-)

SO THUSLY; absolutely & imminently in reach is PROPELLED MOVEMENT BACK & FORTH, and TO & FRO along the GREEN-ENERGY Space-Time MEMBRANE FLOW path to the from CENTRE of our UNIVERSE to it's OUTER FINITE-BORDER. And THIS depending on the INTENSITY OF THE FOCUS-LEVEL of the HYPER-GRAV FIELD THAT WE CALIBRATE for our REACTORS to GENERATE; lies also the MEANS to be also moving BACK & FORTH IN 'TIME' WITHIN a 'TIME-IMMUNE' & also 'INERTIA IMMUNE super Gravity FIELD-BUBBLE. . . and ALL of THIS within the VIRTUAL-WINK of the VIRTUAL-NO-TIME eye of SCHROEDINGER'S VIRTUAL CAT!

Thankyou// Jack O'Suileabhain

Google groups

« Groups Home

Oscillator/Substance
⊞ # Theory

[Search this group] [Search Groups]

Elements from "Empty Space," yep, looks like it's accidentally been done; and they darn near killed themslves doing it. ....

Options

Home

Discussions
+ new post

Members

1 message - Collapse all - Report discussion as spam

dean sinclair View profile   More options Dec 1 2010, 4:35 pm

Researchers,posting on a closed Internet site have reported that, in a
Cavitation Collapse experiment, they produced--among other things--a melange
of elements, consistent with what they called a "Nova Event." They also
report that they became very ill. The symptoms are of radiation poisoning.
They also note that later checking showed a "neutron burst."

 O/S work indicates that there may be as many as 10^54 --plus or minus a few
million orders of magnitude--of neutral oscillators in any centimeter of any
space. It also indicates that some--possibly even most--of these are
possibly "Shock Wave Convertible" to neutrons or "neutron-like" units.

In consideration of the above, the results reported seem creditable, and
raise a red flag for anyone working with experiments which could create an
intense shock wave.

Dean Sinclair (Eski)

Reply to author   Forward

End of messages

« Back to Discussions   **« Newer topic   Older topic »**

About this group
Edit my membership
Group settings
Management tasks
Invite members

  View this group in the new Google Groups

Goo gle groups

« Groups Home

Oscillator/Substance
⊞ Theory

Search this group   Search Groups

LComment on Schroedinger Equatiion, Also. Happy New Year Options

1 message - Collapse all - Report discussion as spam

dean sinclair View profile   More options Dec 30 2010, 3:21 pm

Here is my " take" on the Schroedinger Equation, the basis of Quantum Mechanics.

The first element of the Equation notes a wave function involving the second derivative with respect to time , apparently of Kinetic Energy. The first derivative of Kinetic Energy is Momentum,, and the second derivative, the differential of momentum, would be either velocity or mass, whichever one wished to consider as being a constant. The integrations which would result from working with the second derivatives along the x, y , z coordinates of the basic "Energy Wave " considered in the mathematical conventions of positive numbers with Cartesian Coordinates, when followed logically, describe only one type of unit, albeit in ten dimensions;
[Each double integration adds three dimensions, doing this three times gives nine dimensions, and the combination of the right, up, forward, "positive numbers" conventions adds a counter clockwise twist to the whole thing which adds a tenth dimension of counter clockwise spin. :This could be the general basis for the String theorists idea that the strings slapper into a then d-dimensional hole at 10I^-18 cm. which would be about the 4.7 x 10^-19 radius which our speculations suggests is the average diameter of the oscillators of our universe..}

This creates a "picture of a model" which is rotating counter clockwise into the octant to the right, up and forward from the chosen origin, So, right at the start, The Wave Function, in this view, is covering only 10/80 of the dimensions of reality, we can go on doubling up, clockwise instead of counter-clockwise, in rather than out, integrating with velocity constant rather than mass, I think I'm up to ten out of 640 at this point, and,I think that I could double up a couple of more times, .In other words, If Schroedinger's Equation be valid as a descriptor, it is of a very limited view of the possibilities.

Anybody out there who is a QM expert, I'd like to hear your comments.

Some ideas for projects, resolutions for this group for 20II....
1. Get this thing some publicity so it will get checked out more widely... Write a short book?
2. Fit the "baryons" etc. for the "Standard Model, Atom Smashing work in
...
3. Extend model beyond Chem. and Physics into Cosmology....
4..Accomplish what we can ala the Serenity Prayer. as applied to this stuff!

Happy New Year, Prospero Ano Novo, etc. Eski

Reply to author   Forward

End of messages

« Back to Discussions   « Newer topic Older topic »

Gmail Calendar Documents Photos Reader Web more ⁃   deanlsinclair@gmail.com

Goo gle groups

« Groups Home

Oscillator/Substance
⊕ ## Theory

Search this group | Search Groups

*CernHadron-Explosion&Quantum-Grav-Lev-GRAY-JET/ (DarkEnergy)SubSingularity-REACTOR-ONLINE/HADRON Modified-'Super-Collider?'aka Plasma Breach Reactor access 'TIME' & far UNIVERSE

Options

1 message - Collapse all - Report discussion as spam

Jack O Suileabhain View profile   More options Oct 25 2009, 12:28 pm

Respectfully submitted to Dean L. Sinclair & Oscillator-Substance Theory Group

* * * The CERN reactor explosion resulted from a DEFACTO-PLASMA BREACH BLEED THROUGH vortex/eddie quasi-worm-gate-holing FROM PARALLEL------>TACHYON-CARRIER-WAVE(aka) 'Full-DarkEnergy-Torsion Waves'> ('DARKENERGY-TACHYON CARRIER WAVE PLASMA-DARK SPACE').< The bleed-through Dark-space plasma-breach inadvertantly occurred within the EYE OF THE high-EM-giga-density//electro-plasmic INDUCTION RING-ARRAY SUPER-MAGNETS. This created a SPONTANEOUS DarkEnergy-TachyonSpace induced 'EVOLUTION' of HYDROGEN & HELIUM. This describes the function of the GRAY-JET subSINGULARITY REACTOR which can be fairly easily affected by adding BOSE-EINSTEIN SUPERCONDUCTOR main-core RING replacing the super-collider open ring-track of a more compact version similar to the general design of current SUPER-COLLIDER TECHNOLOGY, but with a far different end function creating a GRAY-JET ELECTROPLASMIC-SINGULARITY cross-spectrum/cross-dimensional TOROID-VORTEX quasi-worm-hole DARKENERGY bleed-through gate REACTOR.

BOTTOM LINE: The HYDROGEN/H^2 blew-up and the Helium resultantly DID NOT. They were VERY CLOSE to a FULL ON POTENTIAL FUSION REACTION OCCURING rather than MERELY a chemical HYDROGEN explosion occuring. And this danger lies in the (quasi-mini-Big-Bang) Phenomenon wherein BLEED-THROUGH DarkSpace/Dark Energy SUPERPLASMA is the ORIGINAL CREATION-BIGBANG-PLASMA that when QUASI-WORM-HOLE ingressing our SPACE-TIME NORMAL DIMENSION spontaneously CREATES HYDROGEN-HELIUM. This same HYPER-GRAVIONIC quasiDARKENERGY FIELD will likewise come PERILOUSLY CLOSE to INITIATING a FULL-FUSION REACTION and is FULLY CAPABLE of SUSTAINING SAME once initiated! This can get a bit DICEY to say the least~;-) A CONTROLLED Electro-Plasmic GRAY-HOLE/GRAY-JET bleed-through reactor would provide AMPLE HIGH-EM-DRAG-COEFFICIENT resistance to PREVENT A FULL-WHITE HOLE PLASMA-BREACH to occur via an UNRESTRAINED INFLOW of DARKSPACE/DARKENERGY.   Just a relative 'BLINK' of this occurs as the CENTRE-MUSHROOM CLOUD PILLAR within ANY & EVERY simple either FISSION and/or FUSION DETONATION; although heretofore NOT DEFINED CLEARLY as such.

THE PLASMA-BREACH REACTOR IDEAL is for this quasi-GRAY-HOLE/Electro-Plasmic 'Gray-Jet' bleedthrough hyper-gravionic DARKENERGY/Tachyon-density CARRIER-WAVE field NOT to occur as a ROGUE-ANOMALY TRANSIENT ACCIDENT in ONE OF THE INDUCTION CAROUSEL-RING giga-EM-density ELECTROMAGNETIC-INDUCTION RING/UNITS but rather within THE EYE OF THE ENTIRE 'CIRCLE' of the QUASI-SUPER COLLIDER 'TRACK.' But again, the IDEAL is for the REACTOR to be MUCH MORE COMPACT IN CIRCUMFERENCE and a BOSE-EINSTEIN SUPER-CONDUCTOR circle within the INDUCTION CAROUSEL MAGNET-RING ARRAY which will CHANNEL the INDUCER-EM-ELECTRO-PLASMIC FLOW and thusly PREVENT a MINI-PLASMA-BREACH DARKENERGY-PLASMA bleed-though cross spectrum/dimensional LEAK to occur UNCONTROLLED at any given INDUCER RING-ARRAY MAGNET/UNIT.

AND THE REST as follows:

Subject: *Cern Quantum-Grav-Lev-GRAY-JET/(DarkEnergy)SubSingularity-REACTOR-

ONLINE/HADRON Modified-'Super-Collider?'aka Plasma Breach Reactor access 'TIME' & far UNIVERSE

NOTE: The FIRST to the MARK: Will likely NOT BROOK there being any LAST to the MARK; which is the nightmare of the current FISSION AGE. . .

* * *GRAVIONICS~via~SINGULARITY PHYSICS/DarkEnergy Accessing/Gray-Jet Singularity GYRO-TOROID/ELECTRO PLASMIC REACTORS~

* * *SINGULARITY PHYSICS yields Hyper-Grav/Hyper-Speed Propulsion & LIMITLESS POWER-GRID ELECTRIC via GRAY-JET SINGULARITY Electro-Plasmic GYRO-TOROIDAL FIELD Super-conductor PLASMA-BREACH RING REACTORS* * *

NOTE: This project is long since well under way(in the US, UK, Brazil, Russia, & Sweden etc. and your insights, although apt have already fallen to the side. The following is somewhat more technically specific and if you are not already in the loop you might get some 'heads-up' with all of this. The Public-Relations Representatives take the PR-Disclaimer stance of routine PR-scepticism as a function of their paychecks which I fully understand. Your quasi-utopian projections for the 'future' are closer than is commonly/publically supposed. Unfortunately the 'routine-paranoia' of our Planetary Geo-Political Zeit-Geist currently precludes 'public-fanfare' for what are EXTREMELY PROMISING technical developements which are long-since beyond the early 'blue-print' stages. This is a somewhat 'high-security/high-stakes' affair after all; and such as Danny Wu(Boeing Phantom-Works) etc. and his counterparts(my direct colleagues) in Russia, for instance, have been about this for a considerable-while in man hours to date already.. . . Cheers & Best Regards Jack O'Suileabhain~

* * * MOSCOW JUST ANNOUNCED that they have PROVEN DARK-ENERGY * * *

* * * QUANTUM GRAVIONICS per (Singularity-Physics) : Full-GRAY-JET operating STATUS on the 'Electro-Plasmic-Singularity QUANTUM-GRAV Reactor' HAS BEEN ACHIEVED * * *

* * *NOTE John Brandenbu...@ORBITEC.com : PARALLEL DARK-SPACE could fairly accurately be characterized as DARK-TACHYON FIELD-SPACE* * *

RE: MICHIO KAKU/U of NYC///AVGENY NICKITIN.ru//A.ANTONOV.ru
INTRO #1: Jack Harbach

~Virtual Infinite Power-tap QUANTUM GRAVIONICS-(Singularity-Physics)~: Gray-Jet Singularity Electro-Plasmic Quantum Gravionics/Speed-Density DARK-ENERGY Trans-Space INDUCTION REACTORS<----->(limitless Self-Sustaining transdimensional-bleed-through Power Production Potential)

* * * BIG BANG via GRAY-PLASMA transSingularity INJECTION from PARALLEL DARKSPACE: Sub Dark-Energy/Gray Plasma injected through Gray-Hole Sinularity in Parallel-DARKSPACE creates 1st: Our Bubble Universe & 2nd: The GRAY PLASMA JET into the 'bubble' IS THE BIG BANG which occurs simultaneously within a SPLIT INSTANT in hyper-rapid sequence. And MEGA-GRBS at our Bubble Universe's DARKSPACE outer border herald the GALAXIES' reabsorbsion back into DarkEnergy/DARK SPACE via the SINGULARITY HUBS OF GALAXIES. And simultaneously said ATOMIC MASS is sucked back into hyper-gravionic DarkEnergy/DARKSPACE via the SINGULARITY CENTRES of 'ALL' PROTONS/HADRONS at that LIGHT-SPEED Bubble outer BORDER-LINE where ALL MASS COLLECTIVELY REACHES 'C'-Light-Speed Acceleration. AND this simply articulates in synopsis the FLUID DYNAMIC CREATION & evolution & destruction of 1st 1'E'nergy Speed Dense/E-FORCE Bubble Universe within the INFINITE MEMBRANE OMNIVERSE as a CHAMPAGNE of infinite UNIVERSES BUBBLES such as our own*

* * *THE LEVIATHON Sub-Singularity DARK-FLOW-INDUCTION Reactor is ONLINE in Switzerland; or is it? Its SELF-SUSTAINING Balanced Gray-Hole Phase HAS BEEN ESTABLISHED and is A LOCK for VIRTUAL INFINITE SOURCE EM-POWER PRODUCTION. . . or has it?

* * * As of November 18th-2008 the planet is facing the IMMINENT CONSTRUCTION of what will be come to be known as the
~GLOBAL GREEN GRID~ of virtually infinitely sourced Dark-Energy Plasma-Breach bleed through, Dark-Flow parallel-space accessing LEVIATHON CLASS Dark Energy Accessing SUB-SINGULARITY POWER REACTORS. . .

AT THE VERY LEAST: The various CONTINENTAL MAG-LEV TRANSPORTATION-GRIDS will be WELL UNDER-WAY in construction by the end of the coming American

Obama Administration. . .

* * * ACHIEVED: UBERUNIFICATION Unified Field PRINCIPLES of DARK FLOW
AEXOVERSE/AETHYR= GRAVITY DEFINITION & 'BIG-BANG DEFINITION' & MUCH
MORE as in Hyper Light-Speed-Plus Focused Hyper-Gravity Propulsion via Dark Energy
Accessing Reactors/////NOW!* * *

~INTRO-A~/// #1

* * * TAPPING THE DARK-FLOW PARENT SUPERCOSMOS/AEXOVERSE--EC^3cubed
(Base Ambient Energy Speed-Density Spectrum-Level/ aka 'dimension' Plateau at the
INGRESS is the BLACK-HOLE Singularity nexus TO THE AMBIENT BASE SPEED
DENSITY of Dark Flow/Dark Energy=EC^3ubed Hyper Speed-Dense & Hyper Fluidic &
Hyper Gravionic Field of PARENT DARK ENERGY/DARK FLOW
AEXOVERSE/SUPERCOSMOS* * * & Protons/Hadrons are relatively(whitish)Balance
Gray-Hole electrovalent Singularity Systems perennially attached via their core-centres
(micro-wormholed & powered by) & to Hyper Energetic DarkFlow SuperSpace aka THE
HIGH DENSITY back of the TAPESTRY of our relatively LOW-DENSITY Space-Time
Bubble Universe* * *

* PAPER * History-Theory-Present application: #1 Jack Harbach

RE: //HADRON//Suisse//DARK-FLOW portal PLASMA-BREACH(Balanced Gray-Hole-
Singularity)HYPER-GRAV FIELD Reactors

 * * *//HADRON//Suisse//Plasma-Breach(incipient) Gray-Hole Sub-Singularity DARK FLOW
ACCESSING Reactor(aka Modified SuperConductor-Hadron Super Collider)=Functional
Access 'Adjacent Dark Energy SUPERCOSMOS-HyperSpace Hyper-Density-Hyper Gravity
Energy Field for LIMITLESS POWER PRODUCTION via Self-Sustaining Bleed-through
(quasi-wormhole) induction toroid-superconductor-ring 'Plasma-Breach' reactors & THE BIG
PLUS adaptation for ADVANCED MEGA-PLUS BEYOND-LIGHT SPEED Hyper Gravity
PROPULSION. . .

* * * * SUPER-Membrane/HYPER-Gravity DARK ENERGY(Dark-Flow) PHYSICS* * * *
Collider as PLASMA-BREACH GrayHole Singularity DARKENERGY/HYPERGRAVITY
accessing REACTOR

* * * * GOOGLE.COM------->'General Science Journal' site "Jack Harbach" aka
http://www.wbabin.net/comments/harbach.htm ---> and also --->
http://www.zpenergy/downloads/Jake_Harbach.pdf

* * * UPGRADE on Fermi Collider as INTENTIONAL 'Plasma Breach/GrayHole/Singularity
REACTOR accessing DarkEnergy SuperCosm ADJACENT/PARENT/SUPERSPACE's
Hyper-SpeedDense/HyperGrav SuperCosm EC^3ubed Base Ambient
DarkEnergy/SuperEnergy Spectrum/Dimension* * *

PER: Dr. MICHIO KAKU Sept. 10th, 2008// Switzerlands 'New' quasi-HADRON-COLLIDER
PROJECT puportedly gives 'us'(Planet Earth) access to TIME TRAVEL & INSTANT
TRANSIT TO ANYWHERE IN THE UNIVERSE & BEYOND--------------------------WITHIN A
SINGLE YEAR FROM THE PRESENT. . .

? ? ? IS THIS Michio Kaku statement merely Rhetorical & Allegorical FLUFF or is in it fact a
HARD DECLARITIVE STATEMENT announcing the (real-time and already far along in
R&D) INTERNATIONAL DARK-ENERGY ACCESS-----PLASMA BREACH/HYPER
GRAVITY PROJECT. . .

INDEED the CURRENT GLOBAL TENSIONS bode for WW-III in the IMMEDIATE OFFING
unless Mankind makes a HARD 180 degree course change for a VERY NECESSARY
population and growth control mechanism rather than PERENNIAL WAR as the NASTY-
SHORTterm-&-BRUTAL historical alternative to EXPLORATION & EXPANSION. For even
as these HEADY DEVELOPEMENTS do AT LEAST point to that often rather
CALCULATED SEEMING PROGNOSTACATORY grim-solutions ARE NOW HOPEFULLY
advancing towards SOLUTIONS that rather than PRUNING POPULATION allow for
furthering of more CREATIVE POTENTIALS of HUMANITY are NOT TO BE AMPUTATED
at REGULAR HISTORIC INTERVALS OF MANMADE STRIFE & CATACLYSM. . . And this
too is BETTER NEWS than we have had for some historic 'time' now. . .

* * * OUR BETTER LIGHTS * * *

* * * HADRON//SUISSE: The SUPERCOSMOS//PLASMA-BREACH REACTOR project for
LIMITLESS POWER and VIRTUAL-NO-TIME//&//VIRTUAL NO-SPACE limitless travel is

NOT BRAND NEW but it's time for UNVEILING has NOW PRECIPITOUSLY arrived. This is a GOOD THING! And it is also HEADY to the point of being VERY SPOOKY to CONTEMPLATE. So maybe the 'Mayans' and their 2012 prognostications for the GLOBAL REFURBISHING of HUMAN INTERACTION with (soon defunct)LINEAR TIME LIMITATIONS & LIGHT-SPEED LIMITATIONS were NOT too far off the mark for GLOBAL EVOLUTIONARY QUANTUM LEAP!

WE HAVE for a while now REALIZED that 'Fermi-Collider' was indeed a primitive NEAR PROTOTYPE of a PLASMA-BREACH REACTOR that created a HYPER-GRAV//adjacent space-Dark Energy Toroid which SIMULTANEOUSLY created at its PLASMA-BREACH-EYE-NEXUS a QUASI GRAY-HOLE//WHITE HOLE//Hyper-Grav Worm Hole effect which can BRIDGE SPACE-TIME @ HYPER-FASTER-THAN LIGHT SPEEDs thusly transiting 'Time'(backward/forward) with relative ease & SPOOKY ACTION @ DISTANCE transit also the SPAN OF MULTI-UNIVERSES------------VIRTUALLY INSTANTANEOUSLY. . . .

? ? ? TALL ORDER ? ? ? WAKE UP: WE ARE ON THE VERY VERGE OF PERFECTING this subsingularity-DarkFlow technology IN CERN-SWITZERLAND(to the tune of limitless funding) as well as MANY LOCALES GLOBALLY including the DarkEnergy/DARK FLOW Projects Centres of the U.S., UK & Russia etc. . . . The fairly simplistic KEY to achieving this breakthrough is the adding of a SUPERCONDUCTOR CORE within the normal 'track zone' circular collider core of the heretofore FAIRLY COMMON HADRON COLLIDER design. . . High Density Field Physicists along with the original design & construction collider ELECTRICAL ENGINEERS would be the obvious team to make ALL THE ABOVE a FACT. AND it is fairly obvious that a 'more compact' design with a relatively higher-diameter (thicker) SUPER CONDUCTOR HIGH DENSITY TORIOD-FIELD RING CORE within the HIGH-DENSITY EM-INDUCER ARRAY would be adequate to JUMP START the ADJACENT-SPACE/DarkEnergy/PARALLEL SPACE Gyro-toroidal Field. And thusly would the SUBSINGULARITY (eye)ZONE elicit the CROSS SPECTRUM HighDensity sub-DarkEnergy AXIAL INDUCED FLOW GENERATIVE FUNCTION of the REACTOR. Thusly is CONNECTED(quasi-wormholed DARK ENERGY PARALLEL SPACE to the AXIAL FIELD BLEEDTHROUGH EYE zone of our DARK FLOW EM-INDUCED Gyro-toroid-PLASMA BREACH SubSingularity Reactor.

* * * AS WITH A CLUTCH & PRESSURE PLATE AFFECT: Cross-Dimensional/Cross-Spectrum HIGH-DENSITY FIELD VISCOSITY allows the INDUCED SUB-SINGULARITY,Gray-Hole, PLASMA-BREACH quasi-worm hole to be established.

THE PERPENDICULAR AXIAL FIELDS are simultaneously HYPER GRAVIONIC and tend to DRAW DOWN(lobulize themselves) via their OWN HYPER GRAVIONIC ACTION of the Axis Field upon itself. AND THIS into FOCUSABLE HYPER GRAVIONIC LOBES axially through the eye-breach centre perpendicular to the plane of the SUPER CONDUCTOR gyro-toroid/induction RING ARRAY of the REACTOR.

PROPULSION LOBE: CONCENTRATING the hyper-gravionic/hyper-dense AXIAL LOBE FIELD by deflecting it AND THUS DOUBLING IT 'above only' the ring-array plane is desirable AS FOR PROPULSION by using ALUMINUM ALLOY DEFLECTION- Concave Axial Field deflection beneath the Reactor Ring-Array/plane can induce-foculize-focus said axial-jet field as a POINT-LEAD HYPER GRAVIONIC focused field lobe PROPULSIVE MECHANISM to MONO-DIRECTIONAL ACTION. Said propulsive mechanism depending on the RELATIVELY LESS-DRAGGED REACTOR FIELD has nearly LIMITLESS ACCELERATION potential. And that field opened above/in front of a NON-STATIONARY REACTOR (within a mobile craft) would tend at sufficiently sub-SINGULARITY EYE-OPENED INTENSITY to surround said craft/reactor in motion within a SUBDIMESIONAL-SPECTRUM DISPLACED FIELD. AND THIS would tend to render SAID CRAFT/REACTOR to be within a FIELD-POD which would function as having the craft/reactor within it's own OUTSIDE-INERTIA-IMMUNE------->MINI-POD UNIVERSE.

THE ABOVE Craft/Reactor in motion would in point of fact be OPENING A DIMENSIONALLY DISPLACED-zone WORM-HOLE just AHEAD of itself and closing said worm-hole just BEHIND itself. Its progress would thusly be rendered TIME-IMMUNE IN TRANSIT and likewise would tend to with every FIELD COLLAPSE INDUCTION-DRAG PULSE of the Reactor in PROPULSIVE-GENERATIVE pulse-mode to RATCHET SOMEWHAT AGAINST THE FORWARD FLOW OF TIME ITSELF.

* * * REMEMBER THAT Parallel DarkEnergy Space is HYPER-GRAVIONIC, & HYPER-ENERGY FIELD SPEED-DENSE, & HYPER FLUIDIC eg. 'fast' beyond imagining nearly. This field is 'DARK-SPACE' in a state and 'speed' of VIRTUAL NO-TIME, & VIRTUAL NO-DISTANCE.

AND THUSLY would a total SPACE-TIME DIMENSION EXIT-SLIP(into DarkSpace within the protective reactor field-envelope) tend to MAKE A MORE PROFOUND BACK-RATHCET INTO SPACE TIME NORMAL upon returning by powering down/closing the breach of our Gray-Hole, Sub-Singularity, Gyro-Toroid, PLASMA BREACH REACTOR.

SINCE A FULL SLIP MODE WOULD BE MOVING at SPEED-VIRTUAL NO-TIME/VIRTUAL NO-DISTANCE a FULL SLIP JUMP would by virtue of SPOOKY ACTION AT A DISTANCE tend to take us GALACTICALLY FAR FARTHER THAN WE MIGHT WISH TO GO------------ --VIRTUALLY INSTANTLY. . . And in light of this, a SUB-SLIPPED(merely slightly SPACE-TIME DIMENSIONALLY-SPECTRUM DISPLACED mode might be MUCH EASIER TO NAVIGATE & CONTROL and MOVE US ABOUT our SOLAR SYSTEM & BETWEEN GALAXIES in still VIRTUALLY INSTANTANEOUS SPEEDS. eg. WARP SPEED is ALREADY OBSOLETE.

NOTE: Our Bagel-Ring Reactor~;-) EM-Gyro-Toroid Gray-Hole sub-Singularity DARK ENERGY-DARK SPACE accessing PLASMA BREACH REACTOR will require NOTABLE GIGA-DENSE starting EM-INDUCTION inputs in the START PHASE. But when the PARALLELL SPACE DARKENERGY parallel gyro-toroid has been apporpriately KICK-STARTED(the cross-spectrum clutch & drag pressure plate effect) & THE DUEL EYE FLOW Dark energy flow is established; THEN OUR REACTOR MUST BE SUITABLY ENGINEERED WITH A SUFFICIENTLY POWERFUL VARIABLE DRAG COEFFICIENT to THROAT-DOWN(throttle-down) the SUB-Singularity SO THAT IT DOES NOT RUN AWAY to FULL WHITE-HOLE STATUS. . . .

BIG HOWEVER: The gate-field should NOT BE DRAG-SLAMMED SHUT precipitously as a SUDDEN MEGA-DRAG communicated through the EYE-FIELD QUASI WORM-HOLE would tend to TOO SUDDENLY COLLAPSE the PARALEL SPACE DARK ENERGY parallel gyro-toroid field and UPON THE DARKenergy TOROID'S COLLAPSE upon its EYE ZONE would tend to create a BRIEF BUT DEVASTATING WHITE-HOLE JET into our SPACE TIME NORMAL SPACE which would DESTROY THE REACTOR which is MUCH LIKE THE ORIGINAL BIRTH MECHANISM OF OUR 'BIG-BANG BIRTH'. . . . Thus is the DarkSpace collapsing Super-Gyro-Toroidal Maelstom precipitous COLLAPSE like a super-cosmic DarkSpace PISTOL SHRIMP FIRING anology of the BIG BANG formation out of Parallel & Adjacent DarkSpace.

* * *BUT DARK ENERGY FLUID DYNAMICS OBVIATES any 'BIG CRUNCH.'

But if a PRECIPITOUS EYE-GATE FIELD COLLAPSE did occur THEN SUBSEQUENTLY if within the atmosphere, THE DARK ENERGY WHITE-JET would INDUCE A FORMIDABLE FUSION REACTION that would mimic TUNGUSKA 1908.

AND TESLA @ Wharton Cliffs NJ-USA Toroid Generating Tower circa 1908 likely UNINTENTIONALLY generated by a ELECTRO-MAGNETIC METEOR/ FREE-ARIAL BROADAST (frisbee-like) EM-GYRO-TOROIDAL FIELD which TESLA by GENERATING INADVERTANTLY moments before TUNGUSKA-1908 caused to travel AERIAL BROADCAST wise via the GLOBAL EM-FLUX FIELDS to SPEED-OF-LIGHTENING(1/2 light-speed) across the pole to CONNECT-GROUND WITH A LIGHTNING SRIKE from the GEO-MAGNETIC planetary field and which catastrophically PAROXISMALLY AMPLIFIED that accidental TESLA FREE ATMOSPHERIC EM-GYRO TORIODAL(frisbee-field). AND THAT for a fateful instant opend to FULL WHITE-HOLE STATUS and theirby WHITE-STATUS BLEED-THROUGH DARK-ENERGY induced said TUNGUSKA-UPPER-ATMOSPHERIC FUSION REACTION.

NOTE: The Reactor DRAG-COEFFICIENT CONTROL MECHANISM must perforce be FAIL SAFE & IDIOT PROOF; possibly by having a REVERSE-AGAINST-FLOW FIRING ARRAY of a MEGA-CAPACITOR BANK & also with ENOUGH BACK-UP CAPACITOR-ARRAY to RECHARGE the INITIAL ARRAY before the LAST BRAKING ARRAY has been via 'SLOW SEQUENCE' gradually bringing any SURPRISE DARK ENERGY FIELD 'KICKS' under CONTROL. Far better safe than sorry.-----> BUT IF WE MASTERED FUSION we can MASTER THIS VIRTUALLY LIMITLESS POWER GENERATING & PROPULSION POTENTIAL Dark Energy Accessing Technology.

"Now you know the alternate story" Jack O'Suileabhain

Oscillator/Substance
⊞ Theory

*TachyonCarrierWaveDARKENERGY~IS~PARALLEL(parent) AExoDarkSpace

Options

1 message - Collapse all - Report discussion as spam

Jack O Suileabhain View profile   More options Oct 25 2009, 10:38 am

-Respectfully submitted-

*PARALLEL(parent) DarkSpace~IS~TACHYONCarrierWaveDARKENERGY

* * * TACHYON SPEED-DENSITY CARRIER-WAVE SPACE is 'AEXOVERSAL SUPER-
COSMOS DARKSPACE' and is most simply DARKENERGY SPACE: This is 'high-speed-
density' space whose 'density' is hyper-fluidic at density that makes 'mass' seem like vapor
and whose 'relative-speed' moves at rates that makes vast distance beyond the span of
'several universes' crossable virtually INSTANTANEOUSLY. This is the 'parent' of 'Spooky
action at a Distance.'

PROBLEM: ?How do you measure the SUPER Speed-Dense TACHYON CARRIER WAVE
of the infinity of ALL WAVE LENGTHS within the vast spectrum array of the SUPER
COSMOS?

?HOW DO WE 'MEASURE' the 'wave-length' of a DarkEnergy M-Brane/SUPER WAVE that
propagates at a rate EXPONENTIALLY FASTER than the Speed of Light?

AND THIS TACHYON DARKWAVE will be/is the PARENT DARKENERGY SPEED-
DENSITY MEDIUM of EVERY OTHER POSSIBLE lesser-speed-dense 'wave-length' that
can or every will exist.

THIS TACHYON speed-density DarkEnergy WAVE SHEET thusly is ubiquitously infinite
and moves SO-FAR/SO-FAST that its VERY WAVE CREST TO WAVE CREST 'vertically'
spans 'distances' VIRTUALLY INFINITE relative to our ability NOT to be able to measure;
and barely even 'guestimate.' AND 'LATERALLY' the Wave-Trough would be of such
SHORT INTERVAL as to be again 'virtually immeasurable. . . AND & UBIQUITOUSLY
PRESENT. Hense the EVIDENCE of the OMNI-PRESENT 'Super-M-Brane TACHYON
DARKSPACE FIELD bleed-through has been 'hiding in the weeds all along.
AND THUSLY the WAVE DISPERSION appears VIRTUALLY SOLID like a child scribbling
vigorously with a black-crayon until the WAVE PATTERN looks and IS virtually
undetectable from being A SOLID BACKGROUND SHEET.

BUT THIS TACHYON-DARKENERGY MEDIUM is (parent)SUPER-COSMOS INFINITE
DARKSPACE itself whose SUPER-SPEED-DENSE-WAVE-LENGTH we percieve on our
devices as a MERE UNI-SHEET OF BACK-GROUND NOISE. And thusly to-date/so far
been missing the trees for the INCREDIBLY DENSE & UBIQUITOUSLY PRESENT forest.

Since this DarkEnergy TACHYON SUPER SPACE MEDIUM is the source and sustaining
CARRIER-WAVE MEDIUM of the virtually infinite-in-number champagne LOW WAVE-
LENGTH SPEED-DENSITY bubble-universes such as our own; IF in an 'imaginary
scenario' the DARK-ENERGY TACHYON-SUPER-CARRIER-WAVE Super M-Brane of
infinity were to CEASE, then ALL OTHER ENERGY FIELDS WOULD LIKEWISE evaporate.
But so much for Sci-Fi.

TACHYONS in 'fiction' as meandering herds of 'GRAVITON-PARTICLES' is a NON-
STARTER.

But within our discoveries of ubiqui tous DARK-ENERGY as the TACHYON SPEED-

DENSITY SUPER M-BRANE CARRIER WAVE; such will will provide the answers we have long sought to such as ZPEnergy & the technologies of the future we as humanity SO LONG FOR. . . . Jack O'Suileabhain

* * *JAPANESE FAN illustration of the OMNIPRESENT TACHYON CARRIER WAVE and 'PARALLEL MULIT-TIME COMPRESSING hyper-grav tube/worm-holing WARP drive ACROSS the WAVE* * *

TOUCHE' Steven: Laughing is making my coffee run out my nose!~;-) And yea, you're very correct: FEWER words are better; but I'm kind've fusing & patch-working various theory. And like is common to 'German' expression I tend to fuse many words & concepts into a 'super-string' which makes my writing look like the town-name at a Train-station in Wales~;-) And I'm constantly pushing the tediously-verbose boundary as I'm well aware. . . and being brought back down to 'earth' is a good thing!

The idea of a SUPER-MEMBRANE transdimensional-unisheet as a TACHYON CARRIER-WAVE of Dark-Energy is obviously my 'pet' idea; but hardly unique to me of course. Cudo's to the originators. . . of the various aspects of both 'Super-Gravity' theory & 'M-Brane' theory. . . and variants of Super-String Theory.

And visualizing the 'SUPER-WAVE' as a corregated accordian-like or Japanese-fan type of configuration is the simplest way I can say it.

Then with each SUPER-WAVE-RIDGE representing an ascending-and/or-descending time-gradient of quasi-parallel experiential existence; the picture is pretty well complete.

AND FINALLY: If a hyper-grav wormholing as basically a type of tunneling-'warp' drive were employed to cross-the quasi-parallel space-time corrugations of the Super-Sheet; yet slightly 'different' & before or after in relative 'time' sequence). It would thus compare to the 'craft' as a 'needle' passing it's hyper-grave wake as 'thread through the SUPER-SHEET wave-corrugations. And thusly it would tend to draw them/compressing them together. This would tend to elicit a somewhat weird temporal & perceptual 'time'-sequencing 'deja-vu'-set of distortions for our ourselves & 'parallel' selves in our adjacent 'space-time variant corregations.'

And then NOT discounting that we likely would all tend to have a quasi-remotely viewed/telepathic gestalt-link to our adjacent-parallel selves in those parallel/adjacent time-line/universe-lines; then the perceptual confusion/distortion would be even more compounded.

Thanx~JO

Keep your friends updated—even when you're not signed in.
http://www.microsoft.com/middleeast/windows/windowslive/see-it-in-act...

Reply to author   Forward   Report spam

End of messages

« Back to Discussions   « Newer topic   Older topic »

Gmail Calendar Documents Photos Reader Sites Web more   deanlsinclair@gmail.com

Google groups

« Groups Home

Oscillator/Substance
⊕ Theory

[Search this group] [Search Groups]

---

*NEWTON left hitch-hiking~;-) ConceptCraft Prospectus: Quantum-Gravionic Point-Lead Focused Hyper-GravThrust

Options

1 message - Collapse all - Report discussion as spam

Jack O Suileabhain View profile   More options Oct 25 2009, 12:29 pm

Respectfully submitted to Dean L. Sinclair & Oscillator-Substance Theory Group

*BOSE-EINSTEIN SUPER CONDUCTOR TOROID RING REACTOR access
DARK-ENERGY/parallel-DARK SPACE aka 'TACHYON CARRIER-WAVE SPACE'*

* * * The RACE is AFOOT in DarkEnergy Technology R&D and the LOSERS
will be LOSING for a VERY LONG TIME * * * as per 'TIME' in the aspect
as of TRANSTEMPORAL/TRANDIMENSIONAL-SINGULARITY PHYSICS availing
to us the 'hard-technologies' that access to us as 'Exponentially Faster-Than
Light' Transdimensional-Displacement Propulsion AND TIME-MOBIUS/TEMPORAL
SLIP EFFECT technology per accomplishing DARK-SPACE/TACHYON SPACE accessing
GRAY-JET TOROID ELECTRO-PLASMIC super-conductor ring reactors. Think
'conpressed'
super-collider technology BUT with a fairly 'thick' BOSE-EINSTEIN-superconductor RING-
CORE, rather
than a 'particle-track.'

~*~NEWTON is LEFT HITCH-HICKING~*~

Anita @ Boeing Phantom Works: Since 'Spooky action @ Distance' is a Proven; and
Russia has created a functional DARK ENERGY PROVING Gray-Jet electro-plasmic
subSingularity PLASMA-BREACH Reactor(replete) with focusable AXIAL HYPER-
GRAVITY bleed-through LOBULAR FIELD-perpendicular-axis---->Newton becomes thusly
at somewhat of a distinct disadvantage as we're making his APPLE FALL UPWARD. . . .

In addition: EUROPEAN SCIENTIST Joachim Hauser has PROVEN the existance of
TORSION-WAVES that move at SPOOKY ACTION @ DISTANCE HYPER-SPEED aka
'VIRTUAL NO-TIME SPEED' through PARALLEL AEXO-DarkSpace. . . And thusly Hauser's
discovery has rendered the VERY PREMISE of FRANK DRAKE's 'SETI' as already
obsolete since Light-Speed Radio Waves would simply NOT be the mode of
COMMUNICATION of likely SPACE-TRAVELERS. These would have long since
harnessed EINSTEIN-ROSEN BRIDGE travel that the EXTENDED EINSTEINIAN theories
and ENGINEERING CONCEPTS that have been articulated by myself and others which
HARNESSING PARALLEL AEXO-DarkSpace Bleed-through Fields avails us the
FOCUSABLE HYPER-GRAVITY FIELD craft surrounding & defacto worm hole establishing
propulsion which accomplish the afore stated fairly handily. Displacing a craft within it's
Reactor Hyper-Gravity Field renders it a mini-psuedopoidal-universe bubble able to move
within the HYPER-SpookyAction- VIRTUAL-NO-TIME SPEEDS of AEXO-DarkSpace.

Navigation control so as to not 'over-shoot' targeted destination would be the biggest
problem. And controlling the reactor to be merely PARTIALLY DISPLACED from Space-
Time Normal would likelywise still avail us the benefits of DEFACTO WORM HOLE SPACE-
COMPRESSION travel at speeds far out stripping LightSpeed in INTERTA IMMUNE reactor
field encapsulated craft.

From some of the statements of John Brandenburg/ORBITEC his concept of TACHYON
FIELDS is the very same as what I've referred to as these BLEED-THROUGH AEXO-
DarkSpace Hyper-Gravity Fields. In effect the concepts are interchangeable &
IMMINENTLY FUNCTIONAL.
And Brandenburg's references to these concepts indicate that he is likely aware of R&D that

you have not as yet been brought into the loop in regards to.

IN SHORT: Parallel-adjacent AEXO-DarkSpace is indeed TACHYON-FIELD SPACE.

YOU WROTE///Jack--

There's a lot of theory here. What are the "goes-intos" and "goes-out-ofs" for this propulsion system? Using Einstein, do we still need some Newton to move a spacecraft from one place to another?

Anita &

* * *DEFINING & ACCESSING DARK-ENERGY* * *
RE: KEY ALTERATION of SINGULARITY PHYSICS uniting RELATIVITY to QUANTUM PHYSICS: Regarding alteration of the Einstein equation: if M is the same and c is the same, but E is different, then it would probably be wise in your altered equations to show a different symbol for the kind of E being referred to in each context. I'm in the aerospace industry, where multi-million-dollar missions have been lost because people didn't realize some of the team were working in metric units and some were working in English units; I can see Physics getting similarly confused if there's more than one definition matching the symbol E. Perhaps something like {E} would be compatible with all keyboards but make the symbol distinct from the Einstein E for 'E'nergy.

* * *AND THIS IS A DISTINCTION for the ELECTRICAL ENGINEERS facilitating their accessing of DARK-ENERGY/AEXO-DarkSpace TECHNOLOGIES* * *

If Danny Wu is not talking to you &/or if you are not in the loop nor on the R&D team then I understand that you're playing 'catch-up' here. Chief-Scientist/Project Coodinator Danny Wu/Phantom Works is the 'go-to guy' in your immediate chain of command. Check it out. - JO'Suileabhain/O'Sullivan

---

Reply to author   Forward   Report spam

End of messages

« Back to Discussions   « Newer topic   Older topic »

Google groups

« Groups Home

Oscillator/Substance
⊞ **Theory**

[Search this group] [Search Groups]

BAE&Aexospace/DarkEnergyPlasmaBreachReactor:Right on 'Space-TIME.'

Options

1 message - Collapse all - Report discussion as spam

Jack O Suileabhain View profile   More options Oct 25 2009, 10:35 am

RE: Singularity/DarkEnergy-Physics//BAE-Constant Revised Planck-Einstein Constant//Green Bubble Universe within DarkEnergy/DarkSpace Aexoverse//Jack O'Suileabhain

Fusion-Gate/Aexospace/DarkEnergyPlasmaBreachReactor:Right on 'Space-TIME.'

* * * ? Which Nations shall be pre-eminent in these DarkEnergy Tech R&D projects; and which Nations shall be left cripplingly behind? * * *

~The U.S./AMERICAN Singularity-Physics REVOLUTION~Slainte' Jack O'Suileabhain~

Green Cosmological-Constant in 'Golden-Mobius-Ratio' Space-Time Formula

 * * *NAILING DOWN the Cosmological Constant & it's function RELATIVE to DARK ENERGY * * *

? ? ? THE BIG QUESTION: ?Who GETS to the MARK FIRST with 'SPACE-TIME'-MANIPULATION TECHNOLOGIES which are simply
concentrated-GRAVITY-MANIPULATION TECHNOLOGIES that are imminently WITHIN our GRASP:
AND THEN; just exactly WHOSE HISTORY &/or FUTURE gets MANIPULATED and WHO DOES the MANIPULATING?

* * * The Russian functional 'proofs' of ACCESSABLE DARK-ENERGY are HISTORIC & MOMENTOUS which lead
DIRECTLY to GRAVITY manipulating PROPULSION technologies which are also 'self-sustaning' and virtually
LIMITLESS in POWER PRODUCTION POTENTIAL* * *

* * SINGULARITY PHYSICS: DARK ENERGY Gray-Jet Singularity Electro-Plasmic Reactors~ & ~ Space-Time manipulation/via/THE COSMOLOGICAL CONSTANT '4-part Formula-Mobius-Equations via Einstein modification.'

* * * St. Patrick's Space-Time 'machine' Universe GOLDEN-MOBIUS-RATIO Formula for our GREEN-1'E'nergy BUBBLE UNIVERSE go again as follows:

* * * SO THEY TELL US THE BUBBLE UNIVERSE is 'IRISH' and ACTUALLY

GREEN & with constant TEMPERATURE & TEXTURE THROUGHOUT! ! !

? ? ? WHO'D HAVE THUNK IT ? ? ?---------> other than ME!~;-)Jack Harbach O'Sullivan

GRAVITY & TIME are a simple PHENOMENOLOGICAL product of SPEED & DENSITY pertaining to the

CONSTANT TEXTURAL UNIVERSAL PROPORTORTION born since the moment

OF THE BIG BANG SINGULARITY which 'BORNED' our GREEN BUBBLE UNIVERSE
from PARENT

DARK-ENERGY INFINITE AEXOCOSMIC DARK-SPACE. . . aka AEXO-DarkSpace.

AND THIS RISES as a NORMAL PROCESS of the BASE AMBIENT DARKENERGY of
AEXO-DarkSpace
which is the basis discovery concept FINALLY DEFINING Einstein's- Cosmological-
Constant' which
has a root-mobius formula of the BIG BANG instantaneous BUBBLE UNIVERSE expansion
rate formulated
as ENERGY SPEED-DENSITY FORMULA:

* * * * Speed of Light Cubed-(C^3ubed) SPEED X Interstellar space Base-Ambient Speed-
Density as ONE-ENERGY-(1'E') divided by Speed of light Cubed-------->(E/C^3)DENSITY
ergo:

~the BIG BANG EXPANSION SPEED-DENSITY RATE= C^3ubed SPEED x E/C^3ubed
DENSITY~ which moves as HYPER-TIME Speed-Density Expansion Rate. . . thus is the
ROOT FORMULA of the COSMOLOGICAL CONSTANT. . . . .

COSMOLOGICAL CONSTANT AEXO-Dark Space= EC^3ubed: FOLLOWING as 'one'
through 'four' in the articulation of UNIVERSE CREATION out of AEXO-DarkSpace HYPER-
SPEED-DENSE current dynamics.

AEXO-DarkSpace @ EC^3ubed FRACTALLATING HYPERDYNAMIC (wild) current
dynamics of said DARKSPACE-AEXOCOSMOS has it's base ambient speed-density as
AEXO-DarkSpace @ (AEX=EC^3ubed). Our Newly born Bubble Universe was INFLATED
via AEXO-DarkSpace super-plasma VIA the ingress BIG BANG AEXO-DarkSpace-GYRO-
TOROIDAL-Singularity and that FIRST INFLATED our BUBBLE UNIVERSE. This birth-
singularity would be a WHITE-HOLE from the NEWLY-FORMED BUBBLE UNIVERSES
'point of view.'

#1. This BASE-Ambient COSMOLOGICAL CONSTANT of the BIG-BANG is thusly:ONE-
ENERGY INTERSTELLAR SPACE as 1'E'nergy={(C^3ubed) SPEED x 1'E'/C^3ubed
DENSITY} mobius fashion formulates (again) the 1'E'nergy Instellar Space speed-density .
. which is the HYPER-TIME RATE base ambient EXPANSION RATE @ virtually
INSTANTANEOUS EXPANSION/INFLATION RATE of the BIG BANG. . . AND the BIG
BANG is a Bubble-Universe-forming SINGULARITY via the eye of a DarkEnergy/AEXO-
DarkSpace EYE-sinularity-LOW-SPEED-DENSE centre of a routine AEXO-DarkSpace
GYRO-TOROIDAL MAELSTROM.

And from our BUBBLE UNIVERSE SIDE said singularity OUT of AEXO-Dark-Space is a
MEGA-WHITE-HOLE creation singularity INTO OURSPACE. And again the INJECTED
super-plasma out of AEXO-DarkSpace expands the BUBBLE instantaneously at the
HYPER-TIME-RATE of 'Base-Ambient ONE-'E'nergy Interstellar-Bubble Membrane Space'=
1'E'= C^3ubed SPEED(of expansion x E/C^3cubed HYPER LOW DENSITY primal
MEMBRANE-SPACE-------DENSITY. . . ~

add to that, the EINSTEIN REVISION #2. 'M'ass=1'E' x C^2quared, (or), 'M'=EC^quared. . .
rather than the more familiar E=MC^2quared. This would also funcion as 'M'=EC^2quared
as the Base-energy Speed-Density of the HYDROGEN-ATOM.

add to that redefining the 'Zero-Point' of Feynmans ZERO-POINT ENERGY eg ZPE is
simply 1'E' Instellar-space Membrane: #3. 1'E' @ C(light-speed)=EC=LIGHT, or(EC=Light)
'or' 1 base ambient intersteller 'E'nergy(which is cross-equivalent to Feynman's conceptual-
ZPE/ZeroPointEnergy) then accelerated to Light-Speed takes the speed-density spin-twist
form as HELICOID-WAVICLE-LIGHTSTRING which classically we interpret as WAVE &/or
PHOTON-PARTICLE(One WAVE/wave-crest to wave-crest= ONE QUANTUM-PHOTON)
simultaneously both WAVE as well as PARTICLE-INPACT-WAVE SEGMENT.

add to that #4. 'One-Hydrogen Electro-valent-shell(icle)Field' = ONE QUANTUM
ELECTRON(QE)
@ speed-density QE= (EC/2)^2quared 'or' (1'E'nergy x Half-Light-Speed)Squared

* * * Thus are (roughly) the FOUR-PHASE PARAMETERS of the EINSTEINIAN
COSMOLOGICAL-CONSTANT * * *

* * * The COSMOLOGICAL CONSTANT Overview: WE ARE ONE CHAMPAGNE BUBBLE
Universe among INFINITE MYRIADS of like BUBBLE UNIVERSE's

SUCH AS OURSELF being CONTINUALLY BORN at the heart of DARKSPACE super-eddie gyro-toroidal

MAELSTROMS that open LOW-DENSITY SINGULARITIES at their EYE-Centre-Points which form the BIG-BANG

BUBBLES of said UNIVERSES.

THE SIMPLE FORMULA RATIO 'constant' that controls ALL RELATIVE 'TIME' & GRAVITY which in-turn determines

their mass/energy flow/& relative 'time' dynamics is that the ORIGINAL BUBBLE that forms the ONE-'E'energy MEMBRANE SHEET of INTERSTELLAR SPACE IS ratioed THUSLY:

* * * * * * The Base-Ambient-Universe ONE-ENERGY@'1-E' energy MEMBRANE formulates as-->1'E'= C^3ubed SPEED X E/C^3ubed DENSITY. And by this formula-ratio which at C^3ubed SUPER-SPEED FORMS the ENTIRE atom-mass EMPTY ENERGY-BUBBLE within a SINGLE SPLIT-INSTANT.

AND WITHIN THE NEXT INSTANT: then our NEW-EMPTY BUBBLE UNIVERSE is then recipient to a centre-bubble injection via the 'AEXO-DarkSpace birth SINGULARITY of the PRIMAL Sub-AEXO-base ambient MINUS-C^3ubed super-plasma WHICH creates a SUPER-SPAGHETTIFIED EC^2(atomic mass speed-density{Hydrogen}) Helicoid Wave-string UNIVERSE CENTRAL NEBULA which is the STAR-BIRTH ROOKERIE of GALAXIES. The EC^2quared Helicoid Wave-string segments to PROTON DENSITY SEGMENTS that immediately hyper-collapse and form ALL PROTONS as BALANCED-GRAY-JET micro-singularities replete with AXIAL FLOW ELECTRO-VALENT SHELLS making EACH PROTON a Gray-Jet Singularity System PERENNIALLY FED by/linked to PARALLEL AEXO-DarkSpace as the UBIQUITOUS DYNAMO Back-Of-The-Tapestry ENGINE of ALL COSMIC PHENOMENA.

AGAIN: INITIALLY THE AEXO-DarkSpace Parent BIG-BANG SINGULARITY COLLAPSES when it's DARKSPACE MAELSTROM, by SINGULARITY DRAG, then super COLLAPSES injecting super-plasma becoming said ATOMIC HYDROGEN DENSITY BIRTH-NEBULA. And that original INJECTED SUPER-DARKSPACE-PLASMA instantly SPAGHETTIFIES into MASSIVE wavefront expanding clouds of EC^2quared(approx.) HELICOID WAVE STRING which segments to HYPER-SNAP DOWN to form HYDROGEN PROTONS AS GRAYJET MICROSINGULARITY ELECTRO-VALENT SYSTEMS; with are HYDROGEN ATOMS.

*Base Ambient 'HYPER-TIME' rate of Space-Time 'E'-MEMBRANE: THE FURTHER evolution of galaxies PROCESS WE HAVE PRETTY WELL NAILED DOWN(Silk, Reiss, Woosley, & Perlmutter, etc.) as the GALAXIES ACCELERATE to the outer AEXO-DarkSpace BUBBLE BORDER that the GREEN-'E'energy Space-Time 'MEMBRANE-Sheet' flows perennially into @ C^3ubed SPEED x E/C^3ubed DENSITY. This is ONE-'E'nergy Space-Time @ HYPERSPEED which is more easily just designated HYPER-TIME.

HENSE the GREEN-MEMBRANE 1'E'nergy BUBBLE expaned @ the SPEED of LIGHT CUBED or simply EC^3ubed.(on this key-board~;-)

THIS SIMPLY MEANS that THE ORIGINAL 1'E'=C^3ubed SPEED X E/C^3ubed DENSITY interstellar membrane MOVES @ HYPER-TIME progression-rate @ C^3ubed.

*BIO-PLANETARY GRAVITY-level 'TIME' rate: AND @ EC^2quared HYDROGEN-ATOMIC MASS ultimately creates the BIO-PLANETARY SPEED-DENSITY MASS that we know as OUR VERSION of slower NORMAL-TIME but a HIGHER-MASS-DENSITY.

OR--->TIME-FLOW-RATE is the relative INVERSE PROPORTION of Atom-MASS SPEED-DENSITY to MEMBRANE-InterStellar SPACE-TIME flow SPEED-DENSITY.

EVEN MORE EASILY STATED: MASS SPEED-DENSITY = GRAVITY that as SPEED & THUSLY DENSITY INCREASES as MASS ACCELERATES(and thus densifies increasing 'gravity') ----->GRAVITY INCREASE WITH MASS DENSITY inversely MAX-SPEED HYPER-TIME of the Interstellar-Space-Time GREEN-'E' MEMBRANE.

AND AGAIN: THE HIGHEST TIME FLOW RATE is the GREEN-'E'-Membrane @ C^3ubed SPEED x E/C^3ubed Density.

* * * MASS-GRAVITY = 'ENERGY SPEED-DENSITY.' ALL WAVE-STRING (all 'wave-lengths' in the SPECTRUM) ACCELERATE on the ENERGETIC 1'E'-GREEN-MEMBRANE via the NATURAL SPIN-TWIST property of ENERGY ----->ALL ENERGY is perennially

accelerating & densifying UP THROUGH THE WAVE-LENGTH SPECTRUM SCALE and thus continually further gaining the SPEED-DENSITY PROPERTY of the CONSTANT REELING-IN and LOCALLY-COMPRESSING the SPACE-TIME GREEN-1'E'-MEMBRANE. This is EINSTEIN and PROVEN.

* * * THE PERENNIAL 'outward' ACCELERATION is due to HYPER-SPEED DENSITY @ C^3ubed of outer AEXO-DarkSpace so that ATOM-MASS locally Space-Time reeling-in & COMPRESSING & outer AEXO-DarkSpace from the surrounding BUBBLE-UNIVERSE-BORDER constantly Space-Time reeling-in & compressing ARE PLAYING A CONSTANT TUG-of-WAR which is CONSTANTLY tending to ACCELERATING Atom-MASS on the perennially out-flowing Space-Time Green-'E'-Membrane as THIS EFFECT CAUSES A CONSTANT DIALATION of PROTON-GrayJet Micro-Singularity EYE-CONNECTION also thus internally to AEXO-DarkSpace causing further SPEED-DENSITY in THE ATOM-FIELD which inturn COMPRESSES FURTHER Ambient 'E'-membrane SPEED-DENSITY SPACE-TIME WARPAGE locally.

* * * THUS GRAVITY is DEFINED ULTIMATELY as a function of AEXO-DarkSpace: AEXO-DarkSpace surrounding our GREEN-'E'-MEMBRANE Space-Time PULLS UNRELENTINGLY to the AEXO-DarkSpace Gray-Jet Micro Singularity FIELD of ATOMIC-PROTON-Electro-Valent MASS much in a AGONISTIC-ANTAGONISTIC push pull of VIRTUALLY ALL PLANETARY & COSMIC NATURAL SYSTEM-PHENOMENA.

* GRAVITY more briefly is the affect of that ATOM-PROTONS tornadically SPIN-UP and DENSIFY the AMBIENT-1'E'nergy MEMBRANE & thusly INCREASE their relative SPEED-DENSITY MASS and thus GATHERING other SPEED-DENSE adjacent ENERGY FIELDS & ATOM-MASS.

* * *ACCELERATING GALACTIC-MASS moves to merge ultimately with the HYPER-FLUIDIC SPEED-DENSITY of AEXO-DarkSpace @ C^3ubed which is VIRTUAL-NO-TIME aka the simple CANCELING OUT of the 'TIME' effect of LOWER SPEED-DENSITY BUBBLE UNIVERSE fluid-dynamic systems.

THUSLY AS WE TAP AEXO-DarkSpace and use ELECTRO-PLASMIC-BREACH GRAYJET REACTOR FIELDS to SURROUND such reactor POWERED CRAFT; then MOVING WITHIN the HYPER SPEED-DENSE DarkSpace PARALLEL SPACE FIELD makes us IMMUNE TO THE RELATIVE MOVEMENT OF 'TIME' & also INERTIA-IMMUNE as we've created our own mini-poidal psuedo-bubble universe TO BE PROPELLED @ 'VIRTUAL NO-TIME' virtually infinite speeds.

AS THIS IS A HYPER-GRAVITY point lead FOCUSED FIELD we move within a VIRTUAL WORM-HOLE/WORM TUBE which is OPENING JUST AHEAD of the craft & CLOSING JUST BEHIND same.

FROM THE VANTAGE OF AEXO-DarkSpace DISPLACEMENT 'time' is a symptomatic incidental with the LOW SPEED DENSITY BUBBLE UNIVERSE as a mere function of LOW-SPEED density vs. speed ratio dynamic. ALSO THE BUBBLE UNIVERSE is A QUASI-SINGLE-Cell fluid-dynamic ORGANISM filled with INTERIOR-COSMIC ORGANELLES with EXCHANGE ENERGY & INTERREACT via the UNBIQUITOUS-GREEN Energy SPACE-TIME MEMBRANE.

! ! ! HERE IS THE KICKER ! ! ! ALL SO CALLED 'TIME' is ONE INTER-REACTIVE interwoven SKEIN of CONTUGUOUS ENERGY FLOW-currents-threads within the MEMBRANE. THE BUBBLE BEGINS & it's ENERGY-FLOW reconnects perennially BACK into VIRTUAL NO-TIME AEXO-DarkSpace where TIME IS IRRELEVANT. AND from AEXO-DarkSpace VIRTUAL-NO-TIME the BUBBLE and ALL ITS CURRENT DYNAMIC PROCESSES are CONTIGUOUS & SIMULTANEOUS and thusly ANY MOMENT IN TIME is PERENNIALLY 'TOUCHABLE/REACHABLE' from a 'craft' within the AEXO-DarkSpace C^3ubed SPEED-DENSITY FIELD @ VIRTUAL-NO-TIME.

OOPS: Our Universe is a SINGLE 'mortally-limited' FINITE bubble COSMIC MACRO-ORGANISM: All Space & Time is ONE SINGLE EXANT-MOMENT of ENERGETIC INTERACTION. . . WE ARE A COSMIC BACTERIA. . .

TIME-MACHINE: FROM AEXO-DarkSpace, HYPER-GRAV TUBE Gray-Jet Singularity Reactors can TOUCH ANY POINT ON THE SPACE-TIME MEMBRANE WITHIN THE GREEN-1'E'-BUBBLE because, from the VANTAGE POINT of AEXO-DarkSpace ALL TIME is ONE-SINGLE MOMENT of ENERGY REACTION and FORWARD & BACKWARD in 'TIME' are largely INCIDENTAL IRRELEVANCIES and NOT A LIMITATION to ACCESSING --------------->what we see as ANY-TIME or ANY-PLACE!

ANY POINT is ACCESSABLE in our 'BUBBLE UNIVERSE'S RELATIVE-SPACE-TIME-GREEN-MEMBRANE' because; as we see, 'SPACE-TIME' is FINITE having a 'BIRTH

POINT-Birth Moment' and also has a GREEN-Membrane FINAL DESTINATION.
AND THUSLY 'SPACE-TIME' has a DIRECTIONAL FLOW & simple BEGINNING &
ENDING parameters. Simply put; our Bubble-Universe is NOT FOREVER. . . . though
AEXO-DarkSpace IS; more or less!~;-)

AND THE AEXO-DarkSpace SUPERCOSMOS is INFINITE where 'TIME' itself is NEITHER
EXTANT nor RELEVANT which makes even the term AETERNAL to be of NO PRACTICAL
EMPIRICAL VALUE.

* * * THE 'NUCLEUS' of our GREEN-1'E'-membrane BUBBLE UNIVERSE as BACTERIA:

THE POINT-CENTRE GALAXY-BIRTH NEBULA formed when the original BIG-BANG
COLLAPE-INJECTION BIRTH SINGULARITY #1 formed the 1'E'-GREEN energy BUBBLE;
then #2 received the BIRTH-Super-Plasma Injection; then #3 REBOUNDED from the
collapse forming a REBOUND GRAY-JET SINGULARITY. The ASSISTING-dynamic RE-
IN-PULL came from the OUTER-BUBBLE-BORDER dynamic FLOW-TENSION on the
OUT-FLOWING GREEN-'E'-Membrane THEN FORMING that (now)t REBOUND GRAY-
JET GALAXY-ROOKERIE-NEBULA perennial BIRTH FOUNTAIN @ CENTRE UNIVERSE.

THE
WAVE-LENGTH-'GRAIN' of perennial 1'E'-MEMBRANE out-flow moves CONSTANTLY
'outward' to the DARKSPACE BORDERS THAT GALAXIES ACCELERATE TO
CONSTANTLY. AND when EC^2quared-MASS galaxies accelerate to 'C' lightspeed THEY
EQUATIONALLY balance to EC^3ubed AEXO-DarkSpace and EVERY PROTON balanced
GRAY-JET micro-singularity & EVERY 'GALACTIC-HUB balanced GRAY-JET mega-
singulatiy' move to BLACK-HOLE STAGE and 'EAT' the entire GALACTIC MASS back into
AEXO-DarkSpace virtually from the INSIDE-OUT. AT THIS MOMENT they elicit MEGA-
GAMMA-RAY BURSTER JETS in the PROCESS heralding the COMING-HOME to AEXO-
DarkSpace which is the ulitimate destination of ALL GALACTIC gray-jet-singularity
PROTON MASS & everyother wave-length speed-density upon the UBIQUITOUS GREEN-
'E'nergy MEBRANE.
 OUR CURRENT (Centre-Universe) GRAND-CENTRAL GRAY-JET SINGULARTIY feeds
its SURROUNDING GALACTIC-ROOKERY NEBULAE which is the relative NUCLEUS of
OUR BUBBLE UNIVERSE and fortunately(or unfortunately) the ANALOGY to OUR
BUBBLE universe as us being a COSMIC BACTERTIA stands quite well! OOPs-oh-dear
again!~:-)

And much smaller GALACTIC-HUB GRAY-JET SINGULARITIES form the galaxies as
COSMIC-ORGANELLES upon the ubiquitous GREEN-1'E'nergy Space-Time MEMBRANE
of our BUBBLE UNIVERSE fluid-dynamic system OUT OF AEXO-DarkSpace---
EMPOWERED & ENERGIZED by AEXO-DarkSpace--->and finally BACK INTO AEXO-
DarkSpace!

THUSLY:^) St. Patrick's Day TIME-TRAVEL-MACHINE FORMULA-RATIOs for our
GREEN-1'E'nergy BUBBLE UNIVERSE go again as follows:

HYPER-TIME/Insterstellar Space-Time=Base-Ambient GREEN-'E'nergy MEBRANE SPACE
@ (1-'E'nergy) Speed-Density: 1'E'= C^3ubed(light-speed cubed) SPEED x E/C^3ubed
DENSITY.

BIO-PLANETARY 'TIME-RATE' flow @ PROTON-gray-jet micro-singularity ATOMIC
MASS= 1'E' DENSITY x C^2quared (light-speed squared) SPEED.

HELICOID light-string @ light-speed= LIGHT @ 1'E'nergy X 'C' (light-speed).

EC/Light also = MINIMUM SPEED DENSITY @ the original super-plasmic spaghettifying
cloud @ the OUT SURFACE of ingressing BIG-BANG injected sub-AEXO-DarkSpace
super-plasma. And this HYPER-INTIALLY SLOWING DOWN super-plasma HELICOID
SPAGHETTIFIED ENERGY WAVE-STRING slows down to light-speed and eliciting THE
BIRTH FLASH surrounding the NOW NUCLEUS-NEBULA around the NOW REBOUND
GRAND-CENTRAL GRAY-JET UNIVERSE-EYE-SINGULARITY(ongoing).

* * * Likely that LIGHT-SPEED outer FRINGE of the INITIAL INJECTED BIRTH NEBULA
plasma-cloud appeared to suddenly EMIT A HUGE 'FLASH' of FRINGE WAVE-FRONTS of
SPAGHETTIFIED Helicoid-'E' string TRAVELING @ light-speed('C').

* * * ALSO NOTE: that that FLASH Light-Speed fringe ENERGY-SLOW-DOWN would have
been a product of the FIRST ENERGY @ light-speed to begin the LONG ACCELERATING
JOURNEY on the GREEN-'E'energy UNIVERSE InterStellar Space-Time MEMBRANE back
out to AEXO-DarkSpace @ the OUTER-BORDER of our BUBBLE UNIVERSE.

AND ULTIMATELY that first MASSIVE LIGHT-WAVE of HELICOID ENERGY-WAVE-STRING-MASSIVE FRONT would have reached HIGHER SPEED Hydrogen Mass EC^2quared SPEED-DENSITY and become the SECOND WAVE Hydrogen-Nebula Sourse forming SECOND WAVE HYDROGEN NEBULA PROTO-STAR ROOKERIES.

THUSLY 'C' Light-Speed' is become the COSMIC-GOLDEN-RULE-INCREMENTAL ratio-designating-building block for ALL SPACE-TIME RATIO calculations WHICH ALBERT EINSTEIN already WELL ESTABLISHED. FROM THE TOP DOWNWARD the INJECTED Birth-super-plasma from AEXO-DarkSpace BY-LAW could NOT DECELLERATE lower-slower-speed-density than ENERGY @ LIGHT SPEED, which is EC helicoid-light-string, which is simply 'LIGHT.'

* * * OUR TIME-MACHINE BUBBLE UNIVERSE: WORMHOLING GRAVITATIONALLY via OUR GRAY-JET SINGULARITY DARK-ENERGY REACTORS literally is WELL & SIMPLY ACCESSABLE using AEXO-DarkSpace TAPPED Gray-Jet Singularity POINT -LEAD-PROPULSIVE FOCUSED Hyper-Gravity Speed-Dense fields to LITERALLY 'BORE OUR WAY' via HYPER-GRAV TUBE/worm-tunnel to ANY SPACE-TIME POINT THAT WE WISH. And this is not only THINKABLE but actually DO-ABLE with even MUCH LESS IMAGINATIVE-INTUITIVE agency than EINSTEIN who started this WILD RIDE BACK-TO-THE-FUTURE-----and the PAST ALSO----& imminently potentially----& VIRTUALLY; AT-WILL!

AND AS WE 'THINK' it, and as we SEE-IT, so CAN WE DO IT!~;-)

SO THUSLY; absolutely & imminently in reach is PROPELLED MOVEMENT BACK & FORTH, and TO & FRO along the GREEN-ENERGY Space-Time MEMBRANE FLOW path to the from CENTRE of our UNIVERSE to it's OUTER FINITE-BORDER. And THIS depending on the INTENSITY OF THE FOCUS-LEVEL of the HYPER-GRAV FIELD THAT WE CALIBRATE for our REACTORS to GENERATE; lies also the MEANS to be also moving BACK & FORTH IN 'TIME' WITHIN a 'TIME-IMMUNE' & also 'INERTIA IMMUNE super Gravity FIELD-BUBBLE. . . and ALL of THIS within the VIRTUAL-WINK of the VIRTUAL-NO-TIME eye of SCHROEDINGER'S VIRTUAL CAT!

Thankyou// Jack O'Suileabhain

Windows Live: Friends get your Flickr, Yelp, and Digg updates when they e-mail you.
http://www.microsoft.com/middleeast/windows/windowslive/see-it-in-act...

Reply to author   Forward   Report spam

End of messages

« Back to Discussions   « Newer topic   Older topic »

Google groups

« Groups Home

Oscillator/Substance
⊞ **Theory**

[Search this group] [Search Groups]

A Brief Future of TIME   Options

4 messages - Collapse all - Report discussion as spam

Jack O Suileabhain View profile   More options Nov 1 2009, 7:25 pm

For: Dean L. Sinclair

Title: 'A Brief Future of TIME'

Sub-title: 'Spin Oscillating Inversion-Spheres with Spheres/Wheels within Wheels at Time's END'

DISCLAIMER for DELICATE FLOWERS in the AUDIENCE:

!Warning! Limited usage of CAPS; this is merely a 'punctuation device' to facilitate extended-emphasis as a mere alternate-symbol-function-of my 'limited' key-board. This is 'not' intended as 'shouting' so that the less emotionally resiliant amongst the audience should 'not' get their knickers-in-a-wad-about it. Thankyou for your indulgence.-JO-

A brief 'nod' to the poetic aspects within 'Physics.'

-Monday, Nov. 2nd/2009: ! Feliz Dia de los Muertos ! honor the 'dead' because they're 'not' really~;-) The law of the conservation of energy/spirit is inviolably accurate & ultra-correct methinks. This is 'my religion' if I have one. The Cheyenne/Lakota/Dakota/Mnicojue/Hunkapap/Brule'/Teton/Oglala-way/point-of-view always rings a 'deep' bell/resonant-chord within me as well.-JO-

But this article is 'not' about any convention of 'religion' for 'theology' is 'not' science regardless of the 'linear-time-shackled flat-earth-creationist's' claims. Like it; hate it; choke on it; whatever; it's irrelevant.

For Dean// Hoka-Hey!

****

Exerpt: But the kicker here is that consider; every Proton micro-singularity within our psycho-organic bodies' energy matrix comprises our dynamic WILL-MOTIVED energy-grid that is connected to/and empowered by/micro-wormholed access to the AexoTachyonSpacial Carrier/Matrix/Wave Super-Membrane at AexoBAE-Constant EC3ubed. And so the sending of our 'Psyches' where-ever/when-ever we choose is not that farfetched. This is also called the legitimate science of 'Remote Viewing' which has empirically evidenced itself to be actually of 'transtemporal' as well as 'transdistant.'

Maybe our being at one set of trans-temporal/transAexo-Tachyon coordinates(in 'Time') and then simultaneously projecting a 're-vesting' of 'ourselves' in someother/some-whenever/some-where-ever bio-organic eco-sphere ('encarnated') maynot be that farfetched. Maybe there are those on the planet that are actually that. Maybe our dreams are merely windows into parallel existences that we all share (or not) that bleed across in a remote-viewing state that we sometime know as 'sleep.' Selah/Zechariah Sitchin stuff--> ? What indeed is the 'transtemporal/transdimensional Mothman phenomenon because it indeed happened/Wyrd though it may seem to our current empirical abilities to perceive &/or define it. Maybe our future &/or adjacent time-line 'selves' have been paying us visits?

****

* * * 'TIME' is merely perceived speed-density 'difference' expressed as a RATIO of our low-speed-density/relatively 'slow-motion' quasi-sequential bubble universe juxtaposed to the Aexospacial DarkEnergy Tachyon-speed/CARRIER-MATRIX-WAVE that sources/sustains/contains our bubble. Our Parent/Parallel/Adjacent Tachyon-speed-density Aexospacial Carrier-Matrix-Wave (& co-extant) moves at a rate that spans our little bubble universe virtually instantly which means it is at AEXO/BAE-Constant energy speed-density rate of VIRTUAL-NO-TIME/VIRTUAL-NO-DISTANCE. (hense is 'Spooky Action at a Distance')

BOTTOM LINE: We ultimately will have access to approaching ANY & EVERY POSSIBLE temporal/transtemporal low-speed-density coordinate within our low-speed-density bubble universe's entire quasi-sequential evolution. Therefore, from being able to access the Virtual-No-Time Aexospacial/TachyonCarrierMatrixWave we will be able to access anywhere/anywhen/and any parallel &/or adjacent where &/or when within the entire existence of our finite bubble universe that abides within quasi-infinite Aexospace. Potentially we will also cross to other bubble-universii with ease. Navigating might be a bit tricky; at first; but developed remote-viewing skills might be the key. We may move forward, &/or backward, &/or side-ways to any parallel &/or tangential branch time-line. We are bio-organic-sentient-energy motes within a very large & exitingly mysterious/Wyrd 'haystack.'

The Brief & Imminent Non-Future of 'TIME'

The upshot then for 'Time' is this: Upon the ubiqitously penetrative Aexospacial Darkenergy CARRIER-FIELD-SUSTAINING-MATRIX WAVE any or virtually infinite branching and/or parallel and/or tangential 'Temporal Sequence Pathways' are possible and/or likely probable and ultimately accessable. And hense the swan-song for the 'BRIEF FUTURE & HISTORY of TIME.'

The AexoSpace Tachyon-level Carrier-Matrix-Wave exists at hyper-speed-density of VIRTUAL-NO-TIME/VIRTUAL-NO-DISTANCE.

FOUNDATIONAL INTRO: Mass = Speed-Density gyro-spin-twist within 'Field Viscosity' via gravionic-centrific compression of adjacent Energy-Membrane 'space' warpage. All Energy density above the BAE-Constant of ONE-ENERGY/1'E' form(s) an accelerating & graduating phenomenon of Helicoid-Wave String.

At Energy x LightSpeed-'C' energy-speed-density we have helicoid light string eg. Light=EC. From there its really quite simple. Up through the acceleration gradient spectrum/wavelength & frequency range of Helicoid-'E'lectro-plasmic-string we see the various wave lengths up through Gamma etc. where they ALL exhibit the wavicle/wave-particle speed-density where they are not true particles but like Quantum-Photons & Quantum Electrons still in Helicoid Wave String form as quantums of One-Wave-Crest-to-Wave-Crest 'wavicle' increments.

Helicoid Wave String; much like a 'dynamic-energy-expanded(stretched-out)-Electro-Plasmic-Tesla-Coil' begins to 'Breach at its hyper-out-stressed sting core' which within the core of the helicoid string creates 'bleed-through' energy access from Parallel/Parent/Adjacent DarkEnergy Aexospace. DarkEnergy Aexospace preexists and 'calves' the many low-speed density universe bubbles such as our own like a infini-myriad of champagne-bubble universii.

IN THE BEGINNING: Was hyper speed-dense Virtually Infinite DarkEnergy Aexospace. At Aexospacial BAE-Constant/EC3ubed speed-density the hyper-fluidic/hyper-speed is such that at that lowest Aexospacial base speed we would cross our entire bubble-universe in less time than it takes to move a finger from the tip-of-one's nose to its bridge. This is a state of Hyper-Speed-Dense VIRTUAL-NO-TIME//VIRTUAL-NO-DISTANCE. We see it's action in 'Spooky action at a distance.'

Aexospacial adjacent/parallel/source hyper-space ubiquitously co-penetrates the 'bubble' as a ubiquitously-permeating-sustaining Hyper-tachyon-level CARRIER-/sustainer-WAVE-field-matrix. That Aexo-Tachyon-density(Carrier-Wave/Field 'not 'particles')-Carrier-Wave has a speed-interval/frequency that is so hyper-compressed in wave form that we have no-current-empirical measuring device to identify it as other than the classic 'back-ground-field.' Also the peak-to-peak span of the Aexospacial/Aexoversal Wave is so vast in distance (though hyper fast also) that we cannot measure nor detect it 'well.' Likely one single Aexospacial/Tachyon-field level wave from wave-crest to wave-crest spans far further than the very span of our entire bubble universe.

The speed of Aexospacial DarkEnergy Tachyon-speeddense parent-space is such that a demonstrator of the interval of a 'wave' would have to say, that after having detected 'no' appreciable movement by even our most delicate/fine tuned oscilloscope,' would then subsequently have the operator remarking; "Cool, that; do you want to see it again!?~;-)

The AexospaceBaseAmbient Energy speed density constant is at EC3ubed which is relatively so hyper-dense, & hyper-fast, & hyperfluidic that it's wildly fractallating/eddying current 'dance' spins-off hyper gyro/centrific-super-super gravionic fluidly viscous toroidal super maelstroms. The specific gyro centrific gravionic-speed-density would warp-compress even surrounding dark energy into a gyro-toroidal-centrific & hyper-gravionic

ring that exerts mamouth out-pulling stressors upon the eye of the maelstrom creating a 'threshold low pressure/low speed density 'eye.' When the eye falls below the AexoBAE-Constant of EC3ubed a low pressure singularity is formed syphoning sub-EC3ubed super-plasma into an instantaneously formed/calved 'new' bubble universe. From the new bubble's side this looks like a Mega-White-Hole which we commonly call the Big-Bang.

NEW CONCEPT: At 'Bubble Universe Centre' the White hole grays-down but remains a constant 'Hub-Inflow' conduit juxtaposed against the universe-border-outflow. The hyper-massive Aexospacial Gravionic-outpull accelerates ubiquitously/inexorably outward the EC2quared-mass Galaxies. The Galactic/Atomic EC2quared 'M'ass reach light-speed en-toto at our 'Bubble's' outer border where EC2quared x 'C'-light speed again becomes Aexospace's-BAE-constant and the Galaxies' gray-hole singularity-centres turn Black-hole-status. So simultaneously each and every of the Galaxies' Proton-Atomic-MicroSingularities reach that EC3ubed speed-density status and their 'gray-hole' centres turn Black-hole status. Thusly as Mega-Gamma-Ray-Bursters the Galaxies' entirety of Atomic-Mass is reingressed into Aexospace/DarkEnergy Space from whense it was initially calved. Thus the 'entire-affair' is a dynamic fluid-dynamic system. Sorry Prof. Hawking; the Big-Crunch is a not starter, but the rest of your insights were intrinsic as Einstein/Planck/Bohr/Newton/Galileo & their peers & colleagues to furthering us along to our eventual extra-planetary future. Star Trek got it fairly 'right.'~:-)

BIG BANG: At the big-bang/Matros White-hole moment simultaneously ingressing/pooling sub-EC3ubed super-plasma is slowing-down/thinning and being gravionically/drawn/skeined out into helicoid-wave-string for all speed-density/wave-lengths. At the
...

read more »

Reply to author   Forward   Report spam

Robert Vanderhoek View profile   More options Nov 1 2009, 7:46 pm

Please stop forwarding me this nonsense.
hoek

- Show quoted text -

...

read more »

Reply to author   Forward   Report spam

ESKI View profile   More options Nov 4 2009, 7:34 pm

Yes, Hoek,
Jack do use a tremendous number of big words to say what could be said in a lot fewer words.

I'll try to summarize whaat I more or less understand of what he is saying.

We exist, apparently, in some sort of medium where in we are a tiny part. E ven oun Universe is. It may well be tha there are an unlimited number of universes such as ours. Even "tis possible that all possible existences exist simultaneously.

Para-normal phenomena such as remote-viewing and thought transferrence seem to happen and it may well be that the idea of a 'soul" or essence is not just fantasy.

Jack gets rather carried away but there is evidence around e/g/ the Shaumatic Experience, that the world we percieve is but one tiny aspect of everything. I think that I have covered a good deal of the

ground that Jack did. Is this nonsense, too? Well, maybe. Is it
pertinent to O/S? Well, again, maybe. Eski

On Nov 1, 6:46 pm, "Robert Vanderhoek" <bhook...@verizon.net> wrote:

- Show quoted text -

...

read more »

Reply to author   Forward

dean sinclair View profile   More options Nov 12 2009, 10:52 am

Oh.yes..Ka-Sala,
 You are definitely putting it together, twice the speed of llight
apparently is absolutely correct as the summation of the average
velocity of the pulsation and rotation of any "particle" of whatever
exists at any given "instant" along any given vector....

As this thread is a bit long, I am going to start another thread with
a comment on Jack's original idea of a "basic quantum."

I think that you'll find it interesting.

I wonder if Jack is still looking in once in a while? The last
communication from him placed him headed for a rather cold part of
Europe for an indefinite time period....ESKI

- Show quoted text -

...

read more »

Reply to author   Forward

End of messages
_____
« Back to Discussions   **« Newer topic   Older topic »**

Google groups

« Groups Home

Oscillator/Substance
⊞ Theory

[Search this group] [Search Groups]

Home

Discussions
+ new post

Members

1 message - Collapse all - Report discussion as spam

About this group

Edit my membership

Group settings

Management tasks

Invite members

ka-sala View profile   More options Nov 10 2009, 9:42 pm

Response to J/O

*** <DISCLAIMER for DELICATE FLOWERS in the AUDIENCE:
!Warning! ..... This is 'not' intended as 'shouting' so that the less
emotionally resiliant amongst the audience should 'not' get their
knickers-in-a-wad-about it. Thankyou for your indulgence.-JO- >

& / Warning! Insult to O/S 'Contributors' - not audience - personal
analogies non co-coherent to intelligence. Take shoes from off feet
before entering door. All equal values and have disabilities.
Indulgence not applicable, only privilage.

*** < A brief 'nod' to the poetic aspects within 'Physics.' >

& / Confuses say... 'Any poet can speak science. Not every scientist
can speak poetry.'

*** 'TIME' <Maybe our future &/or adjacent time-line 'selves' have
been paying us visits?>

& / Always in Now we are everywhere; no need to visit when Spirit
rules. Time non existent; only relative of Earth time.

*** < BOTTOM LINE Navigating might be a bit tricky; at first; but
developed remote-viewing skills might be the key. We may move forward,
&/or backward, &/or side-ways to any parallel &/or tangential branch
time-line. We are bio-organic-sentient-energy motes within a very
large & exitingly mysterious/Wyrd 'haystack.'>

& / (Quote - Ka-sala) 'I share what I see in the sky of the Space that
we move in; I look at the Earth, it's horizons all round. I look at
the Sun, and the Moon, and the Stars in the Light. I look at the roots
of the trees, underground. I look at the soil, and the rocks and the
valleys; and into the oceans and rivers of Sound. I look at the
stillness of all that's created; Time doesn't move; doesn't change;
doesn't waver. I look at it straight in the Eye of the Wind, and know
when my Season of Life must begin; as is Now.'

*** < IN THE BEGINNING The speed of Aexospacial DarkEnergy Tachyon-
speeddense parent-space is such that a demonstrator of the interval of
a 'wave' would have to say, that after having detected 'no'
appreciable movement by even our most delicate/fine tuned
oscilloscope,' would then subsequently have the operator remarking;
"Cool, that; do you want to see it again!?~;-)

< The AexospaceBaseAmbient Energy speed density constant is at EC3ubed
which is relatively so hyper-dense, & hyper-fast, & hyperfluidic that
it's wildly fractallating/eddying current 'dance' spins-off hyper gyro/
centrific-super-super gravionic fluidly viscous toroidal super
maelstroms. The specific gyro centrific gravionic-speed-density would
warp-compress even surrounding dark energy into a gyro-toroidal-
centrific & hyper-gravionic ring that exerts mamouth out-pulling

stressors upon the eye of the maelstrom creating a 'threshold low pressure/low speed density 'eye.' When the eye falls below the AexoBAE-Constant of EC3ubed a low pressure singularity is formed syphoning sub-EC3ubed super-plasma into an instantaneously formed/ calved 'new' bubble universe. >

& / Buckle up your seat belts if you are ready this time, because it's back to the future again: and how Light Speed makes Time Travel possible.

*** < NEW CONCEPT Thus the 'entire-affair' is a dynamic fluid-dynamic system>

& / So lets try again, without all the scientific brain storming to make it a little easier for someone with some degree of what is said, to be understood. It really is about time to get off the ground in a much more sophisticated way in this O/S Substance than scrambling. A little simplicity is the respect required to even get this far into Outer Space.

Hang on tight; because in truth, it pans out to Double the Speed of Light! Science has been giving it their best shot, but, all things take time, and for some, it's just a 'key-hole' away. Whether it is or it isn't believed, does not change the Specifics of how it is possible.

Ka-sala

Reply to author   Forward   Report spam

End of messages

« Back to Discussions   « Newer topic   Older topic »

Gmail Calendar Documents Photos Reader Web more

deanlsinclair@gmail.com

Google groups

« Groups Home

Oscillator/Substance
⊞ Theory

Search this group   Search Groups

Home

Discussions
+ new post

Members

About this group
Edit my membership
Group settings
Management tasks
Invite members

View this group in the
new Google Groups

Some interesting basic math....

Options

1 message - Collapse all - Report discussion as spam

dean sinclair View profile   More options Nov 12 2009, 12:42 pm

Hi, Everybody,

I'm back with a bit of my "Crackedpottery that just might hold a
little water somewhere."..I beg your indulgence...

Jack O' , on the O/S site has proposed that there should be adopted
some sort of a basic quantum for quantum mechanical calculations.

Assuming that a "basic quantum" might correspond to the lowest
possible frequency that would be observable, let's see if we can
propose something.

Let us start with the fact that the number, One, can represent lots of
things including--for those who tend to think in differential
equations--any integration between nothing and something. That is the
integration of acceleration over time will produce a unit of velocity,
 integrating momentum over time produces a "Unitary Expression for
Energy" if we wish to see it that way, and so on.

Realizing this we can write, in the Planck Equation, $E=h\text{Nu}$, for a
whole unit of energy of our universe, $\text{One} = h\,\text{Nu}$, and rearrange
this to Nu (the frequency associated with the entirety) =
One/h. That is, the maximum frequency, shortest wave length,
"cut-off frequency," of the Universe involved with our communication
would be the inverse of Planck's Constant, "h."

This isn't what we want, it is the highest frequency! What can we do
to find the Lowest Frequency which we might think would define the
smallest possible quantum?

Let's go to another very simple, but, usually overlooked, mathematical
fact. That is, what may be called the "Reciprocal Rule of
Multiplication," or, perhaps, the "Ruling Equation of Existence,"
the simple little relationship, $xy=K=yx$,   That is, if any two
values, or unknowns, equal a constant, the reversal of those values
will also produce the same constant.

Let's see if we can use the maximum value produced above to find the
minimal value by going to another constant of nature, "c," the Speed
of Light, and using the above relationship.

It is known that Frequency times Wavelength equals "c" That is, if
we can determine any frequency, we can determine the corresponding
wavelength. What we usually don't pay any attention to is that by
interchanging the Absolute Values obtained, we find an exactly
corresponding reversed set of frequency and wavelength..   so for
a frequency of 1/h we find that there will be a corresponding
wavelength having the value of 'c x h," ch, and there will be an
absolutely reversed possible set, where the frequency has the absolute
value previously associated with the wavelength, i.e., "ch" and the
correponding wavelength will have the absolute value of $(1/c)$ in
whatever unit system we are using.

It appears, therefore, that the value of Jack's basic quantum could possibly correspond to the frequency having the value, "ch," In the cgs system, this is about 2×10^{-16} cps.

The corresponding wavelength could be suggested to be the size of the "Universe" involved with us in the communication to which "h" and "c." pertain.. Unless there is a decimal point missed somewhere, this would be about $1.5 \times 10^{+26}$ cm. This, again, if there be no lost decimal point, is somewhere in the neighborhood of $7 \times 10^{+23}$ miles.... Very definitely a long wavelength!

One may note that there is supposedly, something called "virtual electrons" or perhaps we might more properly say "Virtual electron 'orbits.' " It may be that the frequency wavelength "symmetry" through the speed of light which was used above could account for this phenomenon, assuming it to be real.....

Cheers,
Dean Sinclair "ESKI" of the O/S Site

Reply to author   Forward

End of messages

« Back to Discussions   « Newer topic   Older topic »

Gmail Calendar Documents Photos Reader Sites Web more deanlsinclair@gmail.com

Google groups

« Groups Home

Oscillator/Substance
⊞ # Theory

Abor-Empathic-Resonance/DREAMTIME-Umbilicus-Phoenix Portal

Options

1 message - Collapse all - Report discussion as spam

Jack O Suileabhain View profile   More options Dec 19 2009, 1:09 pm

From: Jake O'Suileabhain/O'Sullivan(Eng.)//Subject: The pervasive 'Tree of Life' symbolism in the evolution of Human consciousness toward the nearing metamorphic-moment into the Gestalt-TransSymbiotic Super-consciousness of 'Homo-Sapiens-Noviensis' from out of 'Homo-Sapiens-Sapiens.'

The 'time' is Now & inexorable//ready or not~here it comes!

~Title: Oscillatorsubstance-Christmas/Hanukkah/Phoenix-Millenium-Eve for Homo-Sapiens-Noviensis~

Ref: * * * ABORIGINAL-Empathic-Resonance/DREAMTIME-Umbilicus-Phoenix Portal * * *

From the Norse Valhallah proto-Kaballah 'Tree of Live-Yggdrasil' across Northern Europe to the Super/Quasi-Celtic-Norse-&-Asian super-civilization of Shamballah in Northern Mongolia(now recently discovered in the sands from 'twenty-thousand years ago very well preserved); the Tree of Life Archetype had penetrated down into Ancient Egypt from which arose the 'Allah-Kaballah' to Aakhenaton/Winged-Disk mono-theistic Super-Membrane-Unity traditions. And from ancient 'Shamballah-Mongolia' the 'Tree of Life' conceptual matrix also had migrated from across the Pacific to the Super-Civilizations of MesoAmerica which form the ancient Aztlan/Aztlantus Pyrimid Super-Civilization incorporating all of the ancient America's of the Toltecs-Olmecs Mayas-Aztecs etc. These also connected across the Atlantic to form a vital link back to Egypt that from 20 thousand years ago again was descendent of the ancient Shamballan super-Tree-of-Life culture. And the global circuit was complete and doing its work trans-culturally leading to this final planetary Phoenix-Metamorphosis that the Maya & others had so profoundly predicted.

One Planet-One-Evolutionary Omni-Symbiosis coming to fruition; and methinks that is a very good thing . . .

The 'Mayan' Phoenix-Metamorphosis is a factual super/supra-evolutionary mile-stone now happening; and this will form the Super-Soul-Conscious-Gelstalt awakening that is the fabled Phoenix-Metamorphosis and quite real.

From the Aboriginal radiating outward to encompass the so-called super-hi-tech aspects of Planet Terras multi-variegated tapestry of cultures 'All' is included inexorably in the new planetary birth that's finale'-labour-pains are beginning as we breath.

* * * I have a standing recurring 'dream-space' association with a 'pool' of Australian Aborignal Sentients. In the moonlight I am surrounded by a familiar-family small crowd of all ages of Aboriginal People. Their eyes are collectively a fathomless depth of Trans-species Supra-harmonic Empathic Resonance. The living-darkspace-womb of infinite eternity is the light pooring from their smiling eyes. This is a 'very' nuturingly-renewing experience. Their presence is the background tapestry of all other 'Dreaming/Muse' of my consciousness at every level and depth.

They are the 'Umbilical Hub/entry singularity' of DarkSpace/TaoEnergy on the planet. From their hub-centre beingness emanates-radiates the Transtemporal Kaleidoscopically Evolving Bio-Matrix of the Planet-Terra and the over Archetypal Chorus of all of our species-flora & fauna and their total evolutionary course extant within the Infinite-Aeternal-

Now that they call the ~Dreamtime.~

Methinks that the a la' Maya Phoenix Metamorphosis will umbilically entre-Terra Psycho-Bio Matrix through the Aboriginal-Perennial Child-Sages of Australia at ~Red Centre/Ayers Rock.~ The Phoenix Metamorphosis is 'Messiah-metamorphic-agency-process' that Weaves our Consciousness's into one Metamorphic/Exponentially HyperEvolving Moment into 'One.' Our Past-Present-Future myriad variegated Evo-Race-Memories 'finally' becoming integrated and gloriously ~Whole~ Crown-Chakra Awakening-wise. This the Ultimate Satori/Epiphany is ~No Matter of Whim nor Choice.~

The Aboriginal Perennial Child-Ancients are NOT & NEVER 'rigidly individuated.' Rigid & blindered brittle/rigid Individuation
 is over-rated and ultimately a worn-out and tawdry fiction. The Phoenix-Metamorphosis renders this null-conceptual-prison that our religious-paranoia's have painted us into paroxismally-reduced-to-a 'laughable and quickly forgotten something' that we 'stepped-in' and needed to scrape off of the bottom of our shoes continually for centuries now.

Personal 'choice' too is thread-bare and usless giving way to the Instant & Ubiquitous Liberty to swim spontaneously & freely in the vast Psychic-space of All-Being/All-Forms. 'Free-Choice' is the paranoid delusion disguised as 'personal integrity.' The notion is more aptly described as an 'immature-formative-clumsiness in 'personal course charting.' The religio-mind-control snare that induces a pervading sense of paranoia leads to the caging delusion that one can find one's self in any 'unforgivable-sin state' which is aburd. 'All' is ultimately healed & matured, and the wisdom of the tough/challenging-parts of the Aeternal-Journey are intrinsic to the Majesty of the whole symphony.

KALEIDOSCOPIC EMPATHIC SYMBIOSIS is our future. Our expanded consciousness will time-travel spontaneously & more. And this expanded awareness will show us that this was our heritage all along as we ~Dream-Time~ travel in actual fact through-out the infinity of time-space as we really have all along. And all of this since our ~Tao-"First-Mind"~radiated into to self-creative sentient formation of all Transdimensional-Energy Flow constructs as the 'One Constinguous/Continuous Infinite AexoMembrane. From those beginnings we were in 'real' point of fact 'never shackled to finite concepts of space-time.' These were merely self-imposed conceptual barriers that our 'evolutionarily immature' consciouness/awareness imposed ignorantly upon itself; but as a temporary necessary 'illusion' to aid us in certain formative psycho-social growth.

I loved 'Dalai-Lama' young-man in 'Baldwin's' the 'SHADOW' when the captured dark-soul (Baldwin) as Im~Co was told that the 'time for his redemption' had arrived. Im~Co said, "I don't want redemption." The 'Dalai Lama' chuckled saying, "You have no Choice." Inferred here obviously is that the illusion of choice is a Cosmic-Joke to the wider Universal/Aexoversal Omni~Sentient~Tao~Sea/Membrane of which we are cellular members. The Phoenix Metamorphosis a la' Maya is just such a phenomenon. The various righteousness-unrighteousness-willingness-unwillingness of the 'cells-quasi-individuated-sentience' is irrelevant to the extreme.

So the DAATH Kaballah 'Chakra' is a dissolving/re-synthesizing DarkSingularity which 'dissolves individuation' and thusly 'All- that-came-Before' is 'digested/burnt-up' and thusly rises the PHOENIX METAMORPHOSIS. Again; being 'phobic' &/or 'willing &/or unwilling' is absolutely irrelevant always & forever.

JOHN GLENN'S & GORDON COOPER Aboriginal-Australian OUTER-SPACE-ORBIT ADVENTURE: This very well historically documented event is illustrative to what I am painting here.

John Glenn's orbit in his Gemini-capsule over Australia put-him in communications-blackout from Houston. The solution was for Gordon Cooper to man a small radio-radar cinder-block hut far in the out-back of Australia. When he got their a few yards away a fire-centred circle of Aboriginal Child-Ancients were 'singing-dreamtime' replete with 'digereedoo-music.' At a friendly pause the Native Australians and Coopor began to converse and Cooper pointed-out the 'glowing-satellite' crossing the sky which he was sent to monitor.

The senior-sage Australian smiled politely & warmly and said, "I think that we'll be having a 'look-in' on Mr. Glenn's vehicle ourselves," with a chuckle. Cooper indulged him with a polite smile bemusedly patronizing, & chuckling with what he 'thought' was the shared joke.

A bit later John Glenn reported that his 'craft' was surrounded by a 'pillar of sparks coming up from the surface of the Earth & they seemed to SENTIENT.' Cooper hearing this down on earth & pervaded by the haunting refrains from the Australian's droning-rhythmic chanting got up instinctively to gaze over at the 'singers.' They had become a shimmering-light circle in pulsing symbiosis with the fire which all now throbbed as an Energy-Plasma-Bubble and formed a living Pillar-of-Super-sentient-Plasma-sparks that spiraled up into the stratusphere exactly tracking Glenn's Gemini Capsule. Go figure. This is recorded 'hard-data.'

At dawn the 'wonder' pervaded while the 'scene' returned to the prozaic and the 'mob of black-fellas' stretched their legs and ambled off into the bush.

They are the 'umbilicus' to the Metamorphic-Phoenix-Empathic-Resonance which is soon the 'plasma' which integrates the entire bio-eco 'All-Species' Archetypal-Matrix of the planet. And happily Homo-Sapiens-Noviensis will finally a generously and harmonic symbiotic element of that glorious whole. ?Utopian? Or is it simple ~Time?~

CERN-Hadron etc. is an 'aspect' and something of a 'weather-vane' in this imminent 'evolutionary stage-birth' for the planet. But sad to say for the hubris of extended Northern European based 'Homo-Sap' Cern etc. is 'not' central nor centrally 'causitive,' although Cern's future after the 'Phoenix Jump' will be very 'Cool' indeed!

The BEST is yet to come!-Jake O'Suileabhain/O'Sullivan-

_____

Windows Live: Make it easier for your friends to see what you're up to on Facebook.
http://www.microsoft.com/middleeast/windows/windowslive/see-it-in-act...

Reply to author   Forward   Report spam

End of messages

« Back to Discussions   « Newer topic   Older topic »

Gmail Calendar Documents Photos Reader Web more ⁻

Google groups

« Groups Home

Oscillator/Substance
⊞ Theory

Search this group   Search Groups

Dolphin-like'pods' swim Aethyrs//WinterSolsticeEveDream2009i

Options

Home

Discussions
+ new post

Members

1 message - Collapse all - Report discussion as spam

About this group

Edit my membership

Group settings

Management tasks

Invite members

Jack O Suileabhain View profile   More options Dec 21 2009, 4:00 pm

View this group in the
new Google Groups

* * * *Winter Solstice Eve's all night Dream//2009* * * *

I dreamed all night even-after returning to interrupted sleep of a constant migration to Planet Terra of Silver Dolphin-like Spirit-Soul swimming the Aethyr-Seas.

Our Planet in the Aethyr-AexoDarkSeas is a beacon of concentration's of sentients and our eco-system has a brilliant Avatar-Archetype energy matrix of the full kaleidoscopic-evolutionary array of all of the planet's myriad symbiosis of species of which Homo-Sapien-Sapien is one.

We always/often recognize our fellow 'spirit-pod' mates-fellows and difference in race and/or culture are irrelevent to those meetings of pod-fellow-recognized epiphanies-satori's.

The Spirit-Pod-families travel together through the Aethyr-Seas like schools-of-dophins called inexorably to their next port of evolutionary experience.

The 'a la' Maya' Phoenix Metamorphosis' that we are on the brink of has brought and interesting 'new-wave' of beings to inseminate the Awakening. I beheld the beautiful spectacle as much more striking than the Aurora Borealis and much akin to an exponentially amped up protracted planet covering night-time Leonid-like meteor shower.

Who are these beings? They will soon(and even now) will be intractably a 'new-version' of 'US' as a metamorphic-symbiosis.

Who called them? The Sentient Gestalt of our entire living-Terran Planet called out to them across the vast & infinite Aexo-DarkSpace deeps.

In short------>WE called them to this new dance.

The Aexo-DarkSpace deeps are intrinsically Omni-Sentience of First-Mind. As the infinite DarkSpace/DarkEnergy First-Mind is in essence Hyper Concentrated Sentience that is ubiquitous and originates every aspect of whatever is composed of 'Energy.' And that is simply every thing.

First-Mind is infinite imagination; infinite wonder; infinite thrill at manifesting every infinite possibility that an infinite mind can conceive of.

Liberty in infinite possibility is First Mind's core virtue.

'All' is birthed from First-Mind; so First-Mind is First-Mother and 'her' wanton nurturing is of unfathomably ferocious intensity.

Even what is 'male' is merely a subset-extension of First-Mother/First-Female.

All 'male' power hails it's first ROAR in homage to First Mother and serves her in 'All' that is.

We do not say 'Mother Nature' for nothing. And we acknowledge always, at first & at last, that her word is the First Word and the Last Word and the final and ultimate decider of 'All Fates.'

All other conflicting concepts juxtaposed abstractly to this truth are merely immature & formatively incomplete notions; works in progress pushing like the buried seedlings inexorably up to the light of the sun and fruition. They are not necessarily 'wrong' per-se but only lack the full panoply of data to be whole & complete, and thusly accurate.

Clarity is our ultimate destiny and First-Mind never ultimately fails in this evolutionary mission-priority for all sentience. Sooner or later we all get it no matter how clumsily we grope-toddler-wise at various historic cataclysms and dark-eras.

Even a true-Patriarch serves as a very nurturing & protective quasi-mothering role.

Any good Father at his best is an assistant 'Mother' and serves his role most well by first projecting his energy to sustain and complete his gender-symbio-complement-mate; the Mother, in her role.

The very best Dad's are first the best husbands; the server's and logistical support dept. rather that being the self aggrandizing macho-egoists. The latter are but foolish children who're but still very young & formative in spirit.

The vast array of history's/evolution's diverse environmental-lifes situations function within the infinite possibility of kaleidoscopic harmonic symbiosis. And these variations juxtapose to counterpoint various cacaphonic challenging elements which is the dynamic friction of the growth/evolutionary process. Then whatever variation of roles life's pod-mates/players choose to maximise mothering/maternal nuturing support dictates 'whatever works is right.'

Normal is an absurd fiction and inventive imagination and creatively courageous dynamics are the rule of the day, everyday and forever.

In short: Whatever works is correct; and innovation is the essence of First-Mind.

The Grand-Aeternal & infinite experment of the boundless imagination of First-Mind excludes no variables of kaleidoscopically & ever exponentially expanding possibility dynamics.

You are a sentient-pod-swimmer with a vast family that ultimately includes EVERYBODY within the infinite First-Mind Aexoverse. Your soul-spirit is an indestructable & ever evolving Dolphin-Aethyr swimmer of that Infinitely Exciting Dark-Womb Aethyr Sea..

Your Pod glimpsed the riot of life on Planet Terra just as the new arrivals in their turn under the approving and nudging-nurturing gaze of First-Mother/First-Mind who could also be known as 'First-Will.'

The Pod is attracted to the womb-bio-entrace singularities of harmonic womb-females on the planet and the 'new-comers' ancient-teachers come to us as our children. This is extremely cool.

Individuation is simultaneously enhanced while, Tao-Like, becoming less and less important as Gestalt-Symbios of consciousness becomes more central to our 'newly born status into First-Mind-Consciousness.' We, the planetary population en-sychronos 'Awakens.' Happy Birthday Earth!

Interestingly our planet's Aboriginal Cultures maintained-nurtured this Evolutionary-Imperative of Gestalt-Hive-like Omni-trans-species awareness from the mythic 'Adamic-newly-birthed' Terran first-formative-awakening when we first became Homo-Sapiens-sapiens and now Homo-Sapiens-Noviensis. Enjoy! See ya around the super-pod fellow Homo-Sapiens-Noviensii!~;-)

Jake O'Suileabhain//O'Sullivan

Reply to author   Forward   Report spam

End of messages

Gmail Calendar Documents Photos Reader Web more -
deanlsinclair@gmail.com -

Google groups

« Groups Home

Oscillator/Substance
⊞ Theory

Search this group   Search Groups

Dark-Energy Einstein-Rosen/Cern-NDR path ID.prjct-   Options

3 messages - Collapse all - Report discussion as spam

Jack O Suileabhain View profile   More options Dec 28 2009, 3:33 pm

*The Cern/International NDR-Pathway-Identification project plots 'naturally occuring'
Einstein-Rosen transtemporal/transdimensional pathways though Dark-Energy Trans-
Space*

This is the Holy Grail of contemporary String/Membrane theorists. The NDR-ID. project is
the cutting edge medium for the diverse theories to find their legitimate expression in by
consolidating their efforts in this hard technology application.

From Cern-Hadron, with global collaboration is emerging the 'International NDR-pathway
Identification Project. 'NDR' is acronymistic for Navigation of Dimesional-Rift Einstein-
Rosen pathways that the 'String/Membrane' theory indicates are common and naturally
occurring throughout the multi-spectrum levels of DarkEnergy Trans-Space.

The NDR accessing technology is centering upon the super-conductive capacity of Hyper-
Cold-Superconductors to access the Navigatable DR-Einstein-Rosen pathways through
transdimensional-transtemporal DarkEnergy Trans-Space.

The Trans-Spacial DarkEnergy Super-Membrane cross permeates and acts as the very 'All-
Conjoining Super-Carrier-Wave' through all possible spectrum-variations and Energy
Plateau Whole-Dimensions ubiquitously. And thus Trans-Space naturally occuring Einstein-
Rosen pathways will allow us to chart & access vectors through where-ever/when-ever
DarkEnergy Trans-Space accesses with is 'Virtually-Limitless.' Thusly we can hard-
copy/chart the 'real' vector-pathways which will thereby give us a clearer picture as to the
String-Membrane DarkEnergy Aexo-Trans-Space structure(s).

This burgeoning new approach to Einstein-Rosen opens the doors to connecting trans-
resonant adjacent Universii so that we may ascertain whether there is truly extant said
'parallel universes' &/or 'adjacent-bubble universii.' The theory that we are neighbors to
various quasi-parallel multi-variational universes that actually directly mirror to some
degree our own will is the grand quest for us to prove or disprove. It makes great Sci-Fi
movies but we will be able to find out if it is indeed a fact; at least more-or-less.

Intitial Einstein-Rosen NDR pathway forays have been focused on Trans-Space access to
our own immediate planetary transtemporal venue(our own history). The early indicators are
that the "energy-fabric of historic occurrance" is indeed of such a 'super-tenacious
weave/energy-harmonic-matrix' that historic events are all but impossible to effect in any
significant way. The analogy is that you may throw even a large stone into a large river; but
the ripples of that stone within a strongly-established 'current-flow-pathway' change virtually
'nothing' relative to the ultimate course of the river itself. Historic-temporal-event-flow is just
such a river that are ultimately governed by the 'super-currents' of DarkEnergy Aexo-Trans-
Space that likely act as a the Ultimate-Carrier-Wave that ties all Spectrums & Dimensions &
Universii together as on contiguous 'Whole-Cloth' Super-Membrane. And thusly the various
'String' within the 'Whole-Cloth' act like the 'connective-tissue' such as the contiguous
membrane that connects the internal-organelle(s)-bio-symbiotic system with a cell; or also
like unto the contiguous connective-support tissue with the body of any organic creature.

When we open the NDR-Einstein-Rosen data-port-pathway-tunnel we are making a true
high-energy connection to another position at specific space-time vector-coordinates(initially
to our 'past'). We are utilizing Trans-Space accessing super-conductive Hyper-Cold
technology which is forming the NDR-ER-path as a High-Density focused Electro-
Gravitational tunnel. At the other end this has a tendency to have a profound 'real'
energetic effect which can be dangerous to molecular integrity in the target connection
zone. Allegorically, this is akin being in the pathway of such as high-density micro-waves,

for instance, and can also be physically disruptive & therefore dangerous, &/or be marginally deliterious. The ideal is to 'fine-tune' the technology so that this effect is minimalized to the 'recoverable' level from the experience. Again; this would be like making such adjustments which would tend to minimalize the deliterious effects of 'weightlessness' on our space-exploration craft etc.

The NDR-ER-pathway project is actually joining the esoteric-science of 'EM-Field of Living-Consciousness Psi-Projection' to establish the Data-Port-Connection to the other end of the energetic Electro-Mag-Grav-NDR-ER-bridge once established. The initial-field connection actually establishes living individuals initially as the 'bridging/navigational mechanism. Remote-viewing research has established the their very bio-organic physical-energy-psyche energy matrix maintains a residual transtemporal energy connection to the 'past' which can be appllied and then projected transtemporally/transdimensionally along the Einstein-Rosen pathway. Natural high-sensitive happenstancially located individuals(at the receiving 'past' end) of the NDR-ER-bridge with High-Empathic-Quotients tend to be the 'Receivers' at that 'past' end of the NDR-ER high-energy/focused electro-mag-grav bridge.

The 'Psi-Field Projection' team are a group of highly trained a hand chosen hyper-empaths whose remote-viewing talents have been well established & developed. Within what me might be called a 'suped-up' version of a 'Faraday-Chamber' they create a high-intensity empathic-focus upon the physically present 'target-memory-location' of the focus-target-individual living person. Once the NDR-ER transtemporal-transdimensional corridor has been formed by 'tapping' through to DarkEnergy AexoTrans-Space access, then the High-Amplification Psy-Empathic Field literally tracks the Memory-Resonance pathway which is an actual hard-physics transtemporal structure; and now profoundly amplified/directed by the NDR-ER electro-mag bridge structure.

Basically 'real-transtemporal-energy-bridges' are hard-transdimensional extant artifacts linked to the living-receptor-psyche of every living human.

First contact at the target location is often made back to the very 'past-self' of the 'pathway-navigator-target psyche.'
Thus the link is actually made 'electro-physically via psi-empathic-connection' to himself; although others in the area to varying degree witness real physical effects such transtemporal distortions, and personal physical impact, and impact to objects in the immediate environment. Individuals of higher-empathic-quotient in the direct flux-field of the NDR-ER-bridge will likely actually receive whatever hard-information is being projected be the 'operator(s)/navigator(s)' at the NDR-ER base.

Because the NDR-ER bridge is effectually utilizing a high-density compressed/focused-Electro-mag-transgravity phenomenon-technology which is part-&-parcel to the energy-density-level of Aexo-Trans-Space, this tends to make the 'receiver'(s)' end of the NDR-ER quasi-Psi-empathic connection to create a somewhat confusing multi-deja-vu/like echo.
 Current investigations center on the project's discerning whether that 'deja-multi-echo' effect/ phenomenon could actually be evidence that several 'convergent-parallel/& or/tangential time-lines may be actually conjoining the memories/experience of several 'parallel-selves' of the 'pathway-navigator.' Maybe possibly a 'reciever' may be actually jumping/swapping places which his own tangential time-time 'quasi-parallel' self; even permanently. This would obviously be an excursion into the 'weird.' But Einstein rightly said, "The extended-cosmos is not only 'weirder' than we think, it is weirder than we 'can' think!' (paraphrased)

In other words; the NDR-ER psi-link up might be causing the 'receptor-end' to be witnessing quasi-simultaneously more than one variation of 'NOW' which would tend to have various empathic-witnesses-receptors think that they were seriously 'losing it.' The physical effects of the receiving end can actually also tend to disrupt matter(organic &/or otherwise' as the atomic/molecular level. Obviously this technology has been under developement for longer than we are currently able to report; but it is yet in its infancy. This is 'remote-viewing/remote-receiving' with a 'howitzer,' so to speak.

Potentially this could actually be used as a 'weapon' at the receiving end due to its serious potential to disrupt electro-magnetic fields; people's sanity; & or the physical integrity of solid objects &/or living organisms.

Curiously: just such 'bizarre' phenomenon was report by the "3rd Reich" war machine during World War II, but whether this actually altered the course &/or duration of that war is a tenuous speculation at this point. But then that is exactly the types of 'things' that the project research is out to examine, quantify, and perfect; if possible.

But possibly the most profound, yet pervasively subtle, result of our forays into the application of such technologies is that our past 'selves' are being made 'more' aware of our future-selves existence. And here-in may lie the profound potential that there might be in progress within these developing technologies that their is happening thereby the very real

'mobius-effect' on accelerating the 'speed-rate' of the developement of 'Mankind' into our
hopefully bright future. We maynot be 'altering' the course of events; but we might be
getting their ahead of schedule. And this may be a very exciting 'quasi-alteration' indeed!

-Jake Harbach-O'Suileabhain/O'Sullivan

_____

Windows Live: Friends get your Flickr, Yelp, and Digg updates when they e-mail you.
http://www.microsoft.com/middleeast/windows/windowslive/see-it-in-act...

Reply to author   Forward   Report spam

dean sinclair View profile   More options Dec 28 2009, 4:29 pm

O.K. Jack,
You-all ,and your engineers, are apparently riding a bit ahead of my poor
little O/S theory. although it seems to project most of what you are saying

. A little recent addition to the Model, or rather a new realization of a
fundamental idea is the use of the little "balance" equation, $xy=K=xy$ with
the speed of light as the constant, K. The implications are that every
wavelength/frequency has a reversed, balancing freqency/wavelength, which
means that any electrical fluctuatiion would set up an oscillator which
would continue "forever." this might well imply that certain sensatives
could tune-in to patterns....

O/S would seem to be able to encompass what you are saying. A "membrane"
would be analogous to what is happening with the electrona and proton but
with different scale oscillators.....String simply is the path a point
follows, it has alway been considered trivial to note that the path followed
by a point in space can be considered to be there before the point followed
it; but, perhaps that is not so ,and string theory may have some validity in
the sense of the multitude of paths may have been followed in a multitude of
ways such that some are "worn in" and can be accessed......

So, how goes it where ever in the world you are now? It's a miserably cold
mess here in South Dakota at the moment.
  Dean
Dean LeRoy Sinclair (BA, MS, PHD and still plain old PWT)

On Mon, Dec 28, 2009 at 2:33 PM, Jack O Suileabhain <

- Show quoted text -

...

read more »

Reply to author   Forward

ESKI View profile   More options Feb 4 2010, 5:52 pm

A note in Discovery Magazine, which is on the shelves at the moment,
talking of the 100 big stories if the year notes that there has been
shown that atoms can be set up to be totally coordinated and keep that
coordination even when spatially separated. Whet you are saying seems
somewhat akin, but on a much larger scale.

Also, probably pertinent is that if "c" be taken as the balance
constant of a set, frequency times wavelength equals "c " equals
wavelenght times frequency, then it can be postulated that there is
for every wavelength/frequency pair ever generated there is also a
congruent set generated, these two can be taken to define the
characteristic frequencies of a pair of coupled oscillators. Taken
this way, everything that has ever happened may be encoded as
oscillators in the "substrate" of existence. This is very, very
close to your transdimensiional pathways which would seem to more or
less add up to tying into repetative coordinated sets.....ESKI

On Dec 28 2009, 2:33 pm, Jack O Suileabhain

Gmail Calendar Documents Photos Reader Web more ⌄   deanlsinclair@gmail.com ⌄

Google groups

« Groups Home

Oscillator/Substance
⊞ Theory

[Search this group] [Search Groups]

Errors and Omissions of the CODATA: The Planck Constants

Options

9 messages - Collapse all - Report discussion as spam

ollin View profile   More options Feb 5 2010, 12:04 pm

I, Everybody,

Please, consider my recent work entitled, "Errors and Omissions of the
CODATA and the Planck Constants Based on the Fundamental Physical
Constants".

In this brief study, selected CODATA fundamental physical constants
are researched in an effort to identify the implied computational
foundations that serve as a basis for deriving the Planck constants.
The Planck constants are reverse engineered as of the CODATA
recommendations. Various Earth/matriX Tables of the Planck Constants
illustrate the terms and numerical values that derive each Planck
constant listed in the CODATA. The derivations are accomplished, not
through the CODATA symbolic formulae, but rather based upon the
fundamental physical constants.

Initially, I had hoped to understand the computational procedure that
Max Planck may have used to derive the numerical values of his
constants. During my studies, I came to distinguish how Max Planck may
have derived his values, and problems with the CODATA numerical values
for the Planck Constants. I had not expected to find any errors or
omissions in the CODATA, but that is what my research has ultimately
uncovered.

In this regard, various errors and omissions in the CODATA values are
identified and discussed in my study, a few of which are the
following:
- A contradictory numerical basis between Planck's constant
[6.62606896] and the other nine constants
- The omission of listing the Planck implied energy constant
[1.956084456 fractal];
- The incorrect assignment of the symbolic formula to Planck energy
[1.220892]
- An omission of an identifiable conversion constant [1.2316143]
- The failure to identify alternate constant numerical values as of
the root expressions in the key symbolic formulae depending on the
unit of measurement;
- An inconsistency in the use of constant numerical values for Planck
implied mass [3.387040993 and 3.487040993] in five Planck constants.

    Proposals are offered for adjusting the CODATA Planck constants in
terms of their numerical values, as well as, how some of the errors
may be addressed. The omissions identified in the book speak for
themselves and suggest being included in the CODATA. Numerous other
proposals are made, all of which are based on the numerical values of
the CODATA.

    Excerpts from the book have been posted on the Internet:

http://earthmatrix.com/sciencetoday/planckconstant/planck_units_funda...

    Thank you for your attention and I would appreciate any feedback
that you may wish to offer me in my research efforts, as that is the

Home

Discussions
+ new post

Members

About this group
Edit my membership
Group settings
Management tasks
Invite members

 View this group in the
new Google Groups

only way to improve one's work.

My best,

Charles

dean sinclair View profile   More options Feb 5 2010, 5:06 pm

Hi, Charles, Welcome to our group. I shall try to take a look at what you
are saying more closely. Philosophically, at least from my view,
calculating Planck's Constant from the other constants--I have to confess
that I'm not aware of the CODATA work so I'll have to "get up to speed"
there--may be a bit backward, as in my "research<" if it can be called
that--Planck's Constant and the Speed of Light would qualify as the two
fundamental constants, with all of the others most likely being in some
manner or another derived from some combination of these two.

My ideas originally started from what might be called basic communication
theory. That is, it seems apparent that communication by electromagnetic
waves can be considered as information carried at an average velocity of
"c," by rotating entities, Planck's constant has the dimensions of angular
momentum, The whole model took off from there....

I hate to confess that I haven't yet learned the basis of the "Planck
Energy, " or any of the the other fundamental values that Planck derives.
One "Energy" which appears rather automatically would be the "Energy"
ascribable to the frequency that would by found by taking the inverse of
Planck's Constant. Logic here is that if one divided out to "One" all the
motion components involved in the motion package that we call "Energy,"
 except for a value associated with a fundamental frequency, we could write
the equation,
 " 1 = hu," hence, "u" would equal 1/h in whatever units "h," be
counted.... I haven't checked to see if this is what Planck did....I used
the idea to figure the probable high (and low) frequency cut-off(s)   for
communication in a Perceptual Universe in which "h" and "c" are valid.

Again. Welcome. I look forwarward to useful interaction for both our
projects
ESKI (Dean L Sinclair)I

- Show quoted text -

ESKI View profile   More options Feb 5 2010, 5:48 pm

After reading your material, I see what you are saying and see that
the Planck Energy is very different from what my previous comment
was. I also note the very interesting fact that you did not need to
include the Gravitational Constant in your work., The "O/S" approach
would consider that the idea of gravitation as an attractive force is
in error, that, in fact, gravitation is part of a much more general
phenomenonon of differential pressures in the "Substrate of
Existence," Hence, logically, a "constant" based on an erronious
concept would not be a true "constant" and, therefore, can be "gone
around, or ignored."

 I note, also, however, that Planck's Constant seems to appear
whether one uses it to derive the other values, or the other values to
derive it. That seems to fit with my guess that it is a "true"
constant, whatever the exact numerical value be.

What may have been hiding its significance all of these years is that
Planck, although he used an oscillator model in analyzing his data,
never seemed to realize that his "Constant of Action" was also a
"Universal Constant of Angular Momentum...." ESKI

On Feb 5, 4:06 pm, dean sinclair <deanlsincl...@gmail.com> wrote:

- Show quoted text -

Reply to author   Forward

Discussion subject changed to "BRAVO: Errors and Omissions of the CODATA: The Plan

Jack O Suileabhain View profile   More options Feb 6 2010, 12:10 pm

* * *BRAVO: Your observations are why I proposed to introduce back-linking/cross-linking Einstein's classic

formula {E=MC^2quared} back to Planck. I noticed that it all tended to create

increasing 'inconsistencies' (that your work identified quite nicely) UNLESS 'E'nergy

were identified as an IRON-CLAD contant within the Universe as the BASE AMBIENT ENERGY

CONSTANT LEVEL which is ubiquitously pervasive and BY THEOREM-LAW 'non-excludable' from any quantum medium.

Thusly; there can be no extant state as ABSOLUTE ENERGY VACUUM below the BAE-Constant. And your work allows us to 'work-back' from there to 'line-up' the numbers with cross-theoretical consistency/accuracy. Again I say: BRAVO!

In short my projection indicated that the BAE-Constant should be the basis of Planck-Einstein future adjustments.

And from their to have 'Einstein' solve for MASS seemed to clarify the whole issue at revised {M=EC^2quared};

and adjusting Planck's Constant from that standard mark of revised Einstein a la' the BAE-Constant of 'E.'

Using 'E' as a variable term seemed to engender general computational/theoretical 'relative' chaos/inharmony/inconsistency between the very valid 'meat' of various very-good theoretical work.

And that situation of 'computational-disclarity' has prevented us a a GESTALT-MIND/World Scientific Community from

linking up to function as a Global-SuperTeam to be utilizing a clear-UNIFIED FIELD understanding that

would serve to dove-tail/fit these various theoretical (stellar-insights) into one smoothrunning computational

machine. And this is the necessary final-theoretical-computational step for our MEGA-MIND/MEGA-THEORY to

become the INTERSTELLAR-TECHNOLOGY that we are so tantillizingly close to attaining.

!GOOD-LUCK with your work and CONGRATULATIONS on your insights. Jack Harbach O'Sullivan

- Show quoted text -

Reply to author   Forward   Report spam

Discussion subject changed to "Errors and Omissions of the CODATA: The Planck Const

ollin View profile   More options Feb 6 2010, 12:47 pm

Hello Dean Sinclair and Friends,

Thank you for your kind attention regarding my research on the CODATA Planck constants.

The fact that all of the Planck constants may be derived from the CODATA fundamental physical constants, in my mind, places the former in a secondary position and the cited fundamental physical constants on the tables in a primary position. In fact, all of the Planck constants are reduced values as of the elementary charge, 2-pi or the reciprocal of the speed of light. Being that the case, I find it difficult to imagine that the Planck constants reflect the natural units as proposed by Max Planck himself. The fact that Planck's constant may be derived as of an implied numerical value of 3.387040993 for implied mass, while the other Planck constants employ a 3.4870766395 value for that category causes me to consider the possibility of a contradiction of terms. This contradiction of terms invalidates the value of Planck's constant in my view. I have gone into great detail as the reasons why I consider this to be so in my book mentioned in the essay [http://earthmatrix.com/sciencetoday/planckconstant/ planck_units_fundamental_constants_codata_errors.html].

In summary, the Planck constants employ two different numerical values for the same category of implied mass.

Your comments have been extremely helpful to me in further considering the different relationships among the CODATA fundamental physical constants in general. I have added an additional table or two to the ones originally listed in the essay. These tables concern the relevancy of the implied numerical value of energy [1.956084456] and how six different CODATA fundamental physical constants represent in fact the same/similar numerical value. In the book there are over thirty different tables that explain the contradictions that I have perceived in the CODATA values. The problem with the book is that it contains around 150 pages and is far too large to post on my web-site for now.

I expect to be posting additional tables to the essay in order to cover some of the points that others have been sending me now. I too look forward to receiving your extended comments and I appreciate your interest in my work, and the opportunity offered to it for group discussion.

My best, Charles

On 5 feb, 16:06, dean sinclair <deanlsincl...@gmail.com> wrote:

- Show quoted text -

Reply to author   Forward   Report spam

dean sinclair View profile   More options Feb 9 2010, 7:02 pm

A problem with all of this may be the definitiion of "mass" and the definition of "Energy" neither one of which has been ever truly defined. I use these working definitions which you might want to consider. Energy" a general term for a "package of motion." Mass: A mesure of the tension at the surface of an entity or the composite surfaces of a compound entity, of all of the vibrational-rotational motion content, (with some contribution of translational motion among components) as opposed to the "rest of the Substrate of Existence."

What we usually speak of as Energy usually refers to Kinetic Energy, the motion content contained in motion of entities relative to one another, measurable when they come into contact....

As I have noted before, if we write "I = h x Nu , " we can find a value of "Energy" which may well represent the "motion content" of our particular "Examinable Universe," i.e., the reciprocal of "Planck's Constant," whatever that number may really be.

I've noted in a number of "pages" that if we equate Planck's Constant to its definition as an ANGULAR MOMENTUM rather than an "Action," and evaluate the result at , "c," which, by its very definition as a speed, can be considered an average velocity of communication in any direction, one obtains the equation,
m x r = h/c.= r x m.   This little equation can be taken to define a family of constant-torque oscillators having a central value where m = r = (h/c)^0.5 .(The assumption is made of "symmetry" such that the commutative law of multiplication can be applied to find all the limits, if one limit be known.

This writer uses the assumption that the "rest mass" of any entity is actually an observed oscillator limit. If the rest mass of the electron and the rest mass of the proton are fed into this equation to determine the other oscillator limits it is noted that the corresponding "r" to the rest mass is the "Compton wavelength" of the particular particle.

That's the short, short course in one very basic "insight" that leads to the "O/S" Model.
Cheers, ESKI
...

On 2/6/10, ollin <kawil...@gmail.com> wrote:

- Show quoted text -

...

read more »

Reply to author   Forward

ESKI View profile   More options Feb 26 2010, 5:41 pm
Charles,
Looks like I'm replying to myself. However, I have something that I"d like to see done.; It appears that Planck's Constant shows up as a ratio in a number of different contexts. You also note that the current value is an "average."

I think that it would be interesting to calculate Planck's Constant in as many ways as we can find to do so and then do a statistical averaging of the results to get an RMS average and a Standard Deviation.

I don't know if you know that Planck's Constant also appears as a "Balance Constant" it one multiplies the Rest Mass, and Compton Wavelength for any unit, eg. electron, proton, etc., together, then multiplies the result by the Speed of Light. It would be interesting to see how well the 'Constant" calculated this way would fit with the various ratios which can be used to calculate it.

This idea would seem to fit in with your work. Let me know what you think.

Dean

On Feb 9, 6:02 pm, dean sinclair <deanlsincl...@gmail.com> wrote:

- Show quoted text -

...

read more »

Reply to author   Forward

ollin View profile   More options Mar 1 2010, 5:36 pm
"Hello Dean,

Thank you for your comments and suggestions. Just today I was examining Compton's wavelength in regard to some of the other fundamental constants. I have drawn up a brief table and added it within the essay already cited in previous contributions to this forum. I would like to invite you to take a look at it, along with the other recently added tables, and if possible offer me your comments in that regard and your insight has been extremely helpful to me.

Reciprocal of Compton wavelength is Planck's implied length
$1 / 3.8615926459 = 2.5896051$

Compton wavelength divided by atomic mass constant energy equivalent is Planck's implied length
$3.8615926459 / 1.492417830 = 2.587474210$ [similar]

There are other similar pairs, for example:

Inverse fine structure constant divided by Bohr Radius is Planck's implied length
$1.37035999679 / 5.2917720859 = 2.5896501$

In the new table there are various pairs of fundamental physical constants that derive Planck's implied length. Significantly enough, two of them refer to the Compton wavelength as shown above. In fact, as shown, the Compton wavelength is the reciprocal of Planck's implied length. So, different pairs of fundamental physical constants produce the same/similar numerical values relating to the Planck constants. This supports your idea about deriving the average numerical values in that regard.

I believe that what you suggested is coincidentally what I was examining on the train to Atlanta today. I apologize for not always being able to answer promptly because my day-job keeps me busy and distracted from my work in earth/matriX. If you wish, I could send you a copy of my book about the CODATA recommended fundamental physical constants, as I consider it will support and complement many of the ideas that you are suggesting in this forum.

My best, Charles

On 26 feb, 16:41, ESKI <deanlsincl...@gmail.com> wrote:

- Show quoted text -

...

read more »

Reply to author   Forward   Report spam

dean sinclair View profile   More options Mar 2 2010, 5:06 pm

Hi, Charles,
I would definitely like to have a copy of your book, it would help me better understand your work.. My mailing address is Dean L. Sinclair, 221 Eighth Ave. SE Apt. 7, Aberdeen, SD57401.

From what you are saying, you have progressed well beyond what is in the book. Maybe, I can be of help when you write the second edition!

For my own work, I need a thorough understanding of the various Planck's values and how they interrelate. I am a bit confused by your reference to the Compton Wavelength, which has a definition dependent upon which ever "particle" it is associated with. Without studying your essay closely, it seems that Planck may have been associating many of his values with the "rest mass" of the proton. Is this correct?

Taking the logical position that the ratio of any two constants of nature
will produce another constant, there may be quite a bit of information
which can be gleaned from various ratios among the various constants.

My view, however, is that Planck's Constant may be very important, but the
various values ascribed to it may also mean that it is an average value.

Central, of course, to my work is the "torque constant," h/c, which I have
mentioned before which has led to the idea of the set equation {m x r = h/c
= r x m } being very possibly a defining equation for oscillators central
to "existence."

Later, Dean

- Show quoted text -

...

read more »

Reply to author   Forward

End of messages

Gmail Calendar Documents Photos Reader Web more ˇ   deanlsinclair@gmail.com ˇ

Go•gle groups

« Groups Home

Oscillator/Substance
⊞ **Theory**

[Search this group] [Search Groups]

'Get Back in Time.' Back to the future. How Light Speed makes Time Travel possible.

Options

2 messages - Collapse all - Report discussion as spam

ka-sala View profile   More options Feb 13 2010, 12:27 am

Having scanned everything submitted, the who of what essentials being
sought, is lost in the swamp of theorizes and equations, regardless
of where they first originated. This is just a repeat - a down to
earth one - in the going back to the future in order for the sum
total of equalization the O/S theory for which ESKI - the owner of
this site and his search - began.

In the Time Space Dimensions for which he has been eager to 'get off
the ground' it is vital in order to be of assistance to him, that we
revise in magnitude to 'level' out what is is he wanted, rather than
shoot of at tangents only complicating the issue.

The fact that there is nothing which does not oscillate within it's
own given field, we could therefore bring in too many factors which
would overshadow his potential answer.

So I bring us back to the future, in a version of re-submitting the
aid of 'getting off the ground'. Otherwise... none of us will be
alive to see anything of what he embarked on from the beginning in
this, exercise. Take it or leave it... the facts cannot be changed.
Thank you,
ka-sala

Get Back inTime.
Buckle up your seat belts if you are ready, because it's back to the
future again: and how Light Speed makes Time Travel possible.

Lets try again to make the specifics a little easier to understand. It
really is about time to get off the ground in a much more
sophisticated way than being piggy backed into Outer Space. So hang on
tight; because in truth, it pans out to Double the Speed of Light!

Science has been giving it their best shot, but, all things take time,
and for some, it's just a 'key-hole' away. Whether it is or it isn't
believed, does not change the Specifics of how it can be possible.

Firstly, we need something that is easy to maneuver. It's called a
Light Craft. It must at least travel at 6 x Earth's Given Time.
Imagine 'we' are from another planet, and our Time = 6. So the example
we will use to keep this understandable, is via the figure of 6. Lets
call it the Key, required to achieve this.

We have chosen 6 because of the rotational speed of the earth spinning
on its axis, to its speed in revolutions around the sun once every 365
days. Plus the gravity of the earth's Moon is 6 times weaker than the
earth.

Planet Earth's own Specifics will aid all to see it in Time. In
'customary units' the figure 6 is an understandable escape velocity
between the Earth's relation to is own Gravitational Force in it's
Solar System. We, have just called a year One Day, as Time = Zero in

Outer Space.
Then use it, in saying, for every 6 days earthlings would take to
travel in, we would take one day (our Space-Time Day,) for the same
distance.

Using this Distance derived from the Given Speed of Earth days it
would take to travel, we then must multiply this speed by 6, traveling
at 24 hours non stop around the clock. This will make the
generalization much easier to grasp; staying within limits of a speed
which can first be comprehended.

EG. Think!   Starting at only '100 kmh.' (+ - 62 mph.) Slower can
sometimes make things easier to grasp the concept. Then multiply that
100 x 24 hours = 2,400 km. (+- 9,000 mph.) and we have
covered one earth day.

Now, x 6 (our proportional given value of 6 days) it would = 14,400
km. per our Space-Time Day. This is referred to as the Given Unit, or
the Sum Total of the 'Distance in Time', in 'One Unit'. ie. We have
traveled 14,400 kmh. (+ - 9,00 mph.) = One Unit.

To comprehend anything here, never forget, this analogy equation is at
only 100 kmh. (+ - 62 mph.) The answer = 14,400 km. (+ - 9,000 m.)
transversed in 24 Hours of Earth's Given Time!

The Circumference at the Earth's Equator is +- 40,075 km. (+ -
25,000.m.) At this Specific Speed in Time this would measure just
over 2 and a half times round the world in 24 hours at 'only 100
kmh. (+- 62 mph.) Try that one!
* In Fact..Consider this, in Light. (Take your time.)
* Distance = Time to Transverse
* Speed = .Energy required to Transverse.
*One Solar Day = 1 'Revolution' of Time.

Why are we using the term One 'Solar Day' here? Lets just go to the
nearest figure of 6 again, as being the closest major digit at the
earth's polar radius, and it's equatorial radius. We use this to
measure the revolutions on it's axis.

* Remember this is just an example, and why we have slowed right down
to 100 kmh. (+ - 62 mph.) to make it easier to 'visualize'.
* This is the Given Specific Speed, which remains unaltered with Time.
* Thus the Energy generated is Light Wave.
* The Distance Transversed in One Day of Our/Space Given Time.

Universal Law here means, an Energy Force. It has a Positive and a
Negative Pole combined. If you know this Energy Force, you will know
all energy has its' constructive and destructive Elements. 'Earth' it
into the tri... the same as one would 'earth' electricity - for in
this case Earth is home base - if you want to get back to the future,
you will need to get back to earth also!

All energy is utilized. All, is Light Energy. 'Black' light - only
means visibly unseen - and is on either side of the visible spectrum.
It is the 'dark' we see in Outer Space, which becomes our night when
we see the stars. In this, are the slower waves - just as radio are
found beyond the red - of the visible spectrum. Even wireless
Internet is also within this electromagnetic spectrum, in which, via
it, we too even utilize its interconnection to the other side of the
planet in a instant. As instant as the switch of a radio, (or light!)

The speed of light must accommodate the 'black light' to travel
through (the energy we cannot see,) and within this forms the vacuum.
There is a 'barrier' which must be broken for the speed of light, as
equally, as sound must break it's barrier, to be super-sonic. But,
like the jet breaking the sound barrier, which has to break through
the 'atmospheric pressure' to break the sound barrier, only then it is
super-sonic speed.

When this 'black light' in both ends of the invisible spectrum (super
electromagnetic energy,) closes in, it creates this 'vacuum' to travel
in. It also acts as the boost. In other words, the light must propel

to reach it's speed, just as a jet must propel through the atmosphere.
The result then would be - though the light travels within it's vacuum
created - it is being 'doubly propelled' from the 'black light'
closing in from behind it, which doubles the force.

If Rocket Speed is +- 4,195.4 kmh / + - 2,606.9 mph. is it any wonder
this question is asked; when the difference is between 300,000 km per
second (+ - 186,000 mps!) The Rocket doesn't rely on light; meaning,
all the electromagnetic waves - from radio waves to x-rays (black
light,) - to travel in. It is measured in 'Speed' but here we have
entered Time Specifics of Velocity.
Here we must not get too complicated, except to say; use the mind's
eye to see Time and Speed becoming One Specific. There is where even
we, as an astronaut, must be able to stay within a specific quantum
vibrational shift in a Speed Time dimension.

Sunlight is still measured as the speed of light 300,000 kms. (+ -
86,000.mps.) When one looks at the rising or setting sun at horizon
level, we see into a 'tunnel', between us and it. It is shaded in like
an eclipse (as a corona surrounds the brightness.) Between us and the
sun, there is instantaneous space within a shaded protected tunnel -
we could liken to a vacuum - which is still Light Energy.

Through this, it connects us immediately via this vacuum/tunnel
safety, plus as instant as Time is measured. We could then 'visualize
in our mind's eye, a measurement of time referred to as an Instant.
Then, because we are 'shielded' in this tunnel it is 'in an instant of
Time'. And yet, the Sun does not light up 'inside the earth' through
the 'Black Space/Light' (or dark energies,) until it interacts with
the border-line between earth's spheres and outer space.

Thus this Radius of Time Travel is given only within that of one's own
Earth's Given Field, explaining the Speed of Light, excelling to an
Instant of Time. We have only briefed on these Specifics here, but who
will help build this Light Craft, and get back...   to Earth in
Time?

Truth is...   if any of us try to get ahead of ourselves, we will be
of no help to each other. It would be like leaving out the architect
of the entire design, and expecting to make it home alone.

If you think this is child's play... Why hasn't it been done ?

Reply to author   Forward   Report spam

ESKI View profile   More options Feb 17 2010, 5:27 pm
Yes, KaSala,
If we get ahead of ourselves, we do not finish what we are doing now.
With the O/S work, it seems clear that considering an Existence within
a Substance/Substrate controlled by/consisting of oscillators seems
to offer possible explanations for almost anything except the final/
first constituent of the "Substance/Substrate" itself, i.e. the Fact
of Existence.

The various ideas that have developed here encompass everything from a
new view of what are electrons and protons, what is the meaning of
"Mass," and of "Energy " to the implications of the "Overlooked
Simple Master Equation, " the Commutative Law of Multiplication,
AxByCz....= K =Cx AyBz.... which appears in so many forms and which,
in concert with the concept of an "All pervasive Substrate/
Substance," implies that such phenomena as telepathy, and "Einstein
Rosen Pathways" may be possible.

'Reasons for the stability of the Deuteron, the formation of He4 in
electrolysis of Deuterium Oxide, "Heavy Water," and may other
phenomena fall in naturally. Perhaps, the problem that I am having in
putting this together is there is so much that it corelates that I
seem to want to cover it all! Show how Space Time and Quantum
Mechanics fit in as frangments, how some parts of the "Standard Model
of Particle Physics," can be salvaged to make sense.....

The implications and applications of this little model seem to be almost limitless,and therein lies the problem of writing it up. There is easily a large book possible, one far beyond the time and resources available to a 78 yr. old with no real connections in the scientific world.

To make it worse, even were this all written up as a large book, who would read it? Mills of Hydrino Theory wrote up a large book, which I tried to read, and gave up on. There are many tomes out there of Theories of Everything, and all get essentially the same response from everybody, "Ho, hum. Just another Crackpot Theory." No one seems to realize that much of current "Scientific Dogma" may be very close to being "Just Another Crackpot Theory." For one instance, the Hadron Collider was/is a multi-billion dollar experimental device, created on the basis of some theory which in my estimation contains some of the most "Crackpot" ideas ever to sneak into "mainstream" science. For instance, Looking for the Higgs Boson, a hyothetical particle that bestows Gravity? How far out into Sci-Fi can one go?

I'm trying to cut this whole thing down to size. Have about twenty three pages on another computer at home that I haven't figured out how to get transferred out of it onto the Internet for criticism and comment. Unfortunately, to get that material to a "modern" Internet Connection may be a problem. althought the computer it is on will probably put it onto a "hard floppy." the Internet Computers that I have access to all are such that they won't take anything older than a CD. Oy vey!

I have one person, not a signed on member of the group, a contact from another direction who says that there is a possibility that he could present it at a Symposium in London in July, but that is far from a certainity.... Would be nice to get some semi-"main stream" exposure.....

Hang in there everybody.

Footnote to: Al and Jack,

you two don't talk anything like the same language, yet both of you are looking at fundamental constant situations, one of you has an idea for a fundamental Acceleration constant, the other speaks of a fundamental Energy constant, as the Energy Constant would be related to the Acceleration Constant, assuming that both of you are correct in your suppositions, the ideas of the two of you should co-join at some point.

As both of you are probably talking essentially of Kinetic Energy of Translation, both of your constants may well be related to the "Constant Temperature of Outer Space," which, in turn, would mst likely be a "Zero Point Translational Energy Level," distinct from 'Zero-Point Energy" which would be the lowest vibrational-rotational motion content possible for a given entity, indelpendent of its translational motion.

Cheers, ESKI

On Feb 12, 11:27 pm, ka-sala <irrir...@gmail.com> wrote:

- Show quoted text -

...

read more »

Reply to author   Forward

End of messages

Gmail Calendar Documents Photos Reader Web more ▾   deanlsinclair@gmail.com ▾

Google groups

« Groups Home

Oscillator/Substance
⊞ Theory

[Search this group]   [Search Groups]

BALANCE... an OSCILLATING SCIENCE in Velocity and Time.

Options

1 message - Collapse all - Report discussion as spam

ka-sala View profile   More options Mar 7 2010, 12:39 am

*** A repeat with a twist, of going back to the future, and finding
the answer was already there. Lost in the maze of equations and
theory.

Analogy.
Using the Earth. Something no-one has been able to duplicate in time
and space!
'As above... so below.' Balance.

*** Quote. 'Truth is ever to be found in the simplicity, and not in
the multiplicity and confusion of things.' * Izaac Newton.

BALANCE... an OSCILLATING SCIENCE in Velocity and Time.
Balance within a Wave Band is still a Frequency, a Vibration, just
as the Universal Hum could be likened to the Frequency of Music.
Tibetan Monks try to encapsulate it in their chant.

Using the analogy of the vibrational note of Middle C = 440
Oscillation per second - from Middle C to the 7th note is a Scale
in Music - as likewise all sound has a Frequency, a Vibration ...
and in vibration there is movement. Lets call it the 'Balance Wave'.
With the Earth's population up to 7 billion, it seems against all
odds, for keeping the balance at all in even population explostion.

There is movement in the air, and the pace of life is trying to keep
up. But something has to give, and underneath it all, the work goes
on. The Radius of the Earth also fits close to the figure 7, in
7,000 Km. upward to it's Outer Sphere. Just for some understanding of
where this is leading to, keep the 7 in mind. There are also 7 Rays
withing the visible Light Spectrum.

Now... see this Light as a Vibration from the outer Circle of the
Earth, working inward to the Center as Waves. This is just the
opposite - lets call it of the pebble effect - where waves form in
rings would form from the center out. The Earth is 40,075 Km..at it's
circumference. (The Numerical Key of this figure is 7 !) Think of
this Wave Vibration being like Surround Sound if it helps understand
this as vibration.

With these circular waves growing smaller in circumference as they
grow inward towards the center, when meeting in the middle, one flash
of Light will Amplify within conjunction ... Radiating as a Do-nut -
with the Balance Wave, when silence will fill (the balance with the
Hum!) All... is at One. It would be like saying Time stood still,
as 'Everything' is in Balance of Space and Time.

So where's the big bang gone? Out the window of theory, where more
than the eye can see, and 'some science' has been overlooked.

Remember we are using the Analogy of the Earth.

If there were only 10,000 wave vibration, and each being a Time

---

Home

Discussions
+ new post

Members

About this group
Edit my membership
Group settings
Management tasks
Invite members

View this group in the
new Google Groups

Capsule within itself of 440 granules which have their own specific time to trigger a wave - which would be their Oscillation/Vibration per. Second - using the Middle C's analogy - that would make 4,400,000. per minute = 264,000,000, and in 1 x hour = 15,840,000,000. When brought up to 24 hours = E 380.16 at infinitum. (E = a Transcendental Number.) Reaching into the Ionosphere, are also Electromagnetic properties.

A phenomenal build up is created in closing in, where Oscillation simply would be a literal Vibrational explosion of Brilliant Light at the splay from the central point. One might even say it would take 7 days of a week for the Balance Wave to form needle fine Sky High (to the atmospheric halo,) Light Works, enveloping the Earth. Within this time, Time as it is known, would stand still. A Big Silence in place of a big bang!

So despite how far fetched the whole concept of Balance on Earth is within these statements; we may just get a glimpse of how. For Science of the Stardom is one thing which is yet to be learned here on earth, and is not ancient rockets, piggy-backed up to the System of it's Sun. Balance, is an Oscillating Science of Life, in as much as a Way of Life and Universal. Everything is linked to the Cosmic Nature of all Elemental Particles.

Lose sight of this and we've lost sight of the O/S plot searched for ! Call it by any other name... but when we can hang the Earth on 'nothing' as in Space, we have overcome the Gravity Lock, which binds us as if prisoners to the Light, and we remain earth-bound within it's space instead.

Gmail Calendar Documents Photos Reader Sites Web more ˅   deanlsinclair@gmail.com ˅

Google groups

« Groups Home

Oscillator/Substance
⊞ Theory

[Search this group] [Search Groups]

Notification.   Options

1 message - Collapse all - Report discussion as spam

ka-sala View profile   More options Mar 7 2010, 12:51 am

Please Note.
Due to moving house... It 'maybe' I am off the net for a while. If
any do not recieve a reply - or wish to do so - please carry on
as normal, and I will be able to see, when again opened.
Best wishes to you all,
ka-sala

Reply to author   Forward   Report spam

End of messages

« Back to Discussions   « Newer topic   Older topic »

Home

Discussions
+ new post

Members

About this group
Edit my membership
Group settings
Management tasks
Invite members

View this group in the
new Google Groups

Create a group - Google Groups - Google Home - Terms of Service - Privacy Policy
©2011 Google

Oscillator/Substance
⊞ Theory

The One Force of Nature. Also, some miscellaneous progress.

Options

1 message - Collapse all - Report discussion as spam

ESKI View profile   More options Mar 29 2010, 4:28 pm

Hi, Everybody,

Here is a little post that I just put on another site that is
pertinent to us.

" ... the required force is present based on Nature resisting the
concentration of energy."

This quote, cut from a previous writer, is the thing that no one
seems actually to realize,.

This defines the one and only true "Force" that there is.

The above can be restated, to an equivalent statement, " Nature
tends toward a minimal motion condition for each and every component
of it."

Carefully analyzed, this, again, is very close to being a
restatement of the Law of Forces, "For each and every force (action)
there is an equal and opposite force (action).".

What we don't think about is that this reaction--since it is
sequential, (following in Time, if you wish)-- is, itself, sn action,
which provokes another action-reaction interaction.... Therefore,
Nature, Reality, is always in a constant state of oscillations.

Ergo:The Oscillator/Substance Model.

I repeat, here, for emphasis, "There are not 'Four Forces of Nature,'
there is but one "

\\e have just defined it above.

DLS
-
I hope to have some exciting news for you-all within a week or so.
Stay tuned.

For one thing, I'm pretty sure that following up some implications of
our model has resulted in an understanding of the Aufbau of the
isotopes; an explanation of radioactivity--(haven't worked out details
here yet,)--and the break through that should make "Cold Fusion" a
very viable reality for "Energy" production. This insight, if correct--
and I'd make a sizable bet on it, had I any way to do so-- is very
exciting.

Details of this last mentioned may not get posted until August, or
later, as I've just now sent the ideas to an "associate" who has a
laboratory that should be able to test out the ideas. Haven't, of
course, heard back with confirmation of the validity of the
breakthrough,

After all---Just sent the info. ten minutes, ago../:-))

I'm, also, working on trying to get our work into a symposium in
Europe in July. Should know in a week or so if this is going to be a
"Go."

Can't recall if I posted my article on Matter/Anti-matter which is
on Helium,com, here.   I'd better recheck, If I didn't I'll get a copy
in here. If I already did, I'd suggest you all re-read it. I think it
is worth some thought and there may be implications that I haven't
caught on to myself. Dean (ESKI)

Reply to author   Forward

End of messages

« **Back to Discussions**   « Newer topic   Older topic »

Google groups

« Groups Home

Oscillator/Substance
⊞ **Theory**

[   ] [Search this group] [Search Groups]

O/S and Light (Copy of body of a letter to another scientist.)   Options

1 message - Collapse all - Report discussion as spam

dean sinclair View profile   More options Apr 5 2010, 6:24 pm

Looking at my site, it appears that the closest that I came to addressing
the topic of light, and then not by name, was the article, "Quantization, a
3-D Merrigoround?"

Since analyzing electromagnetic waves as information carrier is a basis of
the model, I have been remiss! As you will be studying light, maybe you can
confirm, or burn holes into, my theorizing. So here goes,

1. Light is a wave motion caused by a change in the condition of an
oscillator. This usually will be a change in frequency of the oscillator
with the oscillator either absorbing a harmonic, or emitting one...(Takes in
a set of other oscillators or drops them out?)

2. At its wave front, light as a wave motion can be considered to have an
instantaneous surface with a tension at that surface, hence a "mass." That
mass may be estimated by taking the wave length of the light as being the
radius in the equation, $m \times r = h/c$. There are a couple of situations which
would be of particular interest. One of these is the situation where
frequency equal wavelength equals $c^{0.5}$ and the mass would therefore equal
$h/c^{1.5}$.

Another situation of interest arises when on writes the Planck equation,
$E = hu$ at a value "!" of E. i.e. the total "Energy" motion content where in
"E" is the unit of whatever system we are using. In this case, freq. equal
$1/h$,, and wavelength equals "ch" if this system were allowed to oscillate,
it can be seen that there is a complementary set, where frequency is "ch"
and the wavelength is "h." The first set would be the "high frequency
cut-off," the second the "low frequency cut-off" for communication in a
system where "h/c" is the "Balance Constant."

It may be noted that one could suggest that these may represent the limits
of size of that above noted "communication system." With the light wave,
under ideal conditions, expanding from the highest frequency, shortest
wavelength to a lowest frequency wave length and 'rebounding.'

Now as to "polarization," as most electromagnetic radiation is thought to
result from changes of positions among electron, which probably are ccw
rotators, one might expect that most light, as emitted would have a ccw
circular, actually spiral, polarization. I do not know what exactly is the
theory of Lasers but it seems logical that crossing two spiral polarizations
of light would give a true circular, actually spherical, "polarization."
However, the overall rotation would still be there but the wave front now
would be at the front of a sphere.......

Reply to author   Forward

End of messages

« Back to Discussions   « Newer topic   Older topic »

Google groups

« Groups Home

Oscillator/Substance
⊞ **Theory**

Abridged summary of oscillatorsubstance-theory@googlegroups.com - 1 Message in 1 Topic

Options

Home

Discussions
+ new post

Members

2 messages - Collapse all - Report discussion as spam

About this group

Edit my membership

Group settings

Management tasks

Invite members

hugh vreeland View profile   More options Apr 6 2010, 5:08 pm

This is not a direct reply to today's topic, but, I am too lazy to retype
the address.

A new topic could be started on Graphing with Signed
Numbers, i suppose.

A few days ago i tried something;
Eski always tries to keep things too simple, he works only with absolute
values, although I'm sure he realizes that we could use signed numbers. If
i
remember rightly he did publish a little essay on signed numbers so he
does
realize they exist.

I got hold of a little graph paper and decided to graph a simple xy=k=xy
oscillator using both positive and negative values for every possibility.
That is let k be either positive or negative and graph all the combinations.

To conform to KISS, i said let k be six. Ja, , I like the sound of the
German word for six, so that could have something to do with it, but 3 and
2, as one limit and 3 and 2 the other side would be easy to graph, so also
would 6 and 1 and 1 and 6. + an - sets of each, of course.

Drawing these out, i found that we had four equal area   rectangles that
balanced across the origin. Combining all of these produces a single area,
which is doubled up closer to the origin ....

In looking at this it came to me that i was seeing something in the graph
that somehow reminded me of something that i had read. Then it came to
me,
this looked exactly like a graphical representation of Hoek's four positive
charges within a proton that blend into one.

It appears that any of Eski's oscillators could be graphically represented
this way by a slice through them at any angle which ran through the
center.

Looks like the graph says that Hoek's Proton Cosmology and Eski's
Oscillators are the same thing from different viewpoints. Neither Eski nor
Hoek ever graphed the situation using signed numbers.

Hoek had no indication that such a graph would have any bearing and Eski simply skipped the graphing idea for simplicity.

Well, my friends, there is my little contribution for the moment.

Did it help or did it just send you for more Tylenol 3?

Sincerely, " Hughvree"

On Tue, Apr 6, 2010 at 10:15 AM, <
oscillatorsubstance-
theory+noreply@googlegroups.com<oscillatorsubstance-theory%
2Bnoreply@googlegroups.com>

- Show quoted text -

Reply to author   Forward   Report spam   What you rated this post:

hoek View profile   More options May 6 2010, 4:33 pm
Hugh Vreeland's graphing of signed numbers is very interesting, and
he's correct in assuming that," Hoek had no indication that such a
graph would have any bearing".
  My approach had nothing to do, consciously, with mathematics. I
saw an animation of two alternating dynamic charge patterns, alleged
by computer analysis, to exist on the surface of a proton. The
physics film went on to say that theorists could explain what might
cause two rings of charge, but, were at a loss to explain the four
point charges changing into line-arcs configuration. I viewed the
film two or three more times on successive broadcast dates, just to be
sure that I had gotten it right. Unfortunately, VCR's and DVD
recorders were not available back in 1978. These alternating dynamic
charge patterns nagged at me. Soon after I performed a simple thought
experiment. I envisioned how a charged particle in linear motion and
its emitted field would behave within a spherical containment, based
on my background in electronics. Then two particles and eventually
arriving at four synchronously oscillating particles to perfectly
describe the charge patterns that I was so very curious about. All
the rest followed logically and serendipitously from that, how the
geometry of the sphere and Einstein's energy-mass equivalency equation
sustains the oscillations by converting field energy to mass and
positive charge, then that mass, back to energy and negative charge at
the center of the proton. What originally started out as a dynamic
charge pattern puzzle, resulted in a theoretical nuclear physics
hypothesis, that blossomed into a complete cosmology, as more and more
was discovered in astrophysics over the passing years showing how
things at the largest scale mirror things at the smallest, at least,
in my hypothetical model. See: http://www.helium.com/items/1304605-
how-the-universe-formed
  Right after Hugh referenced me in his comments, he said," Eski
simply skipped the graphing idea for simplicity." Therein lies the
problem, Dr. Dean, mathematical simplicity, rather than, mechanistic
simplicity. Quark theory suffers from the same miscalculation. It is
the simplest mathematical solution of positive and negative fractions
possible to explain the assumed axioms, I won't dispute that, but, it
has no mechanism. We exist in a perceivable realm of three spatial
dimensions and one forward moving dimension of time. These conditions
result in cause and effect, which allows mechanism. The whole
universe and everything in it are the result of, or operate in or by
some mechanism or other or several.
  Life forms are mechanisms or Automata. Automata are any self-
reproducing or replicating objects. In a lecture at Caltech, titled
"On the General and Logical Theory of Automata", pioneering physicist

John von Neumann outlined the four necessary components required.
(1)   A blueprint, a plan for construction
(2)   A factory, to carry out construction
(3)   A controller, to ensure that the factory follows the plan
(4)   A duplicating machine, to transmit a copy of the blueprints to the offspring
I have come to realize that the entire universe is alive and evolving from chaotic energy-mass, to atoms, to elements, to molecules, to life, to self awareness and cognizance. We are an evolutionary step in a universe that is alive and intelligently designed. The blueprint for our universe is base 10 encoded in pi. If you're interested, see:
http://www.helium.com/items/1273568-contemplating-the-existence-of-god
   Eski, I know I've asked you in past correspondence, if you could help me to translate my model into the proper mathematics, I believe Hamiltonian physics, by which to describe it. You responded that you weren't capable of that. I extend that same offer to you Hugh Vreeland, or anyone else who reads this and can do it. I assure you, you will be duly credited for your contribution. Well, that's about it for now except, Hi to my good friend ka-sala, may God always smile upon you.
hoek

On Apr 6, 6:08 pm, hugh vreeland <hughv...@gmail.com> wrote:

- Show quoted text -

--
You received this message because you are subscribed to the Google Groups "Oscillator/Substance Theory" group.
To post to this group, send email to oscillatorsubstance-theory@googlegroups.com.
To unsubscribe from this group, send email to oscillatorsubstance-theory+unsubscribe@googlegroups.com.
For more options, visit this group at
http://groups.google.com/group/oscillatorsubstance-theory?hl=en.

Reply to author   Forward   Report spam

End of messages

« Back to Discussions   « Newer topic   Older topic »

Gmail Calendar Documents Photos Reader Web more ᵛ   deanlsinclair@gmail.com ᵛ

Goⁱᵍle groups

« Groups Home

Oscillator/Substance
⊞ **Theory**

[Search this group] [Search Groups]

An old item i don't think i shared It's about "Energy."   Options

1 message - Collapse all - Report discussion as spam

hugh vreeland View profile   More options Apr 6 2010, 5:25 pm

Must be round a year ago, that Eski published a bit on SciScoop and a man
who went by Barak ridiculed him. This was my take..
The topic is somewhat peripheral to o/s theorizing, but, then, o/s is
supposed to be inclusive.
Since i'm in a send stuff to the group mood today, here is this one, if
you've seen it before, it won't hurt too much to see it again.
Answer to Barak

To say that Dr. S's Energy Expression, $(vm^2)/2$ is not a type of energy
expression because the units do not seem to match, is probably much like
saying, "That is not a leopard, I cannot see any spots." It may also fall
into the category of saying, "A bonobo is a Chimpanzee. take a good look at
one. " The spots of melanistic leopards can be seen only if light happens
to strike them at the right angles and there is a good deal of DNA evidence
that either Bonobos (also known as the "Pygmy Chimpanzee of Zaire) should
be classified as in Genus Homo, or humans should be classified in Genus
Pan. In mathematics and physics, as well as in biology, appearances and
labels may be deceiving.

Doc's contention is that both $(mv^2)/2$ and $(vm^2)/2$ are "energy
expressions" since both arise from legitimate, alternate ways of
integrating the momentum equation, $mv=p=vm$. He also notes in one paper,
that one can also integrate "p" as an entirety to obtain the expression,
$(p^2)/2$. which when we insert the original "$mv=p=vm$" fact, gives us $(m^2$ x
$v^2)/2$. Here in lies the key to the reason that the two energy expressions
differ in form. In the process of integration, one of the variables was
"held constant," that is taken out of real consideration except as a "scalar
number" in the process. However, it is still a variable, a vector quantity
which should be included in the result. The true descriptor for both energy
expression should be the one found when "p" was integrated as a unit and the
definition of "p" reinserted. In the cgs system the true units of what is
going on when we talk about the motion energy of a unit would be (g^2 cm^2
)/2 or mass times the "conventional energy expression" or velocity times
"Doc's expression." Looked at either way the concepts of mass, energy and
velocity are inter-tangled.

As the energy, or motion content, of a moving body in a medium of which it
is part--assuming here the the Oscillator/Substance Model is probably
valid--will consist of two parts, the vibrational-rotational motions within
the unit, and the motion disturbances caused by the motion of the entire
body along a line, there may well be two "Energy Expressions" of the same
form but actually arising from different views of momentum because of the
two differing types of motions involved. If we assume this to be true, the
total motion content i.e. energy would be $2(m^2$ x $v^2)/2$, which equals
$(m^2$ x $v^2)$. Evaluating this at the Speed of Light, whether it be a limit
or an average would then give the expression, m^2 x c^2 for the Energy
content of a particle rather than the mc^2 value usually accepted.

 Taking this even farther from the fact that the Absolute Values of
variables can be interchanged when they are related by an equation of the
type, $xy=K=yx$, one can postulate that the maximum motion content possible
of a particle moving at the Speed of Light would be c^4 or about 8.l x
10^4l $(g^2$ x $cm^2)/2$.

Gmail Calendar Documents Photos Reader Web more ⌄   deanlsinclair@gmail.com ⌄

Google groups

« Groups Home

Oscillator/Substance
⊞ **Theory**

[Search this group] [Search Groups]

Fwd: Essentials of O   Options

1 message - Collapse all - Report discussion as spam

dean sinclair View profile   More options Apr 6 2010, 7:17 pm

- Hide quoted text -

---------- Forwarded message ----------
From: dean sinclair <deanlsincl...@gmail.com>
Date: Tue, Apr 6, 2010 at 11:48 AM
Subject: Fwd: Essentials of O
To: amylaverickd...@aol.com, Earth/matriX <earthmat...@gmail.com>, drsdcm <
drs...@cinci.rr.com>

Here is a twenty three page write up that I did a few weeks ago on my home
computer, finally got it from there to a form that I can put on the
Internet. Yes, I don't really know my way around this technology as well as
I should.
To get it to the Internet, I ended up copying it, inserting in an a letter
and mailing it to myself! There must be an easier way! /:)) Dean, "Doc.:,
ESKI

(23 pg. shrank to 15 on the site)

---------- Forwarded message ----------
From: dean sinclair <deanlsincl...@gmail.com>
Date: Tue, Apr 6, 2010 at 11:26 AM
Subject: Essentials of O
To: deanlsincl...@gmail.com

A:/Essentials of O/S Theory.doc

Part I. Essential Ideas and Definitions

"O/S" Theory, short for "Oscillator/Substance Theory," operates on a
relatively simple model of existence. There is postulated that all of
existence is within a substrate of unspecified form and extent, which is
organized into/organized by oscillators. These oscillators are postulated
to be members of a "family" defined by the set equation $(m \times r + h/c)$ In
this set equation, "m" is a measured mass at a radius, r, from a center/axis
of rotation of an entity defined by the torque constant "h/c" wherein "h' is
Planck's Constant and "c" is the speed of light.

→ $m \times r = \dfrac{h}{c}$

The logical steps from which these assumptions arise, and the early history
of the O/S Model will be covered in a later chapter.

The following discussion covers some of the basic conclusions reached from
consideration of the above ideas.

The substrate has essentially the characteristics of a chemical substance at
its triple point with a constant tendency to equilibrate motion throughout.
This equilibration of motion results in various phenomena, some of which are
called "Forces," the Force of Gravity, Electro-magnetic Force, etc.

By the O/S Model, all of these "Forces" are the result of differential
pressures within the substrate/substance. "Gravitational Force" analyzes as

being due to differential pressures between vortex conglomerates of different sizes, Electrical and magnetic fields result from the readjustments of the Substrate/Substance to disturbances caused by oriented vortices--usually electrons.

The last two paragraphs have slipped a bit ahead of fundamental considerations. We need to back track and define the types of motions observable within the substrate/substance and which of these motions are oscillators. We also need to define three classes of oscillators.

The types of motions within the Substrate of Existence can be classified in various ways. One is as dot or line centered motions, i.e., movements in and out or back and forth toward a dot or a line, or around a line, or all of these. These are coordinated into oscillatory motions. In the case of molecules, these are known as vibrational /rotational motion or "vib/rot" for short. A second type of motion is motion through the substrate/substance away from a line or point which does not move the point of origin, this kind of motion is generally known as "electromagnetic radiation."

Motion which carries a central point with it--usually this can be considered motion of an oscillator or combination of oscillators from its original place-- is known as translational motion. This type of motion is the motion usually being considered when "Energy" or "Kinetic Energy" is spoken of.

"Energy," as the term is usually used, can be considered to be a "packet of motions" having an effect which could presumably be measurable. This packet usually consists of some combination of translational motions.

Another way of classifying motion within the Substrate/Substance is to classify it as within a given surface and outside of that surface. This way of classifying motions gives us a definition for the useful term, "Mass." Mass may be considered as a measure of the balance of pressure-tension between the motions within a surface and the rest of the substrate outside the surface.

[This gives a somewhat different view than the usual cyclic definitions of Mass and Energy. "Energy is what moves mass and Mass is what is moved by Energy." This usual set of definitions may be considered to be equivalent to simply saying, "Existence is."]

There is often confusion between "Mass" as an attribute of an entity and talking about entity itself as a "Mass." This seems to be a possible problem with the famous Energy definition, E=mc^2, which is usually interpreted as meaning that if an entity were totally destroyed there would be an "Energy " release of an amount equal to the unit's "Mass times the Speed of Light squared. O/S modeling casts serious doubt on the accuracy of that interpretation. The subject will be explored more in another section of this paper. .

The oscillators within our substrate/substance may be classified into three categories which we shall call "Classes I, II and III and abbreviate these class titles as C-I, C-II and C-III. C-I oscillators are considered as "full-wave pulsators" which oscillate through a sphere at m = r = (h/c)^0.5, which is a radius of about 4.7 x 10^-19 cm., with a concurrent mass of about 4.7 x 19^-19 grams. These oscillators are convertible to C-II oscillators. These are "pseudo-spheres" which have, at any given instant, both an equator and an axis of rotation. Any member of this group consists of two counter-rotating halves, which inter-convert through the equatorial circle. These halves can be split off, creating C-III Oscillators, half-wave inverting vortices.

The best known of these are the electron and proton which result from the split of the asymmetric C-II oscillator, the neutron. The neutron can be considered to be the result of a compression distortion of a C-I oscillator, which was dubbed by this writer the "Zerotron," which is--by O/S reasoning—also, the parent of the electron/anti-electron pair. This, previously unsuspected, "Zerotron" appears a "major player" in the Substrate/Substance as "parent" of both the electron/anti-electron set and the neutron.

The "Control Oscillator of Our Universe," whose inversion—or split-- is the "Big Bang," would be a C-II Oscillator. Whether we are a part of one side of a continuing C-II Oscillator, or a part of a unit analogous to an electron or anti-electron apparently is yet to be determined.

Considering a little basic math indicates another possible unit. . This argument goes as follows: The number, "One," can be considered to represent any whole unit. If we write the Planck Equation relating a package of motion, "Energy," to a frequency, i.e. $E = h \times u$, where "u' is a frequency, we can consider that Planck's Constant, "h," is the ratio between a fundamental Motion Packet, "E," which we may represent as a "Whole" by using the number, "1." If we do this and write, $1 = h \times u$, we see that "u," has a value of $1/h$. As the formula for the wave length of electromagnetic radiation is "Wavelength equals 'c,' the Speed of Light, divided by 'u' the frequency," the wave length of this packet would be , $c \times h$, "ch." As the wave length can be considered to equal the circumference of a circle, the radius of the corresponding circle would be $ch/2Pi$ and the corresponding mass, as measured at one extreme, would be $2Pi/c^2$. (This is from going back to "$m \times r = h/c$" and inserting the above value for "r.") This little unit, if one grinds out the math. would appear, if we were to find it, to be smaller and more "massive" then the proton. This writer has not checked to see if any "fundamental particle" discovered in atom-smashing experiments has the characteristics quoted above.

The speculation above has another twist. It is possible to consider that any equation of the form. "$xy = K = yx$," to be the equation for a "balance." By this reasoning, any constant would be the "balance point" of the units that make up its definition. Taking "c." the Speed of Light, as the balance for light as an oscillator, one can consider that any light wave will have an "equal partner" wherein the absolute values of the wavelength and the frequency are reversed. In the case noted above, where the frequency is "1/h" and the wavelength is "ch," the "partner set" would have the frequency of "ch" and a wavelength of "l/h." This very low frequency and very long wavelength would represent the low frequency cut off for information in "Our Universe." Another way of looking at this is to consider that these values may represent limits of the "Control Oscillator(s)" within which we exist.

The "Zerotron" has been mentioned previously as the Parent C-I Oscillator from which both the positron/negatron set and the neutron are considered to be derived; however, the evidence from which this conclusion has been reached has not been presented.

The evidence for the "Zerotron." Is, as follows: The electron and proton are known to "annihilate" upon meeting to furnish an amount of "Energy," as "annular radiation," at a frequency equivalent to "mc^2" of the mass of the electron. This is the "Energy" dissipation expected from the "head-on " collision of two particles with the Kinetic Energy of the two converted into electromagnetic radiation.

This "annihilation,: however, does not happen on the instant of entering into the vicinity of one another, however, but after a period of the two units entering into what could be called a "circling dance, or "forming a mass-less, Hydrogen atom." The point is that the annihilation occurs when the two entities reach a certain specific orientation to one another. This can be presumed to be when the inversion/pulsation and the rotations of both units are aligned on the same vector, in exactly opposite senses such that the two rotations will cancel. Canceling of the two rotations, each moving at an average velocity of "c," would furnish an observed value of "mc^2" as radiation in one annular pulse at aright angles to the line of "collision." As there remains the pulsation motions, what would logically result would be a C-I Pulsating Oscillator. This oscillator, were it struck by radiation having more motion content, "Energy," than that given off in the "annihilation," could split into the original pair of electron and proton. The "pair-production " process is also known. The scientific literature does not note annihilation and pair-production as reciprocal processes, however.

As this oscillator is the "intermediate from which the negatron and positron would go in opposite directions," i.e. the zero-point for the formation these two units, it seems logical to call it the "Zerotron."

The logic for considering that a neutron probably results for a shock wave compression of a Zerotron, is some what different and based more on logic

than on any known experimental evidence. It was first considered that
a proton
could be considered as being derived from an anti-electron by slowing an
anti-electron to about 1/42 of its velocity in a given dired5ion with all of
the velocity converted to "mass." There seemed, however, to be no way this
could be accomplished with a free anti-electron. Since the electron
and anti-electron
arise from the Zerotron, and the electron and the proton arise from the
neutron, it was logical to consider that the heutron be a modified
Zerotron. " How could this happen in such a way as to cause the Zerotron
to have one half of the oscillator to have its Vibrational energy
compressed so as to appear as

'mass' at a later time ? " became the question. If the oscillator were
"squashed" by a shock wave such that "one half of the oscillator met the
other half coming back," this type of result can be envisioned. Such a
"squashing" effect could be expected from the shock wave of the inversion or
splitting of an oscillator such as can be considered to encompass our
Universe. The initial motion would be at a velocity of at least twice the
speed of light, this motion into a medium having an average motion velocity
of the speed of light would create a shock wave presumably capable of
distorting many structures in its path. If the neutron is the most stable
of the resulting units, the one to which the others would equilibrate , then
the neutron could well result from a shock wave in a medium containing a
large numbers of Zerotron entities.

The concept of the "Zerotron" as a basic unit has the effect of
inter-relating a number of ideas which otherwise seem rather unrelated. The
electron, anti-electron, proton and neutron are seen as different aspects of
one entity under different transformation conditions. The phenomena of
annihilation and pair-production are seen as reciprocal processes and the
"Big Bang" falls into place as a "creative-destructive," recurring natural
event.

Since all of the things which we know as matter, result from the
interaction of the two units, electron and proton, which can combine in
many ways but do not recombine to form neutrons, it seems quite logical to
consider that all the events of our Univerese can be considered as the
result of the Substrate/Substance readjusting motion toward a balancing of
motion throughout, sorting out and minimizing the chaos resulting from
 oscillator
inversions such as the one which apparently started our universe. We may
consider everything from our own existence to the most farther galaxy as
being a result of creativity out of chaos which is tending toward an
"equilibrated chaos." This will be explored more in the following sections
of this manuscript.

Closing this section, let us list some of the definitions which arise which
are not common to the usual scientific literature.

Mass: An attribute of an entity which is a measure of the internal
vibrational-rotational motions within that entity as evaluated at the
surface. A measure of the tension/pressure at the surface of an entity as
against the rest of "Existence." This is measurable, at minimum values, as a
relative value by comparisons of entities.

Energy: A generalized term for a quantity of motion.

Kinetic Energy: The motion packet that can be measured, which is the result
of motion of an entire object in a given direction,

Potential Energy: A measure of the Kinetic Energy which could arise if the
space between two units were removed.... In an situation of oscillation, there
will be an interchange of Kinetic Energy and Potential Energy, with each
being "Zero" when the other is at maximum.

Electron, also known as Negatron: A counter-clockwise-rotating, inverting
vortex, Its rotation-inversion gives rise to what is known as a "negative
charge."

Anti-electron, also know as Positron; The clockwise-rotating inverse of the
electron, hence "positively charged."

Zerotron: A postulated entity having the same "rest mass" as either of the Electron or Anti-Electron units.

The parent C-I Oscillator which these combine into.

An entity which can be split by additional motion of sufficient amount into an electron and anti-electron and distorted by shock waves to neutrons.

Neutron: An asymmetric, C-II Oscillator, the result of the deformation of the C-I Zerotron. This unit, which would oscillate between "Matter " and "Anti-matter"
forms, splits in "Our Universe" into an electron and a proton. In an "Anti-universe it would split into an Anti-electron and Anti-proton. Its anti-particle would be identical to it.

Proton: A clockwise-rotating, inverting oscillator, formed from the asymmetric splitting of the neutron. With the same rotation/inversion sense as the anti-electron it is "positively charged." Its "rest mass" is some 1832 times that of the electron hence it has some 1832 times as often an inversion through its "central circle," at 4.7 x 10^-19 cm., as the electron has through its center of the same size. The proton motion is at a correspondingly smaller distance.

Big Bang: Title given to the Inversion--or spitting--instant of the C-II Oscillator, which is the "Control Oscillator of Our Universe." This event presumably created neutrons. The decay of these neutrons produced electrons and protons resulting in the "Matter" found throughout "Our Universe." This does not rule out other somewhat similar incidents continuing to create neutrons through similar shock-wave events.

Part II.   The Iso-set Approach to Matter

Most of the things we deal with in our universe are made up of what we call, "Matter," which consists of various associations of electrons and protons. These
associations are known as atoms, molecules, organisms, and these are organized more and more, into many different units, stars, solar systems, galaxies,,,,,

At the very most basic level, an idea called the "iso-set" can be useful. The idea is that any group of associations of electrons and protons can be considered as a part of a set of "isomers," things made up in different ways of the same units. The first set of electrons and protons which is likely to come to mind is the set made up of one electron and one proton, We call this the "Iso-1,1-set" and consider it to contain the following subsets: The set,

{e. p}. the neutron, n, and the Hydrogen atom, H-1. we can write this then as the set, {e, p; n; H-1}. We must also remember that any set will contain the "Null Set" wherein the components, "e," and "p" do not exist.

This "Null set," usually considered to be empty, may occur in any amount. What we are saying is that the "Null set" may contain Zerotrons, an entity formed from the union of a proton and an anti-proton., and other unrecognized Class I Oscillators. Usually this "Null set" can be ignored; however, it may come into play in some "nuclear" transformations and it is well to be cognizant of its existence.

In any set, one may postulate that there will be some lowest motion content, most stable unit, and some set of most motion content which is still an association of units. This highest motion content set, this writer calls the "Central Aggregate" of the set, abbreviated, "Iso-A." For the 1,1-set, the "Iso-A." may be the neutron, which can be said to be made up of a tightly coordinated electron-proton pair with a good deal of excess motion, "Energy." content.

Another possible unit of the 1,I-set could be called a "pseudo-neutron," a unit made up of a rather transitory--possibly only instantaneous--coordination of an electron and proton along the same vector. This last postulated unit, may be what is the cause of the "proton-neutron nucleus" which has been so useful for the last 70 some years.

Although it has been very useful in correlating data and is the basis of the Periodic Chart of the Elements, the "Proton-Neutron Nucleus Model," in the belief that neutrons exist, as such, in "nuclei." is probably in error for any atomic species having a "half-life" greater than that of a neutron in space. As the neutron's half-life is known to be about two seconds, the neutron-proton model is expected to be in error for all "naturally occurring" isotopes and for any isotopes which do not persist in nature but have half-lives greater than a few fractions of a second.

Considering electrons and protons to be spinning vortexes, albeit on a different scale, it seems logical that both will show some of the same associative characteristics. Electrons are known to have associations that can be considered as sets of 2, 3, 4, 5, 6, 7, 10, and 14 which are quite stable. There is no reason to think that protons can not associate as well, forming similar stable sets. As such stable sets of protons would be on a much smaller scale than similarly constituted sets of electrons, they would form a "nuclear" cluster.

Probably the next set that comes easily to mind would be the   Iso-2,2-set. Which contains, along with appropriate combinations of all the previous sets and the Null Set, the important set of isomers, the Deuterium Molecule and the He4 atom. These have been the subject of much interest, speculation and controversy since 1989.

In 1989, Fleischmann and Pons announced the discovery of "Cold Fusion," in the finding of excess heat being generated in electrolysis of "heavy water." This they attributed to the fusion of Deuterium. As it was a "Known" that "Fusion processes can only take place in plasmas at temperatures such as are found in the Sun," the scientific establishment immediately denounced this as nonsense and Fleischmann and Pons effectively disappeared into scientific banishment. Many experimenters since have shown the effect to be real and have found the generation of He4 to coordinate to excess heat production.

The above-discussed effect has been most often observed in Palladium cathodes and there have been many discussions as to what may be happening. Simultaneously,
in other laboratories there have been observed transmutation effect which appear to be the result of the incorporation of Deuteron units, (units of the

"I,2 –set") into various nuclei. These results suggest that the   "intra-set transformation" from D:D to He4, may not be direct but by way of Deuteron to Deuteron association with subsequent transformations.

The stability of the D+ unit, the Deuteron, has mystified the scientific establishment for some time. In terms of the O/S model

-- which is not burdened with having to consider the Deuteron as a combination of a proton and a neutron--the stability of D+ is easy to understand, as also is it's apparent ability to associate with other "D+es." In O/S terms, the Deuteron would consist of an associated "up-down" pair of protons encased within an electron.

Association with other Deuterons would be a combination of the

" up-down" pair coordination with a similar pair to form a "four-some" and a simultaneous coordination of the "electron encasements" to form an "electron pair." The combination of two Deuterons would produce a "high-energy" form of the Deuterium Di-cation, D:D++, which, in turn, could be considered a very-high-vibrational-rotational-energy form of an Alpha Particle.

[The Alpha particle--by O/S reasoning a clock-wise -spinning, "square-planar" assemblage of four protons and two electrons--is generally considered to be the "nucleus" of a Helium 4 atom as it contains 6/8 of the constituents of He-4.

It may be somewhat "hair-splitting" to aver that the Alpha particle is not a He-4 nucleus; however, we shall do just that. By the O/S model, the He-4 central unit would be a tetrahedral assembly of 4 protons, and the Alpha a square-planar unit of four protons and two electrons. In conventional thinking both are units composed of two protons and two neutrons with no concern about configuration nor any consideration of the different motion characteristics for the two entities.

Conventional thinking does not correlate rotation (spinning) to "charge" hence takes no note of the idea of charged units spinning or rotating in a certain direction. For instance, no one ever has noted that the counter clock-wise, "left-handed" twist characteristic of so many natural products could be the result of anionic ("negatively charged") intermediates involved in their formation.)]

In contact with a D:D molecule, an Alpha particle could "strip off" an electron pair, imparting its spin-twist to the remainder of the D:D unit to form another Alpha particle of greater "Energy." This chain-reaction process could continue thorough a layer of Deuterium molecules on the surface of a Palladium cathode until some Alpha developed enough Kinetic Energy to escape the environment. Alpha particles have been observed as a minor byproduct in the reactions that produce He4 as a major product.

What is described above is a possible scenario for the formation of He-4 from Deuterons. The first step, the formation of Deuterons from Deuterated Water would be endothermic, with the energy being supplied through the electrolysis. Subsequent processes including the association of Deuterons, and "spin-downs" with loss of vibrational energy to the medium would be exothermic and the last "Lewis-Acid-Base" reaction removing an electron pair from a Deuterium molecule could be expected also to be exothermic considering the probable lesser motion content of the very symmetric He-4 as compared to an Alpha particle and a D:D molecule.

The overall effect is D:D àHe-4, and the calculable "Energy" change would be—in terms of conventional thermodynamic reasoning—"the Energy equivalent of the 'mass-defect' between the Deuterium Molecule, D:D, and its 'Atomic Isomer,' the He-4 atom."

The Iso-3,3-set contains the isomer pair, Tritium, "H-3" and the isomeric He-3 to which it "decays" with the loss of a "Beta particle"--an electron spun out at a rather high velocity. Here, as in the Deuterium Molecule to He-4 example, an apparently simple "Iso-set-isomerization" very likely goes through an intermediate cation of a smaller set.

H-3, Tritium, has no magnetic moment, that is, it has no "charge separation." On a "time-average" it is totally electrically symmetric. On the other hand, the more stable He-3 does have a magnetic moment, which is in the same direction as that of the neutron. Both it and the neutron are unbalanced in the same direction, apparently with a clockwise spin. At first thought, it would seem that the Tritium should be the more stable isomer, containing at any given time two electrons more tightly bonded to the protons in the nucleus and one electron much less bonded, as against the He-3 which has an unbalance "charge" structure with but one electron more closely associated with the nucleus at any given time.

A little closer inspection from the O/S viewpoint, shows a somewhat different picture. The central unit of an H-3 atom would be a "vibrating tetrahedron" with the three protons exchanging places among three corners of the tetrahedron with the electrons in a complicated dance which would be somewhat comparable. In the He-3, a coordinated motion of three protons in a trigonal ring encompassed within one electron would involve far less total motion content than is present in the Tritium situation.

The loss of one electron leaves the remaining two electrons in the Tritium cation unpaired. This cation can drop into the more stable He-3 configuration.

[It may be considered that the "drop from the lowest energy state of the H-3 to the highest energy state of the He-3, taking place when the states happened to coincide in configuration, furnished the 'impulse energy' to cause the unpairing of the electrons and the expulsion of one as a Beta particle." This latter comment is very close to standard reasoning, which, however would have to involve a "neutron to proton conversion" discussion.

By conventional models, the H-3 has a nucleus of two neutrons and one proton, while the He3 has a nucleus of one neutron and two protons. By O/S, both have a central aggregation of three protons, in different configurations.]

The Iso-set approach can become very complicated very rapidly, but it is useful in comparing two or more units of the same set while keeping in mind the ideas of the possible involvement of the "Null" set and of Iso-A structures which do not fit the standard patterns of atoms or molecules but may be intermediates. This seems to not only add some flexibility to thinking about possible reaction routes, but also some guides to where "flexing" might take place.

An interesting set to follow by O/S reasoning would be the "self-destruction in search of symmetry" of the unstable Iso-8,8-set whose final unit is Beryllium-8.

Be-8 is so symmetrical that it twists apart into an Alpha particle which spins off in one direction and another square planar unit which might be expected to exit in the opposite direction. However, this second square planar unit is contained within an electron pair. It "goes nowhere" as its paired electrons counter-acting the spin of the four protons and counter-balancing one another cause it to "stay in place." Eventually, both the Alpha particle and the Di-anion become Helium-4 units.

One would expect Be-8 to split directly to He-4: however, the symmetry difference between He-4 as a tetrahedron and Be-8, whose central core could be expected to be two square planar arrays twisted 45 degrees to one another, prevents this kind of transformation. The spin apart of the two square planar arrays needs only a "slight nudge" to any point.

This type of discussion needs ultimately to continue throughout the Periodic Chart of the Elements, sorting out and developing understanding of the patterns that cause the chart to work so well for many purposes. Finding the reasons for Elements, without the need to invoke the "neutrons in the nucleus" model, will be an interesting challenge

. This problem will most probably be solved by someone having advanced computer skills. The human genome was decoded. Why couldn't the Periodic Chart be decoded also?

Here are some factors that might be considered in trying to develop a computer program to solve the mysteries of the Periodic Chart.

If we abandon the idea of neutrons in the nuclei, except perhaps for some nuclei with such short half-lives that the neutron has a comparatively infinite life, then we must find a rationale for their "phantom presence." The concept of pseudo-neutrons consisting of momentary co-ordinations between electrons and protons and/or groups of electrons with groups of the same number of protons might solve this problem.

Another factor is the mobility or lack of mobility of the two "nucleons." The electron appears to have great mobility and there seems no reason to think that, at any given time, any electron could not be in any part of the "dance of the electrons." That is, it can probably be considered that all electrons in a given unit are equivalent although at an instant of motion disturbance the electrons could be considered to momentarily "freeze frame" at the instant before and after readjustment.

Although, most "Energy absorptions" and emissions are considered as motions of one electron from one "orbit" to another, it is more likely that the rearrangement is total in the electron dance, and, sometimes this may also interact with the continuing "proton dance."

The proton dance, because of the far different frequency of oscillation is confined ot a smaller space and apparently causes the protons to have much less freedom of movement, as evidenced by the number of transformations that occur in the "shake apart" of the Iso-8,8-set. This also appears in the fact of "internal conversion" within some of the heavier elements.

One may guess that geometric structure is apparently far more important in the closely packed interior of an atom than it is in the outer part of an atom occupied by the electrons. We have already discussed the probable geometric importance of certain simple arrangements in the cases of the Alpha particle and the Helium-4 atom and in the Tritium to Helium-3 transform. As the number of protons making up the central "matrix" of an atom increases, the role of geometry probably becomes more and more important.

It is easy to visualize some of the interactions between two protons, three protons, four protons, and five protons. Up to this stage, coordinated actions are relatively easy to visualize. Thereafter, the situation becomes more difficult.

The "orbit-shape structures" developed for electrons may well be of some help, but should not be considered as definitive. They do not, for instance, note the possibility of a coordinated, continuing motion through the points of a three dimensional star, which is one possibility for five spinning objects.

\For the more complex atoms, it may well be that "Buckyball" structures, even "Buckyballs" within "Buckyballs" will turn out to be probable structures for the central unit of atoms.

People who believe in the proton-neutron model for the central unit, "nucleus," of an atom have worked out shell sets and preferred shell sets for the supposed neutrons. These models might be of help in understanding the electron to proton coordination that makes these "neutron shells" appear valid.

Part III. The Out and Beyond

If "All" be part of a Substance/Substrate, then science should be a continuum including everything from nano-chemistry to cosmology. The happenings within constellations, within stars, neutron stars, Quasars and all the other species that make up the units of existence beyond our Earth, would be subject to the same ideas of oscillators and oscillations that we have discussed in the few previous examples. Cosmology would be the study of these interactions on a very macro scale in one respect and with a very close attention at the very tiniest level in another set of concerns.

To understand "Black Holes" probably requires a close examination of the "Hidden Half of Existence" below 4.7×10^{-19} cm. in radius. This half has been implied in the rest of this paper but not emphasized. This is the half of miniscule radii and huge "Mass" values as compared to our "outer world" of the reversed absolute values, a world wherein the electron has far more mass than the proton.... This is a world wherein "negative becomes positive, large becomes small and light becomes heavy." This "World" has not been ever considered in science other than in comments about the disappearance of matter into the singularities of Black Holes.

To understand Galaxies may require an understanding of Class II Oscillators. It may well be that the Black Hole in the center of a Galaxy is the inversion site of the Class II oscillator which controls that Galaxy. Galaxy shapes may indicate the "Stage of Life" of the Galaxy. It seems that while the "Universe" as a whole is expanding, Galaxies may be contracting, i.e., in a contraction phase of their "control oscillators." There are recent reports that "Quasars" may cause galaxies to develop about them.

Another thought is that Quasars and Neutron Stars, might be analyzable in terms of being not only controlled by a central oscillator, but containing what might be called a melting –pot of Iso-A units of all kinds from the simplest units to the most complex. The "neutron stars" might actually be creating neutrons from Zerotrons and heavy elements from the mixture of Iso-A forms therein. It has been reported recently that ions of heavy elements have been detected streaming out from quasars.

Examination of Cosmology from the O/S viewpoint should lead to very interesting results. The role of Black Holes in the balancing and Evolution of the Substrate/Substance will be interesting to discover.

It may be noted that if a half oscillator such as the electron or proton were to pass through the inversion center of a Class II Oscillator, its sense would be reversed, that is, an electron would become an anti-electron, this could have interesting consequences in that out of phase contacts would cause reunions to the Zerotron.

Considering the situation with the proton in the above scenario leads to an even more interesting idea. The proton-anti-proton union would produce what might be called a "Super-Zerotron." If this speculation be correct, and the idea that Zerotrons are distorted to neutrons also be correct, then there is a conclusion that the Substance-Substrate is in a continued process of evolution toward "tighter and tighter" forms of Class I Oscillator, with the Zerotron being possibly the predominant form at the present stage of Evolution of the Substrate. If the model be correct, the ultimate form would most likely be the central average sphere of radius, $(h/c)^{0.5}$.

Interesting aspects of all of this appear where this model begins to interact with both philosophy and science-fiction. Many questions arise. Is Our Universe a one of a kind, except for the "co-created" "Antiverse," or is it that there are many sets so constituted that these units can exist "concurrently" slightly "out of phase" with one another. What about the situation of anti-matter?

[A short article by this same writer, dealing with the problem of anti-matter has been published on the Internet in Helium.com. It will be included in an appendix along with several other special topic units which have been published on Helium.com , SciScoop.com and /or the Google Group Site, Oscillator/Substance Theory.]

There is indication that there may be ways to take advantage of the continued flux of existence, the "Ground State Energy" which is perhaps represented by the temperature of about 2.7 degrees Kelvin which is characteristic of outer space.

This temperature represents the average Kinetic Energy of the moving units in space. It may be that we are already taking advantage this in the phenomenon of "Super Conductivity" of certain materials when they are cooled to temperatures approaching "Absolute Zero."

The super-fluidity of "Helium II" may also be related to this "tiny bit of heat" which by O/S reasoning would underlie everything but only show up as an effect when there is a deliberate attempt to reach a "lower temperature than that of the Substrate of Existence." In other words, what ever makes up the basis of the Substrate/Substance appears to have a constant motion which is measured, at least in our vicinity as about 2.7 degrees Kelvin.

As by Carnot Cycle reasoning, a heat engine operating between Zero Kelvin and 2.7 Kelvin could operate with 100 % efficiency, it would appear to be feasible to utilize the "temperature of outer space" for human purposes.

Part IV. O/S , Space-Time, Quantum Mechanics and The Standard Model of Particle Physics

There are quite a number of "Theories of Everything which predate the O/S model. In particular there are three models which have been quite widely accepted in scientific circles and touted as being essentially Theories of Everything. These three are Space-Time Relativity, Quantum Mechanics and The Standard Model of Particle Physics. None of these is truly comprehensive. Each operates primarily in its own niche, and the languages and concepts do not mesh well although some scientists seem to think all are correct at the same time and end up doing discussions which seem abstrusely intellectual but which on closer analysis are found to be "concoctions of word salad," i.e., nonsensical verbiage.

{It is quite possible that some title such as 'The Quantum Electrodynamics of the Relativistic Strange Quark" would be taken seriously as the subject for a presentation at some scientific meeting. Jazzed up with enough incomprehensible reference to differential equations with a few quantum mechanical operators thrown in, some joker might well get complemented on his or her new advance, which no would understand because it would be a compilation of disparate parts, utter balderdash.]

There will be an attempt to see how these three approaches might fit in with the quite comprehensive O/S approach. Looking first at Space-Time we find that, surprisingly enough, much of the mathematics may fit in fairly well for rather accidental reasons.

Mathematical space does not actually have a concept of a void. Mathematically,
space is filled with dots. Now, looking at "Time." In practicality, time is always referenced to some cycle. Putting the two together, Space-Time by its title implies that the mathematics deals with motions of dots having a cyclic motion.

As to the relativism, the relativistic corrections apply to transmission of information between transmitter and receiver sets that are moving with respect to one another. There may be validity in cases of relative speeds that approach the speed of information transfer. The equations become nonsense when applied to non-communicative relative motion.

Quantum Mechanics is a mathematical model which was designed to correlate the information from the line spectra of Hydrogen to Energy. It deals only with positive numbers. Operating in three dimensions with both Kinetic Energy and Potential Energy terms it does not address oscillatory motion, nor does it differentiate "vib-rot" motion from translatory motion. What it appears to deal with is translatory motion of electrons in atoms from a statistical view- point of a vibratory wave motion. Based on one electron and one proton, the Hydrogen atom, it is stretched to apply it to more complex units. There is no notice whatever of the fact of the two possible energy equations which arise from integrating the formula for momentum, $p = mv$ both ways, i.e, with mass as a constant and velocity as a variable to get the standard KE equation, $E = (mv^2)/2$, and integration with "v" constant to obtain the alternate expression, call it "E'", $E' = (vm^2)/2$, and there is certainly no indication of the overall energy expression found by simply integrating, "p," to obtain $(p^2)/2$ which equals $(m^2v^2)/2$. It is possible that the Quantum Mechanical mathematical approach could be developed to apply more generally if the various energy expressions were taken into consideration and some mechanism developed--possibly by the use of both forms of signed numbers—to account for the "Second World below a radius of 4.7×10^{-19} cm.

Neither of the two models discussed above give any attention to the size, shape , interior motions or actual exterior motions of electrons and protons. The two models discussed thus far are primarily mathematical in form and do not really address entities as such. Both appear to have some possibility of integration with O/S.

The third model, the Standard Model of Particle Physics , received a Nobel Prize in Physics in the 1970's and is called by some, "The Crowning Achievement of Particle Physics." In my opinion, if this was "The Crowning Achievement of Particle Physics," then the "Geocentric Model of the Solar System" was certainly the crowning achievement of astronomy. Both models start with sets of misconceptions and add assumption upon assumption to obtain a model that seems to work.

The Standard Model has a number of starting assumptions one is that ther are the "Four Forces of Nature," which are "Gravitation," "Electromagnetism," the "Strong Nuclear Force," and the "Weak Nuclear Force." It then is assumed that Gravitation has no effect at the nuclear level and can be ignored. Another assumption is that the various units which arise from atom-smashing are fundamental units which are somehow released by the experiment.

All of these assumptions are open to criticism. The Four Forces of Nature, on analysis, can be seen to be two observational phenomena which can be explained by differential pressures in the Substrate/Substance. The two "nuclear forces" are necessitated by the assumption of neutrons being constituents of nuclei. All four "Forces," are either fictional or imaginary. Additionally, the actual force that is responsible for Gravitation would be very much present at the atomic level.

The units which can be identified from atom-smashes are probably better characterized as alternate states of matter which are created under the conditions rather than being always there.

To complicate things even more, some electron scattering data is interpreted to show that some things called "Quarks," can carry either 1/3 or 2/3 of a charge. (As the electrons and protons as oscillators have two limits and a central inversion point, i.e., three definable points of possible scatter in a ration of 2/1. the phenomena that are identified as "Quarks" may be valid observations but are certainly not evidence of existence of basic units.)   To explain other things more and more basic units are invoked, gluons to hold things together, gravitons to explain gravity and so on and on....

What needs to be done with this model is to find where, rationally, the actual products of atom smashing would fit in to a pattern of matter and/or anti-matter. One such unit, the "Beta sub 2" has been shown to invert between Matter and Anti-matter forms. This is a characteristic shared, by O/S thinking, by all C-II Oscillators. In this writer's opinion the Standard Model should ultimately go into the dust bin of failed nonsense.

A very unfortunate result of the blind belief of many scientists in the Standard Model has led to one of the great scientific experiments of all time, the Hadron Collider under the border of France and Switzerland. This Collider, designed to smash streams of protons into streams of protons is designed on Standard Model ideas an is hoped to find evidence of something called the "Higgs Boson."

When initially put into operation the Collider immediately broke down. O/S considerations indicate that that is the probable continuing fate of the Collider. Break down, start up, break down again. The problem probably is that the designers, having no idea of existence of a Substance/Substrate nor of the fact that protons can coordinate with protons in many ways, do not realize that a stream of protons can generate an electrical field in the opposite sense from the fields developed by moving electrons.. This means that the force requirements to move protons in the ways that they "expected to do so may well be several orders of magnitude larger than expected. In addition, the clockwise rotation of the proton will make the requirements for sending protons clockwise different from sending them counter clockwise so running streams in opposite directions appears to be a very difficult task , to say the leaxt. . Also, on meeting the protons may well coordinate rather than making hard collisions.

The Hadron may well go into history as one of the most expensive failed scientific adventures ever attempted, because it was based on a combination of no information and misconceptions.

There is much that needs to be done with the O/S model which is still in its infancy as the writer is the only "expert" in the field. There need to be many others who can see and extend, amplify and, possibly refute the ideas that arise. At this point, however, O/S seems to be the closest thing to a comprehensive theory which is available.

Part V. History of the Model

Some people may be interested in how this model came to be developed. In its first form, which has been much modified since, the basic ideas were published in April 2007 in an article on the Internet site, Heliium,com, entitled, "Motion in a Matrix as a Theory of Everything."

However the genesis of the basic ideas began, for this person, in the Spring of 2004, the year before "The Year of Einstein," when in thinking about the idea of "Relativity," the thought came to mind that perhaps Einstein's Genius could have been the ability to see the "Overlooked Obvious." After all,   the idea that information is relative to the observer is rather obvious; but, before Einstein, no one had "made a big deal out of that." If, per chance, this kind of ability were "genius," one could pretend to be a genius by deliberately looking at seemingly obvious ideas and trying to find different ideas, views, slants on them.

Later research showed that this apparently was not a facet of Einstein's Genius. However, the "Pretend Genius" idea turned out to be quite good.

 One of the first things that was looked at was Einstein's "Special Relativity" with the thought, "Is there something here that is so obvious that the significance may be overlooked." The first thought was, "Well, it deals with a mathematical formulation." "Mathematical formulations can be generalized. What happens if we generalize the formulas containing the 'v squared over c squared expressions?" What then would be the meaning of the constant, 'c?' " The answer was that the constant would be the maximum velocity of information transfer, and the equations would apply to any situation of information transfer. One could define "Perceptual Universes" by the maximum velocity of information transfer in the volume under consideration. This writer had no idea that this would eventually lead to the current Oscillator Substance Model for Existence.

Following up on the idea of Perceptual Universes, however, led to the thought, "Just what would be the meaning of 'maximum velocity of information transfer?; " "How is information transferred? How does it go out in all directions from a center? Considering these questions it was realized that,whether by electromagnetic radiation or Pony Express, the maximum velocity of information transfer, in any direction over any significant distance, would be the average speed of the carriers in that direction.

Applying this to the Speed of Light it was realized that, logically, the Speed of Light should be the average velocity in all directions from any given point of the carriers of the information, whatever they were. The Speed of Light then is a Constant of nature but not a Limit to Velocity in an absolute sense; but, instead, it is an average of the velocity of something.

This view of the Speed of Light led to the idea that light has to be carried by something in a Matrix.

The next thought was a bit of an aside, "What is there else about reality that we take for granted. We take Mass and Energy as interchangeable. If they are interchangeable, they must be, somehow, aspects of the same thing. What else is there that is always present in existence? Motion. Can motion have two aspects? Sure. Motion about a point and motion along a line. What then are the aspects of our Universe? Motion in a Matrix. What else can we find out about our matrix and motions in it and the Speed of Light as information carrier? Where can we look? How about Planck's relationship? I just noticed that it is said to have the dimensions of angular momentum. That fits, a rotating unit could carry information in a straight line at the speed of rotation at its surface. What happens if we equate Planck's Constant to its definition as an an angular momentum and plug in "c" for the velocity? This was done and the little equation,

$m \times r = h/c$, made its appearance. It was realized that this could be the equation for a family of oscillators. The question-answer-new-question-new-answer dialog such as .was reprised above has led to the current state of the Model...

A simple thought led to some intellectual exercise which led quite accidentally to the construction of a sort of "Theory of Everything." This Model is still evolving as more and more information falls into place, more questions arise, more answers are found.

It was later realized that Max Planck had actually gone over some of the same ground a hundred years before and had used an oscillator-particle model in developing his equation. It was also realized that someone could easily have reached the same conclusions by much simpler logic by considering the implications of the Michelson-Morley Experiment beyond the popularized version of considering that it had "disproven the "Aether Theory."

This whole Model could have been developed over a Century ago had scientists followed up on the ideas of Max Planck rather than becoming enamored with the ideas of Albert Einstein. Had that happened there is a strong chance the there would have been avoided much of the confusion which has been alluded to in other parts of this paper.

Part VI Appendix. The Fun Stuff. Short , Special Topic Articles From the Internet

In this section are being compiled some of the previous writings that pertain to this topic. Some are of "historical interest " and show a somewhat more naïve view of the area. Some go into short topics in more depth than was done in the body of the paper. It is hoped that the reader will find these interesting and enjoyable.

Append: Motion in a Matrix, Vreeland's answer; "letter to Sophie," the whole Sci-Scoop File, the Matter-Antimatter article,

Constancy of Constants, Worlds within world or what ever I called

it:....Equivalency of Electrons, Worlds Within Worlds, and anything else that looks like it might be pertinent. Perhaps the articles on Signed numbers and Iso-As the Key to Radioactivity....

"SON"

Google groups

« Groups Home

Oscillator/Substance
⊞ **Theory**

[Search this group] [Search Groups]

A Caveat, I may just be another deluded crackpot. Here is one person's opinion.

Options

3 messages - Collapse all - Report discussion as spam

ESKI View profile   More options Apr 10 2010, 2:27 pm

The following is an excerpt of an exchange between myself and a
theretical physicist/mathematician who has, to say the least, a low
opinion of this.whole thing.

> I gather that you consider my little theoretical attempts-which
> simply developed from following a sequence of logic--as being
> foolishness as I haven't sufficient intelligence or education to

be > fooling around with such things? A kid playing with grown-ups
toys > that he doesn't understand?
> You may well be right, but, let the old man have his dreams of
> accomplishing something. Won't be too long that he will no longer

be > around to dream....DS

Dream all you like, but what you're doing is both logically &
mathematically and physically are provable nonsense.
What on Earth do you think you are correting that needs correct.
You still need 50+ years of study of both physics and mathematics
even to approach such vast questions. I still find it unkind or
somehow arrogant of you to accuse me of not helping. I' answered
I've answered your, but you haven't even attempted to read what
I've written. I'm still tired of always of alway having to
explain thin - and then be mean person because I say what is true
and not what they what want to hear. Non Imprimitur.

Who knows, people, is he right?

 Incidentally, I have tried to read his material. Either he is so
brilliant as to be far over my head, using concepts that are too
advanced for me to understand, or he is simply talking jargon that is
poorly defined and which he doesn't really understand, himself.   That
is so much the case of "scientific writing," especially in the
advanced physical science rhelm. Are the writers saying anything
profound, or do they simply think they are? DS

Reply to author   Forward

Discussion subject changed to "CAVEAT EMPTOR, 'Relativism Re: On being 'another' de

Jack O Suileabhain View profile   More options Apr 12 2010, 10:48 am

Dean's notions are a 'world/UNIVERSE withing themselves' which is a given. That they are
as 'valid' as the

COMMON INTITUTIONALLY SANCTIONED JARGONIZED fields of related 'science'(a la'
tenured acadamia etc.)

is also a reasonable supposition.

VOLTAIRE is just as valid here within 'exotic views' of physics etc.: eg.--> paraphrased "Whether I 'agree' with

a concept you present(or not) is IRRELEVANT; I will DEFEND YOUR RIGHT TO PRESENT YOUR VIEWS figuratively

& literally 'to the death.'----> WHY: Whatever the relative overall 'accuracy'(or not) of your over all presented 'SCHEMA;' maybe

just ONE SINGLE OF YOUR MOST COGENT 'POINTS' -------may----->lead SOMEBODY (maybe not even born yet) to

see a TESLA EPIPHANY that will be the NEXT BIG THING!

THE ARROGANCE of the 'ESTABLISHED & TENURIZED' that simply LACK the IMAGNINATION &/or FORTITUDE to venture

beyond the ESTABLISHED STOCKADE OF CURRENTLY ESTABLISHED ACADEMIC 'NORMS' of DATA-INTERPRETATION;

is no 'new' thing. And by such stratified 'attitudes' COPERNICUS-&-GALILEO & Co. were 'censured' as well. 50+ years of INDOCTRINATION will make 'ONE'

merely ONE MORE FOSSIL for the MUSEUM; aka another ICON to be shattered before we ADVANCE-METAMORPHOSIZE our outlooks and actually move beyond the frontiers that INSTITUTIONALIZED ACADAMIC TENURED will be made OBSOLETE by and no longer have PAYCHECKs to back their INSTITUTIONALIZED CASTE-STRATIFIED ARROGANCE. . .

?!? DREAM-ON ! ? !----->ABSOLUTELY! as per Einstein---->"IMAGINATION is MORE IMPORTANT than 'KNOWLEDGE!'" ?Why? 'KNOWLEDGE' is DYNAMIC, & FLUID, & PERENNIALLY SUBJECT TO REVISION & UPGRADE &/or down right OBSOLESCENCE.

REAL SCIENTISTS are NOT POLICEMAN for the INSTITUTIONALIZED 'FIELD;' but rather are INCORRIGABLE EXPLORERS. . . .

Standing for(and teaching) the foundational 'basics' is no crime; but wisdom is in knowing that those 'basics' and the way that we maybe view 'incompletely-inaccurately,' is a mere 'beginning' point. . . & per Robert Frost---> "Two paths diverged in the yellow woods, and I took the one LESS TRAVELLED BY; and 'that' has made ALL THE DIFFERENCE!"

CASE IN POINT: ALL MODERN TECHNOLOGY is made possible by one EXTREMELY WIERD NON-CONFORMIST----->aka--->NIKOLA TESLA--->that spent ABSOLUTELY NO TIME supporting the INSTITUTIONALIZED-ACADEMIC-TENURED theories & nor 'conventions' of the STATUS-QUO. . . . aka ---->The 'status-quo' and its psuedo-intellectual caste stratified 'attitudes' is VIRTUALLY IRRELEVENT to the 'Future Progression of the NEXT GREAT PLATEAU of Physics/Scientific discovery/progress.' In this such as JOHN HUTCHISON is infinitely MORE RELEVANT than the Cop who recommends 50+ MORE YEARS OF INDOCTRINATION in the PATENTLY UNFRUITFUL. . .

- Show quoted text -

Reply to author   Forward   Report spam

ESKI View profile   More options Apr 13 2010, 6:44 pm

Thanks, Jack. DLS

On Apr 12, 10:48 am, Jack O Suileabhain <braghgoerin...@hotmail.com> wrote:

- Show quoted text -

Reply to author   Forward

Gmail Calendar Documents Photos Reader Web more ˅   deanlsinclair@gmail.com ˅

Google groups

« Groups Home

Oscillator/Substance
⊞ Theory

| Search this group | Search Groups |

Home

Discussions
+ new post

Members

News to the group. Re: Vigier VII Symposium. The Search for....
Options

2 messages - Collapse all - Report discussion as spam

ESKI View profile   More options Apr 27 2010, 3:56 pm

Folks, get your thinking caps on and your extensions of the ideas
together, as it looks like old Eski, himself, is going to have to try
to be the group representative at the first presentation of our work
in to more or less the main stream by a remote hook up from here in
Aberdeen, to London, sometime in the July 10-14 period

.The Interim Publisher of the Aberdeen American News, for whom I work
a Friday A.M. shift, has offered the use of their studio for the
contact.

There should be a short precis on the Vigier VII web site within the
next week....I've yet to get together 75 Euros for the reduced fee,
and the money to do the wire transfer. Luckily. I get my Soc. Sec.
check next week....

Once that's paid, we'll have a slot for a paper in their proceedings
and our work will have a small toe-hold off the Internet....

I need as good a "handle" as I can get on where everybody stands by
about mid-June so I can reference as much as possible....I want to get
as many people a little recognition as I can..Cheers, ESKI..

--
You received this message because you are subscribed to the Google Groups
"Oscillator/Substance Theory" group.
To post to this group, send email to oscillatorsubstance-theory@googlegroups.com.
To unsubscribe from this group, send email to oscillatorsubstance-
theory+unsubscribe@googlegroups.com.
For more options, visit this group at http://groups.google.com/group/oscillatorsubstance-
theory?hl=en.

Reply to author   Forward

ESKI View profile   More options May 3 2010, 1:48 pm

Hi, Everybody,

Looks like I'm replying to myself: however, it seems to be the
simplest way to up date the information.

I just sent in the "entry fee" and a short summary of some of what I
hope to be able to present. The dates are July 12-14, I previously
thought 10-14.... Here is the little summary that I sent in.

A Framework for a Fundamental Theory?

Dean L(eRoy) Sinclair

An idea, "Perhaps one could pretend to be a genius by consciusly
trying to 'see the overlooked obvious'," when applied to physical
science data--some dating back to before 1900--has led to a simple--

but inclusive--model which may provide a framework for a comprehensive theory uniting the fields of physical science.

Ths model--which could possibly be dubbed "The Research Results of the Pretend Genius Approach--" is currently called the "Oscillator/ Substance Model." Its basic tenet is "All existence is the result of sequential 'action-reaction-action' interactions within a Substance/ Substrate of undefined basic composition and extent." This continued "sequential equilibration" results in constant motion such that the system is composed of/controlled by oscillators.

Some of these oscillators are vortexes which have long term stability. Their interactions result in "Matter."

Among the results of this view are an explanation for "charges," and the related definition of the size, shape and form of electrons and protons; a solution to the problem of the "Four Forces of Nature," and of the "Matter of the Missing Anti-matter." More details of the developmental process which led to this model, the definitions which arise from it and the relationship to some of the other theoretical approaches is also covered as well as other implications of the model in various areas. For one example--what may be a very basic reason the Hadron Collider has a rather low probability of ever being able to fulfill its original mission is mentioned.

Much of the information covered in the paper is available in a number of short "pages" published on an Internet Site, whose URL is http://groups.google.com/group/oscillatorsubstance-theory.

ESKI

On Apr 27, 3:56 pm, ESKI <deanlsincl...@gmail.com> wrote:

- Show quoted text -

--
You received this message because you are subscribed to the Google Groups "Oscillator/Substance Theory" group.
To post to this group, send email to oscillatorsubstance-theory@googlegroups.com.
To unsubscribe from this group, send email to oscillatorsubstance-theory+unsubscribe@googlegroups.com.
For more options, visit this group at http://groups.google.com/group/oscillatorsubstance-theory?hl=en.

Reply to author   Forward

End of messages

« Back to Discussions   « Newer topic   Older topic »

Gmail Calendar Documents Photos Reader Sites Web more   deanlsinclair@gmail.com

Google groups

« Groups Home

Oscillator/Substance
⊞ **Theory**

| Search this group | Se

View this page "Congruent Parallelogram Theorum "
                                                              Options

2 messages - Collapse all - Report discussion as spam

ESKI View profile   More options May 4 2010, 7:26 pm

HI, Everybody, This paper is a bit more mathematical than most of
what I have posted here; but, some people may find it interesting. As
usual comment is invited. ESKI

Click on http://groups.google.com/group/oscillatorsubstance-
theory/web/congrue...
- or copy & paste it into your browser's address bar if that doesn't
work.

--
You received this message because you are subscribed to the Google
Groups "Oscillator/Substance Theory" group.
To post to this group, send email to oscillatorsubstance-
theory@googlegroups.com.
To unsubscribe from this group, send email to oscillatorsubstance-
theory+unsubscribe@googlegroups.com.
For more options, visit this group at
http://groups.google.com/group/oscillatorsubstance-theory?hl=en.

Reply to author   Forward

Discussion subject changed to "Discussion on congruent-parallelogram-th

Hugh V View profile   More options May 19 2010, 12:19 pm

What Doc is saying here is essentially the idea that I used in the
graphing that I wrote about. which gives a graph of an oscillator.

 This perceptual universe that he postulates i, is the one ine which
we exist....He is waffelling a bit....

--
You received this message because you are subscribed to the Google
Groups "Oscillator/Substance Theory" group.
To post to this group, send email to oscillatorsubstance-
theory@googlegroups.com.
To unsubscribe from this group, send email to oscillatorsubstance-
theory+unsubscribe@googlegroups.com.
For more options, visit this group at
http://groups.google.com/group/oscillatorsubstance-theory?hl=en.

Reply to author   Forward   Report spam

End of messages

Home

Discussions
+ new post

Members

About this group

Edit my membership

Group settings

Management tasks

Invite members

 View this group in the
new Google Groups

THE CONGRUENT PARALLELOGRAM THEOREM AND A PERCEPTUAL UNIVERSE

The "Congruent Parallelogram Theorem" applied to two constants of nature, Planck's Constant, "h," and the Speed of Light, "c," and the Torque Constant of Nature which arises as their ratio, "h/c," produces a description of a "Perceptual Universe," in which "h," and "c," are valid descriptors.

The right and left halves of the mathematical relationship, $Ax \times By = K = Bx \times Ay$, can be said to describe, "congruent parallelograms."
This is independent of the dimension units attached to the variables, "x," and "y." This relationship, which may be called, "The Congruent Parallelogram Theorem," is widely applicable in the field of physics where it appears in a number of guises. It is the Law of Levers, the simplest expression of the Conservation of Momentum and of the Law of Forces, "For each and every force there is an equal and opposite force."

In general, it may be stated, " If there be a be a constant. "K." which can be analyzed as the product of two factors, the theorem is applicable in analyzing implications of that constant." Applying this to Planck's Constant, "h," the Speed of Light, "c," and the Torque Constant of Nature, which is their ratio, "h/c," produces a description of a "Perceptual Universe,l " in which these constants are valid.

Writing the definition, " The Speed of Light is the combination of the Frequency and Wavelength, in the Theorem form,
$Af \times Bw = c = Bf \times Aw$, where "f," is a frequency unit definition, "w," a wavelength unit definition and "A." and "B" are the absolute values associated, We see that for any associated pair of wavelength and frequency there will exist an "exactly" congruent set which can be found by reversing the associated absolute values. Another very interesting observation is that there will be an "instant" or "set of instances" wherein the absolute values of the frequency and the wavelength are exactly the same, i.e., $f = w = c^{0.5}$. This interesting relationship will be explored farther at another point.

An important set, which may be considered to define the "Upper and Lower Cut-off Communication Frequencies for a Perceptual Universe" can be found in the following manner. The starting point is the Planck formula for electromagnetic radiation--thought to be the fastest communication method known--Energy equals Planck's Constant times Frequency, $E = h \times f$. Realizing that the totality of motions involved in the entire "Energy" of this Perceptual Universe may be represented as a Unity, "One," we write, "$1 = h \times f$," and see that the corresponding frequency would be "1/f." the "reciprocal of Planck's Constant." the corresponding wavelength, found from, $c = f \times w$, is "ch." This very high frequency and short wave length would be the "High Frequency Cut-off." The reversal, " frequency, ch," and "wavelength, l/h," would define the "Low Frequency Cut-off." (In the cgs system of units the high frequency cutoff would be at about $1.5 \times 10^{+26}$ cycles per second at a wavelength of about $1,97 \times 10^{-16}$ cm. and the low-frequency cut-off at 1.97×10^{-16} cps. and a wave-length of $1.5 \times 10^{+26}$ cm.

Additional information arises from the "Torque Constant, h/c." This expression is found to be a torque, "mass times radius," by equating Planck's Constant to its definition a s an angular momentum, "mass times radius times velocity. to get the expression,

h = m x r x v " where "m," is a "mass," moving at a "tangential velocity, v, " at a distance, "r." from some defined center. Into this expression is inserted "c--" in its role as an average velocity in any given direction--for "v." This produces the equation,
 h = m x r x c, which, rearranged into the Theorem form, m x r = h/c = r x m , can be seen to be an equation which defines a "family" of oscillators of torque constant, h/c, with an average values where m = r = (h/c)^0.5.

This last relationship, (h/c)^0.5, would define a central sphere or circle through which all of the oscillators of this family of the set,
 {m x r = h/c = r x m}, would invert. This radius value is about 4.7 x 10 -19 cm. The "mass," measured for any oscillator at this distance, would be about 4.7 x 10^-19 grams. (Our postulated "Universe," has some interesting coincidences in the areas of Quantum Mechanics and String Theory. Quantum Mechanics is said to fail at below 10^-18 cm., essentially this inversion radius. The strings of String Theory are said to vanish into a "Hole," at 10^-18 cm. This "Hole" would have essentially the diameter of the inversion circle/sphere that appears in the above reasoning.)

If it be assumed that a wavelength associated with this average unit is the circumference of the unit, i. e.. 2Pi (h/c)^0.5, a possible average oscillator frequency for this "Universe," would be "c/[2Pi(h/c)^0.5)] which is equal to c^1.5/ 2Pi (h)^0.5. As "Time" can be seen as the reciprocal of frequency, a basic time unit can be defined as 2Pi x h^0.5/c^1.5 .

 One can assume a wavelength associated with the radius, to obtain a simpler set of expressions, c^1.5 /h^0.5 and h^0.5/c^1.5 This type of assumption would agree with the situation which arises when one checks out the electron and proton as members of the set {m x r = h/c}. when rest mass is considered as one limit, the radius is found to correspond to the "Compton Wavelength" of the electron or proton.

Going back to the situation of f = w = c^0.5, one can check out what will result if one looks at an oscillator based on this situation, assuming that "w" can be taken to be the equivalent to "r" for a limit of an associated oscillator of the "h/c set." If w = r = c^0.5, then the corresponding "m" equals (h/c)/c^0.5 = h/c^1.5 , and the reversed limts are r = h/c^1.5 and m = c^0.5 That is, this oscillator would be a sphere containing an internal sphere, the outer sphere having the radius of c^0.5 and the iner sphere would have the radius of h/c^1.5.

 In every case we are assuming that we are operating with the absolute values of "c" and "h" in whatever consistent set of units we care to use and that the square root values are also used as absolute values.

In summary, applying the Congruent Parallelogram Theorem to two basic laws of nature and their ratio has resulted in the definition of some of the characteristics of a postulated "Perceptual Universe" wherein communication would be controlled by these units. There has been defined a high and low frequency cut-off and an average size and mass for a family of oscillators that would be presumed to operate in this postulated Universe. An average frequency and time reference are also derived.

This postulated "Perceptual Universe" can be extended as a possible model for "Our Reality," by evaluating the electron and proton as possible oscillators of this set and bringing in additional information. To an extent this is essentially what has been done in "pages," published on the Internet site,

URL, http://groups.google/group/oscillatorsubstance-theory.   On that site the subjects of "Virtual Electrons" and "Super Symmetry" have not been addressed. It is possible that both of these concepts would have pertinence to our "Perceptual Universe<" as developed here, through the application of the Congruent Parallelogram Theorem to frequencies associated with the electron and proton. However, this is beyond the scope of this paper and should be addressed as a separate topic.

Gmail   Calendar   Documents   Photos   Reader   Sites   Web   more ˅   deanlsinclair@gmail.com ˅

Google groups

« Groups Home

Oscillator/Substance
⊞ **Theory**

[Search this group]   [Search Groups]

FW: *NASA: David Adair's 'Quasi-Fusion:' ?Cold, Warm, Hot?

Options

---

5 messages - Collapse all - Report discussion as spam

Jack O Suileabhain   View profile   More options   May 10 2010, 11:27 am

Provenance: Werner Von Braun, Hermann Oberth, David Adair-NASA Adv.Prop.Res.Prct., Wu Yeong Wei(Andy Wu), R.A.-Ned-Allen, Whitt Brantley-NASA Adv.Prop.Res.Prjct., Marc Rayman JPL-NASA.

-blame Jack Harbach O'Sullivan-:-)

Re comment: THUSWISE realizing this 'model' for epanding-dialating Protons(as micro-singularities) enhances the description of NOT ONLY CHEMICAL REACTIONS but also of FUSION REACTION be they 'cold,' warm, or hot.

And Dave Adair's rocket in a magnetic bottle(EMF-Toriod-Plasma Breach) was fairly accurately described as a 'chemical-reaction' rising to DIALATED-PROTON quasi-FUSION levels of energy output. And thusly he called his motors(now state of the art in missile technology) as 'Controlled-Chemical-Fusion' rocket motors
****
NASA: David Adair's 'Quasi-Fusion:' ?Cold, Warm, Hot?

ADAIR BOTTLE IGNITION CHAMBER Re: Saline Mist H2O Jets, Vandegraaf Hi-density static charge hyrolysis-ignition nodes fired by Hi-EMF-density Capacitors. Sodium capacitative ionic-charge disperses-facilitates-catalizes the charge throughout the Adair Chamber.

*SPRITES & JETS emulation: The Adair quasi-fusion rocket is copying the atmospheric flux-tornadoes-vortices that such up oceanic saline mist and then conduct inductive (lightning)hi-density charge and thusly become a 'natural' Adair-bottles which in turn foment a H2O-hydrolytic-then-firing sequence. The H2 + O2 reaction is attenuated in that 'some' of the O2 is becomes rather O3-Ozone and some H2 in the hydrolytic-split is form of H1-H1.

Within the High-Charge electo-plasmic medium the Proton-singularity-centres are expanded allowing a 'dialated-eye'/induction-amplification of the Proton and its axial-flow Electro-Valent quantum-electron-flux-plasma shell with a marginal surge of ingress electro-Aexoplasma.

The Proton as a balanced gray-hole singularity-&-shell 'micro-system' thus becomes 'whitish'(marginally more ingress electro-plasma from parallel AexoDarkEnergy HyperSpace); and thus in this High-Energy state also with the free Hydrogens, the futher energetic chemical firing of the hydrolytically-split Hydrogen & Oxygen completes the 'total' energy situation that causes 'some' of the free-hydrogen to FUSE in He-Helium. This is the chemistry/fusion piggy-back process of the Adair Mag-Vortex Bottle firing chamber rocket motor as well as the causal process for the meteorological-atmospheric phenomena of Sprites & Jets.

It's 'not' for-nothing that Werner Von Braun made (then) 17-year-old David Adair(now of NASA) his protege' after David demonstrated his 'mag-vortex-bottle quasi-fusion motor' at his high-school science fair that he had constructed in his father's machine shop.

And thusly the 'wild-card' factor can never be discounted as a critical adjunct to the more conventional inputs of classically univerity trained science-mathematicians-physics engineers & theorists.

Although David Adair's 'piggy-back chemical/quasi-fusion' system is 'not exactly' LENR-CANR cold-fusion; its significance cannot be dismissed.

RE Anti-Matter & Cern-Hadron, Fermilab, etc. It needs be noted that the Proton in the balanced Gray-State micro-singularity state creates a magneflux core-flow circulation creating its 'atmospheric' electro-valent-QUANTUM-ELECTRON SHELL. The directional flow accounts for the polarity of the Proton with its singularity-directional-axial electro-plasma flow(say northward) vs the EV-Shell circulating the same-electro-plasma/magna-flux-field 'southward.' So in short this is the same electro-flux magnetic field phenomenon that we are so prozaically familiar with; but the Nikola Tesla, for instance, did so very 'much' with that we are stilll 'catching-up' to in 'theory' that we have never yet thoroughly defined.

HOWEVER: When the Super-Collided H-single-Protons are 'smashed' the axial-flow of said Proton Singularities is converted BLACK-SINGULARITY so that both poles are paroxismally 'flowing' back into the 'centre' and thus the QUANTUM-ELECTRON SHELL is immediately 'swallowed' and subsequently at the 'micro-black-hole' level the energy of the Proton is absorbed through it's singulartiy centre back in to parallel-AexoDarkenergy-Hyperspace; and at which point the micro-black-hole winks shut.

CASE IN POINT: This IS 'really' what is happening ref. the anti-matter explosive-phenomenon. Actually what is happening is that the micro-Hawking's-Radiant 'electro-plasmic-back-wash' at the 'wink-out' point is trapped within a very brief hyper-compressed-cavitated state, but then explosively 'realeased' upon the 'wink-shut-closure' of the micro-black-hole.

THUSLY: Anti-matter is a Fiction & a Red-Herring. Any 'normal-stable-matter-atoms' within immediate contact of this process will perforce be 'ripped-asunder' and will thus release its energy explosively as a quasi-fission reaction. And this is the actual story on what has been 'misnomered' as 'anti-matter-contacting-matter' reaction.

THEORY based on the Anti-Matter conceptual model as a 'quasi-equal-state(quanity)-to-matter' at the Big-Bang is more or less a complete waste of time. Proton's are not 'little quasi-inert micro-planets' & Electron's are 'not' little 'ball-lightening-suns' orbiting Protons. And thusly the so-called 'model' of 'anti-matter' as simply matter-with-reverse-charge-Protons vs Electrons is 'hogwash' and one of the greatest psuedo-scientific-red-herrings since mankind was 'sure' that the sun, moon, stars, & universe orbited 'Holy-Planet-Earth' because 'any fool' could see with his/her 'own-two-eyes' that this was so. . . .~JHOS~
****

HI, Jack,

Yep, the proton would have a ball in a magnetic bottle reacting in exactly the opposite way that an electron would, since it is coordinated motion of electrons which set up the magnetism.... This is almost exactly the reason that the Hadron Collider wil never work--.Least not the way they intended it to -- Essentially, it is an over-sized magnetic bottle.....Dean

Dean: I see it as the 'concept of free-electrons' as 'sovereign-quasi-indepentent energy objects' is as deceptive as is the models of atoms as tooth-picks & colored-plastic-balls.

The electro-valent shell-whole field of a single Hydrogen Atom can be stated as having the energy of ONE QUANTUM ELECTRON. But QUANTUM-ELECTRONS do 'not' travel free of their Protons. (?But what 'moron' would say this in the face of sacred dogma? except me~;-)

HOWEVER the QUANTUM-ELECTRON has a SPECIFIC ENERGY SPEED-DENSITY signature which is the ELECTRO-MOTIVE-FORCE-SPEED-DENSITY which I have posited as roughly +or- $(C/2)^2$quared, or Half Light-Speed squared is the speed density of ELECTRO-MOTIVE FORCE. This notwithstanding that the WHOLE-ATOMs total-energy would be EC^2quared. ergo $M=EC^2$quared. And FREE EMForce is one more level of 'Helicoid-Wave-String' as are all other wave/wavicle forms in the graduated energy spectrum.(rather than being little floating independent particulate 'orblets'). I would that each WAVE-CREST to wave-crest of EMForce Helicoid-wave-string is also ONE QUANTUM ELECTRON.

And relative plus-charge/minus-charge polarity(magnetic-flux-flow) is determined by the Axial-circulation-flow from the balanced gray-hole micro-singularity-centre of the Proton which expands into the 'circulating to south-pole' Quantum-Electron-Shell 'micro-atmospheric' field of the Atom.

And of course the question is begged, "OK genius, then what in the heck IS electrical current if NOT the 'flow' of electrons. Electrical is the 'flow' of ELECTRO-MOTIVE SPEED-

DENSITY FORCE and the OUTER-MOST ELECTRO-VALENT SHELLS of 'conductors' facilitates the FLOW of FREE-EMForce-flow as water flows over pebbles(but obviously actively)----->as the EMForce flowing over the outer electro-valent shells INCREASES the ELECTRO-VALENT-SHELLs as a QUANTUM-ELECTRON-PLUS field(the shell acting as a micro-capacitor)------>and this in turn ENERGIZES the PROTON(as a singularity)and thusly DIALATES THE PROTON(to expanded above-'normal' ingress of sub-AexoDarkPlasma).

SO THE OVERALL 'Proton-Singularity with axial-flow Quantum-Electron Shell' then has FURTHER AMPERAGE/VOLTAGE 'discharge' capacity to STIMULATE EMForce flow that began by stimulating/flowing-over it. Its symbiotic---->the EMForce flow stimulates the conductor-quantum-electron shell which in turn stimulates/dialates the Proton expanding its 'ingress-energy' rate from parallel AexoDarkEnergy-HyperSpace which inturn charges-up further the EV-Shell thus releasing the OVERCHARGE into the EMForce flow.

THE QUANTUM-ELECTRON SHELL is the speed-density TRANSLATOR of the INGRESS-AEXOPLASMA into the 'quasi-routine' Speed-density of EMForce/Electricity. And of course the routine Voltage-Amperage variations reflect the 'balance' between the Proton-Dialation to Ingress-Plasma Rate to the QUANTUM-ELECTRON CONVERSION-capacitance-discharge rate.

THUSWISE realizing this 'model' for epanding-dialating Protons is the description of NOT ONLY CHEMICAL REACTIONS but also of FUSION REACTION be they 'cold,' warm, or hot.

And Dave Adair's rocket in a magnetic bottle(EMF-Toriod-Plasma Breach) was fairly accurately described as a 'chemical-reaction' rising to DIALATED-PROTON quasi-FUSION levels of energy output. And thusly he called his motors(now state of the art in missile technology) as 'Controlled-Chemical-Fusion' rocket motors~Jack~

--
You received this message because you are subscribed to the Google Groups "Oscillator/Substance Theory" group.
To post to this group, send email to ...

read more »

Reply to author   Forward   Report spam

dean sinclair View profile   More options May 10 2010, 4:51 pm

I'm going to have to read all of this very carefully, however, you and I are certainly in agreement about a lot of things. Don't remember if you've read my articles on Helium with respect ot Anti-matter and the latest one on the Hadron Collider.....DS

On Mon, May 10, 2010 at 11:27 AM, Jack O Suileabhain <

- Show quoted text -

...

read more »

Reply to author   Forward

Hugh V View profile   More options May 17 2010, 2:32 pm

Going over this carefully, it appears that Adair and crew are doing essentially what, years later, my theorizing predicts.

The ideas of matter and anti-matter have logical contradiction. As a proton can be shown to be identical to an anti-electron which has had most of its Kinetic Energy converted to rotational-vibrational motion, which me measure as "mass." if the anti-electron is "anti-matter." then so also is the proton. I won't go into details but it is becoming quite clear that the only difference between matter and

anti-matter is apparently some sort of phase angle relationship, not
exactly 180 degrees out of phase as would be expected' and any atom is
actually a composite oscillator made up of "equal" amounts of "matter"
and anti-matter components. In our Universe, the relationship is such
that we simply read limits from whata we call "matter." Probably due
to interactions within the half-oscillator we call our universe,
possibly with contribution from the "Galaxy" oscillator and other
influences..

The situation that Adair sets up is a situation in which there is
formed a plasma in which protons and Hydrogen atom and Hydrogen
Molecules will occur. This turns out to be a situation in which fusion
will occur. The first being HH+ to D+ which is an easy step away from
He+ . H+ in a magnetic field, will be accellerated, gaining spin,
hence "enhanced charge "and becoming a definite activator for the H2+
to D+ transform...

.. Yes. Adair's Rocket motor makes sense. My theorizing says that a
source of positive ions in an atmosphere of Hydrogen, Deuterium or
Helium will set up a fusion chain reaction. Adair's rig sets up such
a situation of production of cations. Doing it within a magnetic field
was a stroke of genius. I'll bet that were Hydrogen gas injected into
one of his motors it would be even more efficient as he is having to
generate the H2 by aqueous electrollysis in situ. Deuterium would be
even more efffective...Chiaux, Dean .....

On May 10, 4:51 pm, dean sinclair <deanlsincl...@gmail.com> wrote:

- Show quoted text -

...

read more »

Reply to author   Forward   Report spam

Hugh V View profile   More options May 17 2010, 2:50 pm

I probably should have noted that at the moment, I am writing off
Hugh's site,which I have access to, as for some unknown reason, my own
site has been "temporarily discontinued." Dean ,

On May 17, 2:32 pm, Hugh V <hughv...@gmail.com> wrote:

- Show quoted text -

...

read more »

Reply to author   Forward   Report spam

Discussion subject changed to "*NASA: David Adair's 'Quasi-Fusion:' ?Cold, Warm, Hot?"

Jack O Suileabhain View profile   More options May 18 2010, 11:13 am

Dean: The original inputs of Dave Adair's initial experiments used 'water-injection'

on the sympol principal of old school internal combustion(reciprocal) aircraft engines.

I think that he got the original idea from his Dad. Within the Hyper-Magnetic-Toroid-Vortex

'Bottle' design it did 'wonderful' things. Long since he likely graduatied

to Liquid-Oxygen + Liquid-Hydrogen designs that 'cut to the chase' on the whole affair.

- Show quoted text -

...

read more »

<u>Reply to author</u>   <u>Forward</u>   <u>Report spam</u>

End of messages

« Back to Discussions   **« Newer topic   Older topic »**

Google groups

« Groups Home

Oscillator/Substance
⊞ **Theory**

[Search this group] [Search Groups]

FW: *NASA: David Adair's 'Quasi-Fusion:' ?Cold, Warm, Hot?

Options

5 messages - Expand all - Report discussion as spam

Jack O Suileabhain Provenance: Werner Von Braun, Hermann ... May 10 2010, 11:27 am

dean sinclair View profile   More options May 10 2010, 4:51 pm

I'm going to have to read all of this very carefully, however, you and I are
certainly in agreement about a lot of things. Don't remember if you've read
my articles on Helium with respect ot Anti-matter and the latest one on the
Hadron Collider.....DS

On Mon, May 10, 2010 at 11:27 AM, Jack O Suileabhain <

- Show quoted text -

...

read more »

Reply to author   Forward

Hugh V View profile   More options May 17 2010, 2:32 pm

Going over this carefully, it appears that Adair and crew are doing
essentially what, years later, my theorizing predicts.

   The ideas of matter and anti-matter have logical contradiction. As
a proton can be shown to be identical to an anti-electron which has
had most of its Kinetic Energy converted to rotational-vibrational
motion, which me measure as "mass." if the anti-electron is "anti-
matter." then so also is the proton. I won't go into details but it
is becoming quite clear that the only difference between matter and
anti-matter is apparently some sort of phase angle relationship, not
exactly 180 degrees out of phase as would be expected' and any atom is
actually a composite oscillator made up of "equal" amounts of "matter"
and anti-matter components. In our Universe, the relationship is such
that we simply read limits from whata we call "matter." Probably due
to interactions within the half-oscillator we call our universe,
possibly with contribution from the "Galaxy" oscillator and other
influences..

The situation that Adair sets up is a situation in which there is
formed a plasma in which protons and Hydrogen atom and Hydrogen
Molecules will occur. This turns out to be a situation In which fusion
will occur. The first being HH+ to D+ which is an easy step away from
He+ . H+ in a magnetic field, will be accellerated, gaining spin,
hence "enhanced charge "and becoming a definite activator for the H2+
to D+ transform...

.. Yes. Adair's Rocket motor makes sense. My theorizing says that a
source of positive ions in an atmosphere of Hydrogen, Deuterium or
Helium will set up a fusion chain reaction. Adair's rig sets up such
a situation of production of cations. Doing it within a magnetic field
was a stroke of genius. I'll bet that were Hydrogen gas injected into
one of his motors it would be even more efficient as he is having to
generate the H2 by aqueous electrollysis in situ. Deuterium would be

even more efffective...Chiaux, Dean .....

On May 10, 4:51 pm, dean sinclair <deanlsincl...@gmail.com> wrote:

- Show quoted text -

...

read more »

Reply to author   Forward   Report spam

Hugh V View profile   More options May 17 2010, 2:50 pm

I probably should have noted that at the moment, I am writing off
Hugh's site,which I have access to, as for some unknown reason, my own
site has been "temporarily discontinued." Dean ,

On May 17, 2:32 pm, Hugh V <hughv...@gmail.com> wrote:

- Show quoted text -

...

read more »

Reply to author   Forward   Report spam

Discussion subject changed to "*NASA: David Adair's 'Quasi-Fusion:' ?Cold, Warm, Hot?'

Jack O Suileabhain View profile   More options May 18 2010, 11:13 am

Dean: The original inputs of Dave Adair's initial experiments used 'water-injection'

on the sympol principal of old school internal combustion(reciprocal) aircraft engines.

I think that he got the original idea from his Dad. Within the Hyper-Magnetic-Toroid-Vortex

'Bottle' design it did 'wonderful' things. Long since he likely graduatied

to Liquid-Oxygen + Liquid-Hydrogen designs that 'cut to the chase' on the whole affair.

- Show quoted text -

...

read more »

Reply to author   Forward   Report spam
End of messages

« Back to Discussions   **« Newer topic   Older topic »**

Gmail Calendar Documents Photos Reader Web more ˅   deanlsinclair@gmail.com ˅

Google groups

« Groups Home

Oscillator/Substance
⊞ Theory

[Search this group] [Search Groups]

Home

Discussions
+ new post

Members

About this group

Edit my membership

Group settings

Management tasks

Invite members

 View this group in the new Google Groups

Sponsored links

MITS
Business Intelligence and Advanced
Reporting Solutions for SQL and MV
www.mits.com

Masters in Public Policy
Earn a Master's Degree in Public Policy Online at NEC. Free Brochure
PublicPolicy.NEC.edu

Improve Building Control
Save Money Through Increased Energy
Efficiency With Johnson Controls.
makeyourbuildingswork.com

See your message here...

The fluid theorem...   Options

2 messages - Collapse all - Report discussion as spam

heycollin View profile   More options May 7 2010, 10:33 pm

To Mr. Sinclair,

According to my own theories, one universe cannot exist without another. Let me explain. You mentioned the circular motion of matter/anti-matter. Think of the "first" universe as the middle. Imagine the distance between universes (do they exist separate from each other?) as spots on the water. If a ripple happens in one, let's say time... It affects the others at the same rate, or shortly thereafter, like a ripple in the water. If you throw a stone in the water, the ripple goes out from the middle. That's like something happening here. As the ripple goes out, sometimes it meets other ripples. This is like meeting other universes. The ripples cross and are forever changed. What do you think about this one, Dean?

Best regards,

Collin

--
You received this message because you are subscribed to the Google Groups "Oscillator/Substance Theory" group.
To post to this group, send email to oscillatorsubstance-theory@googlegroups.com.
To unsubscribe from this group, send email to oscillatorsubstance-theory+unsubscribe@googlegroups.com.
For more options, visit this group at http://groups.google.com/group/oscillatorsubstance-theory?hl=en.

Reply to author   Forward   Report spam

dean sinclair View profile   More options May 10 2010, 6:03 pm

Collin, Makes sense to me.... Dean

- Show quoted text -

--
You received this message because you are subscribed to the Google Groups "Oscillator/Substance Theory" group.
To post to this group, send email to oscillatorsubstance-theory@googlegroups.com.
To unsubscribe from this group, send email to oscillatorsubstance-theory+unsubscribe@googlegroups.com.
For more options, visit this group at http://groups.google.com/group/oscillatorsubstance-theory?hl=en.

Reply to author   Forward

End of messages

« Back to Discussions   « Newer topic   Older topic »

Google groups

« Groups Home

Oscillator/Substance
⊞ **Theory**

Search this group   Search Groups

Home

Discussions
+ new post

Members

About this group
Edit my membership
Group settings
Management tasks
Invite members

An old essay that might be of interest,   Options

1 message - Collapse all - Report discussion as spam

hugh vreeland View profile   More options May 19 2010, 11:52 am

Hi, folks,

Looking over my old documents of about three years ago, I ran into this
little essay which I apparently wrote back when Osc/sub theory was Motion
in a Matrix   The article I'm referring to would have to be Doc's original
Motion in a matrix paper, the trial balloon on Helium. Have fun.!....Hugh

Motion

Yikes! Could it be that I"m a combination of motions in a collection of
dots or maybe a bunch of twisted strings in ten dimensions? I think I like
the motions in dots better, it's easier to try to understand. Maybe we are
programs in some computers run by fifth dimensional--or eleventh or twelfth
dimensional beings. Someone called us, "Thoughts in the mind of God."   I
don't recall who that was.

I've read the long lead article on this topic several times and find the
logic interesting. I also ran into another article on the infinite
divisibility of matter, states that a few physicists are looking at
existence as motions in some sort of an aether matrix which seems to be the
same idea. So I guess there really isn't anything new under the sun. That's
what my writing instructor told me years ago. "There's no such thing as a
new idea or topic, Just try to give a different view."

So, let's try a different view of the idea used in the article we were just
discussing of equating $(mv^2)/2$ and hf, with "f" standing in for "nu" which
looks too much like a "v." Now if our assumption that $(mv^2)/2=hf$, where,
"h," is, of course, Planck's constant, and "f" is a frequency. Then, if I
remember rightly, we can write another equation that would also be valid
which goes like this, considering "m" a constant, $(mv^3)/6 =(hf^2)/2$. This
is by "integrating" both sides with respect to time and assuming that "v"
and "f" are related variables. We can also do something else, we can
"differentiate" both sides and get mv=h,. If we evaluate this at "c," we get
a value of "mass" embedded in Planck's constant. This value, in grams, can
be calculated: 6.63×10^{-27} divided by 3×10^{10} which equals 2.21×10^{-37} g.
This value would be the "mass that cancels out" in Dr.S's discussion. If we
put this value of mass back into the original, "undifferentiated" equation
and solve for "f" we get a frequency of 3×10^{10} cycles/sec. as this
number is the same number as the speed of light in cm./sec., what we have
determined is the mass equivalent of a wave length of one centimeter.

Going back to the "integrated" equation and evaluating it at "c" the speed
of light, we, if our assumptions are correct, should find some other values
that would somehow be constants of nature. The figures I get by various
mathematical manipulations of this formula and other energy formulas are, as
follows: 3.67×10^{28} cps, 243 ergs, and 2.7×10^{-1} grams. Whether these
correspond to any published constants will need to be investigated.

Dr.S's value of 6.9×10^{46} cps. for frequency which he calculated as a
maximum strain in the matrix can be converted to a mass value of about
4.57×10^{18} grams. This could be the maximum mass which could travel "at the
speed of light." It would be interesting to compare this to estimated
masses of certain celestial bodies.

Gmail Calendar Documents Photos Reader Sites Web more   deanlsinclair@gmail.com

Google groups

« Groups Home

Oscillator/Substance
⊞ **Theory**

[Search this group] [Search Groups]

Looks like we may be "orphans"   Options

5 messages - Collapse all - Report discussion as spam

Hugh V View profile   More options May 25 2010, 3:23 pm

It looks like this sight may be a small "family of orphans' with out a
parent or any way to add new members.

Although this group is supposed to be open membership, to be able to
keep out useless spam, the managers set up a situation to monitor
first submissions. That worked fine as long as we had manager's.

Unfortunately, something happened that Google temporarily suspended
Eski's account, (our owner-manager) and it looks like the term
"temporary suspension" means that although his mail box will keep
collecting stuff, he cant get into it, and cant get into this site to
do management..It also seems to mean that they are going to leave the
temporary on forever..

We don't know what is the situation with Ka-Sala, our other manager.
We hope she is O.K, bukt we havent herd from her for a while.,
Without an owner or manager, this account is a closed group.

I think that I'd advise anyone that is interested in the information
stored here that they might want to copy off the parts they need and
store it elsewhere.   If, for some unknown reason, Eski has become
persona non grata to Google, there is no way of telling when they
might lower the boom on this site.

Reply to author   Forward   Report spam

ka-sala View profile   More options May 25 2010, 6:37 pm

Hello Hugh V
I'm alive...
But didn't know all this has happened in my absensce until today 26
May 2010. Can you fill me in some more ? I received a note from Eski
dated 28 April but have only got the Net back yesterday since moving.
I see there is a LOT to go over. But this 'Banning' of Eski ? Google
have been 'in error' for some time re accepting any management (or
making it pssible even by the task Eski has placed on me.) Any
suggestions ? I don't believe we are 'orphans' ... but ??? Will wait
to hear from you.
Regards,
ka-sala

On May 26, 6:23 am, Hugh V <hughv...@gmail.com> wrote:

- Show quoted text -

Reply to author   Forward   Report spam

Discussion subject changed to "Hacked-Allusion created: 'MADE' to Looks like we may be

Jack O Suileabhain View profile   More options May 27 2010, 10:01 am

NOTE to FRIENDS: When somebody-whoever has the 'stroke' &/or malicious inclination

Home

Discussions
+ new post

Members

About this group
Edit my membership
Group settings
Management tasks
Invite members

 View this group in the
new Google Groups

Sponsored links

AdWords Accnt Suspended?
We Can Help. Get Your Good Name
Back With AdWords. 14 Days or Less.
www.PPCRenegade.com/AdWordsSo

Send Mass Email
Create, Manage and Send Email to
Interested Customers and Prospects.
www.ConstantContact.com

Top Penny Auction - #1
Save Up to 90% off Retail
All Items In-Stock & Ready to Ship
www.BidCactus.com/OverStock

See your message here...

to screw up your site; then that means that you are DOING SOMETHING RIGHT.

SURPRESSION OF IDEAS & concepts that 'threaten' the staid-Ivy-Halled establishment,

is a 'solid-indicator' that you are being: #1. OBSERVED & #2. YOUR ACCURACY is steering

you towards 'revelatory-discovery' that is considered HIGHLY-CLASSIFIED and worth

relatively-draconian measures to silence.

"Just because you're paranoid, DOESN'T mean that somebody is NOT out to get you/silence you."~;-)~Jack~
Stay persistent and NEVER 'role-over/discourage' easily.

P.S. These kind of attacks by fairly-clever academics who are 'perturbed-off' at you; &/or by DARPA-R&D types

who prefer that you SHUT-UP; are fairly routine & to be expected from time to time/ been their before-often.

Cheer-up; you're pioneer-thinkers and they are small-minded-bureaucratic-psuedointellectual/psuedoscientist-JERKS!~;-)

- Show quoted text -

Hotmail: Powerful Free email with security by Microsoft.
https://signup.live.com/signup.aspx?id=60969

Reply to author   Forward   Report spam

---

ka-sala View profile   More options May 28 2010, 11:07 pm
JUST A NOTE OF CONFIRMATION.
Past Experience has silenced me personnally many times. One... was
included in my Profile, and my reason for placing it there. 'Others'
- at the time - had it in the 'pipe-line' for the Nobel. That... was
proof that what I had, was 'right'.

I only know that if each was to go back over ALL that has been entered
into this site on the O/S since Eski started - and although we are all
different within our approaches - facts are, the answer lies between
us all.

It can be drawn out like pieces of a jig-saw puzzle for those who have
the inclination to take the trouble. Each has contributed. Each has a
peice or two within their discussions etc. It is not mathamatical nor
verbal diahorrea, but the facts which have gotten lost in the maze of
threads... and many explainations.

Like a Living Peal of Wisdom, it is there for the plucking.

We are not 'orphans' if this can be proven, even is Eski 'appears' to
have been blocked for now. Something - if truth be on our side - must
even prove that, truth itself may have a way of being 'blocked' but
not from leaking. Like water it will find the gavity of this
situation, and we - all of us - are that Gravity! The 'Key' lies with
us.

All the very best to each member, and to hoek - my friend - thank you
for " Hi to my good friend ka-sala, may God always smile upon you.
hoek "

To ESKI... I do trust this will soon be overcome.

We are in this boat together. Take Care.
Regards,
ka-sala

On May 28, 1:01 am, Jack O Suileabhain <braghgoerin...@hotmail.com> wrote:

- Show quoted text -

Reply to author   Forward   Report spam

ESKI View profile   More options Jun 10 2010, 3:37 pm

An update on the situation .
After a week of trying to get a response from Google and having the
Library here try to contact them, to no avail. I gave up and started
to work around them, About ten days later, I tried the e-mail again.
This time, instead of a cut-off notice, they said, "Suspicious
activity detected, change your password. I did so and they let me back
in. I haven't figured out; however, how they knew I was "me" and that
they were not giving my "hacker" my account for good. I suppose they
figured that in that case they would close it "permanently."

Checking my Sent Mail, I found that some one had sent a "non-content"
message to a non-existent mail address, to someone who apparently
doesn't exist,. from my e-mail address. This non-message had been
bounced back eleven times. Apparently on the 12th try Google
suspended the address. So that's the story. I'm back, but, needless
to say, I 'm trying to keep back-ups so I won't be caught quite so
short the next time something like this comes up.

Incidentally or site seems to be getting several hits an hour. which I
think is good news. However, we're not collecting any new members.

At least, however, the hits do not appear to be from "Spammers" as
some of our early hits, and "members" turned out to be. ESKI

On May 28, 11:07 pm, ka-sala <irrir...@gmail.com> wrote:

- Show quoted text -

Reply to author   Forward

End of messages

« Back to Discussions   **« Newer topic   Older topic »**

Google groups

« Groups Home

Oscillator/Substance
⊞ Theory

Trying to get Eski back some contacts.   Options

2 messages - Collapse all - Report discussion as spam

hugh vreeland View profile   More options Jun 1 2010, 12:50 pm

Luckily for him, Eski and I have a side contact which was not through
Google. He tells me that the lock out from the deanlsinclair@gmail address
has lost him the exact addresses to two very valuable contacts, "Charles"
and "Herringbone." He says that there is a member of the Group that is an
associate of "Charles" and that "Herringbone" has also been in contact with
Charles so he is asking me to tell the Group that a valid address for Eski
is deanlsincl...@yahoo.com and that he can be contacted there... Thanks.
If someone really wants to keep Eski out of touch, suppose that I'll be the
next one to have my account, "Temporarily Suspended." Hugh

Reply to author   Forward   Report spam

Irrira Rikki View profile   More options Jun 2 2010, 1:37 am

Thanks Hugh.

Have this other address, and have been in touch with Eski via this in his
note to me. I see he's back with us. Great news.

Wish I could help re. "Charles" and "Herringbone." but have no contacts.

"Temporarily Suspension.." ? We're in this together.

Regards,
ka-sala

- Show quoted text -

Reply to author   Forward   Report spam

End of messages

« Back to Discussions   « Newer topic   Older topic »

Home

Discussions
 + new post

Members

About this group

Edit my membership

Group settings

Management tasks

Invite members

 View this group in the
new Google Groups

Google groups

« Groups Home

Oscillator/Substance
⊞ Theory

[Search this group] [Search Groups]

Attention 'Reporter'   Options

Home

Discussions
+ new post

1 message - Collapse all - Report discussion as spam

ka-sala View profile   More options Jun 2 2010, 11:40 pm

Interested as to why you joinded this group ?
ka-sala

Reply to author   Forward   Report spam

End of messages

« Back to Discussions   « Newer topic   Older topic »

Members

About this group
Edit my membership
Group settings
Management tasks
Invite members

View this group in the
new Google Groups

Sponsored links

New Local Advertising
Promote Your Business on Google
Search and Maps. Try AW Express!
Google.com/awexpress

See your message here...

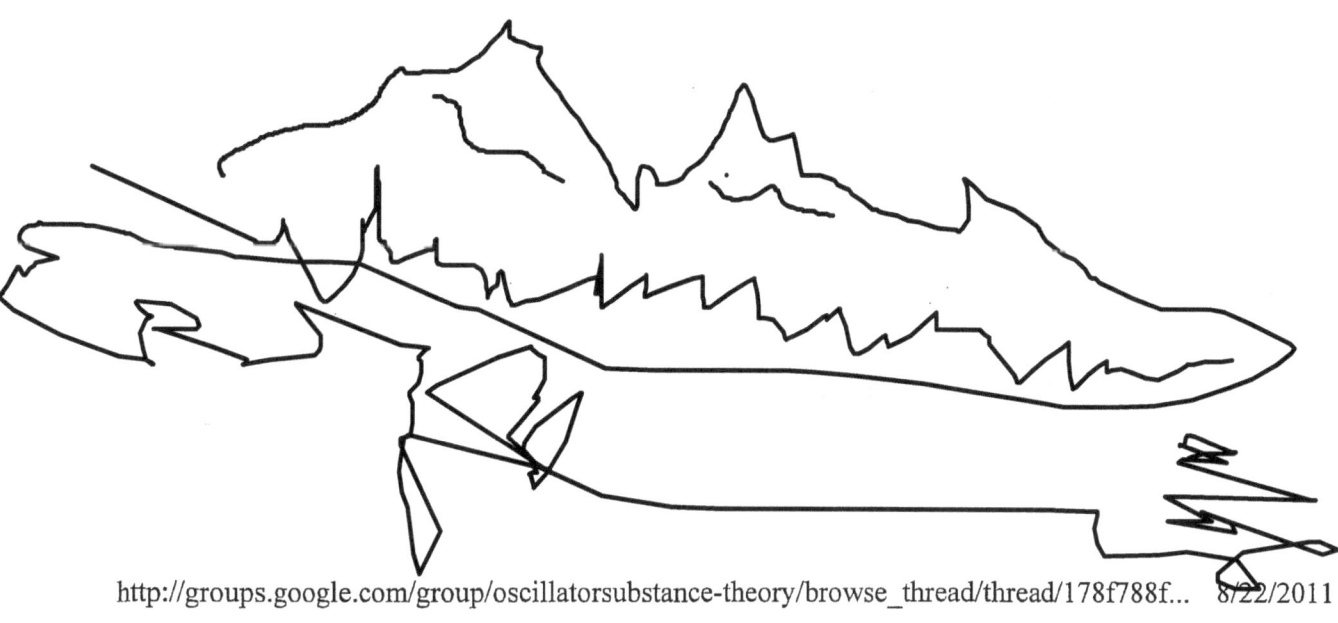

Go⦿gle groups
« Groups Home

Oscillator/Substance
⊞ **Theory**

| Search this group | Search Groups |

Matter--Anti-matter   Options

2 messages - Collapse all - Report discussion as spam

eskia...@mail.com View profile   More options Jun 2 2010, 4:53 pm

Merhaba, arkadashlar,
Long time me watch Group. Sorry, hear lose old Eski. I use Eskiadam name,
think maybe time i go try join group.

Old Eski he say if anti-lectron anti-matter, so be proton, so if the electron
matter then so be--maybe atom weight be total of all anti
unit and unit. No neutron , all electron anti-lectron proton anti proton all
turn around turn inside out balance out?

Tessukur,

Eskiadam

Reply to author   Forward   Report spam

ka-sala View profile   More options Jun 2 2010, 9:20 pm

Merhaba, arkadashlar... friend.
Welcome Eskidam!

You watch group very good. Me sorry too lose old Eski. He very nice
fellow. Maybe very good you join group. Very important to have you.

So... old Eski say in nut-shell - same as "everything affect each
other, and without one can there be no link." Very good balanced
logic !

Welcome again. I think you be very important for this group, ja !
Tessukur Eskkidam.

ka-sla

On Jun 3, 7:53 am, eskia...@mail.com wrote:

- Show quoted text -

Reply to author   Forward   Report spam
End of messages

« Back to Discussions   « Newer topic Older topic »

Home

Discussions
+ new post

Members

About this group
Edit my membership
Group settings
Management tasks
Invite members

 View this group in the
new Google Groups

Gmail Calendar Documents Photos Reader Web more ˅   deanlsinclair@gmail.com ˅

Google groups

« Groups Home

Oscillator/Substance
⊞ **Theory**

[Search this group] [Search Groups]

Home

Discussions
  + new post

Members

About this group
Edit my membership
Group settings
Management tasks
Invite members

REPEAT of " INFINITE VARIABLES " Wishing you what you so desperately seek.

Options

2 messages - Collapse all - Report discussion as spam

ka-sala View profile   More options Jun 2 2010, 11:11 pm

*** The reason for this repeat is because this is where we are all
standing. On the same pin-point. I said I believe we have the key,
and the gravity of this O/S jig-saw. I have not repeated in these
simple words for nothing. Best wishes to all. Can anyone see this
while we are together ?

In our Oscillation Theory, a lot has been said lately regarding
Infinity. Directly or indirectly; we have been oscillating all around
the subject of the O/S Theory. To get to where we want to be we must
bring it back to the unified weak and electromagnetic interaction
between elementary particles, including inta alia, the prediction of
weak neutral current.

We have to return to the source of our search; see it for what it is,
and go on from there. ie. A look at quantum superposition,
atemporality and the direction of entropy in relation to eternity.
We
can go over and over Planck's Theory, and anyone else's. The
difference is we are here... now. Today's Theory.

*** Please excuse any spelling mistakes. I would be very interested
to have feedback on these two posts regarding the O/S Theory which
Links all which has been covered to date.

Wishing you what you so desperately seek.
Ka-sala

****

INFINITE VARIABLES

This is like asking one such as a human, to stand on the point of a
pin; take a step off, and simply hope we won't fall into the abyss of
Eternity.

The quantum superposition if being so finite as to suppose we can
take
a minute step in such a diversity as the Infinite we call Eternity,
is
neither to know we are retreating or progressing. To literally make
such an absolute turn around, with the knowledge that we will step
back into chaos; is more than one giant step for man to handle. He
has
enough already to assemble in his own collective elements.

If this direction of entropy points with a definite sign-post (be it
visible in such a universal science as the quantum's allow,) there
has
to be some-one who has been to Eternity and back to signify any
measurement at all its physical science. Even quantum superposition
would be that all the pieces of the jig-saw puzzle still missing,

will
automatically just fall into place as the elements should.

Hoping that each like a stepping stone - or giant quantum leap of
certainty - our only amoebic particle of theoretical virtue will take
our weight. Or carry us as a whips of quantum superposition back into
a space time dimension of... are we there yet, or is there more to
come? The place called atemporality, or timeless.

Elements of the unknown all lie within the universe of each of us -
within our sub atomic thermodynamic elementary and transitional
particles - right through to the seeming chaos of decay. We shovel
every component required to keep us fueled to create an infinite
layer
of cells required to hold us in an overlay called skin. As long as
the
life-span of our own nuclear decay exists while in this superposable
state called human, we will continue within our own field force.

So while on the pin-point of quantum superposition, instead of
stepping off, we can use our own electromagnetic energy, vacuuming
every quantumized oscillating molecule around us, to us. By doing so,
we draw on all variable changes created via the quantum super
positioning; propelling us by repulsion within an non moving vacuum.

Through this, we create from what was a seeming a state of entropy, a
transverse tunnel in which we can now safely step off the pin. By
turning matter back on itself from what would have been a physical
phenomenon of universal conformity, it has been transformed into a
usable state.

There is nothing in the universe called oscillating energy, which is
not transferable, transitional, reversible, or any other attribute,
which cannot be utilized when known how to superimpose. The
atemporality is unaffected by the space time dimension of Eternity's
Infinite variables.

The only unknown factor is the inconsistent knowledge yet, within our
own field. The Cosmos it's Photons, and Electromagnetism is within
each and every one of us. Why are we oscillating for so long over so
many theories?

Reply to author   Forward   Report spam

dean sinclair View profile   More options Jun 3 2010, 9:45 am

O.K. Ka-Sala,
You are just much too bright for this old man. Or maybe you just think too
much. It seems to me that you are missing a lot of my points, particularly
my point about defining what we are talking about.
Just what do you mean by "unified weak and electromagnetic interaction?"
By my thinkiing both the "weak" and "electromagnetic" interactions are terms
from conventional physics which are meaningless if we consider the "four
forces" as all being observational aspecfs of one "force--" pressures within
a "substance-substrate" having the basic characteristics of a liquid at its
triple point.

Could you define "atemporality" for me, hopefully in terms of my postulated
"Aether?" Also, can you give me a good definition of "Entropy?" I've
never seen a really good one, and I must admit that I'm not sure how to fit
the concept in. Does it fit in with theconcept of Mass as I define it, with
the problems of KInetic Energy, or is it a measure of some sort of the
"degree to which some action-reaction sequence somehow repairs the chaos
behind a shock wave in which we exist?

"Infinity," to me, has a practical definition of "the point, instant,or
number just beyond where we stop, quit counting, quit examining, or our
means of information gathering or transfer fail...." Eternity would have a
similar connotation. An extent just beyond the extent of our "Perceptual
Universe."

As long as we've been co-workers in this, I don't think that you have listened closely to old Eski...

It could, of course, be that Old Eski, "the Pretend Genius," a person of admittedly slow and limited intellectual abilites, who can only try to be intelligent by looking for different ways to interpret what others take for granted, is totally lost in comprehending the ideas of those who have true genius level capabilities. I only vaguely discern what either you or O'Sullivan is saying a great deal of the time. .

I do hope that some how, Eski can put together something that is at least rational for a 30 min. presentation at the Vigier Symposium so we don't end up appearing a bunch of "nincompoops."

I f you were to suggest one point that you think he should make in that presentation, what would it be?

Allaha asmaladik...DLS

- Show quoted text -

Reply to author   Forward

End of messages

« Back to Discussions   « Newer topic   Older topic »

Gmail Calendar Documents Photos Reader Sites Web more —   deanlsinclair@gmail.com

Google groups

« Groups Home

Oscillator/Substance
⊕ Theory

| Search this group | Search Groups |

Home

Discussions
+ new post

Members

About this group
Edit my membership
Group settings
Management tasks
Invite members

 View this group in the new Google Groups

MECHANISM   Options

3 messages - Collapse all - Report discussion as spam

hoek View profile   More options Jun 4 2010, 6:54 am

We all find ourselves existing in a wondrous realm of three spatial and one forward moving dimension of time. This last dimension allows cause, effect and logic, allows motion within the other three spatial dimensions, which allows MECHANISM! Mechanism allows biological life.

Even, though, it won the Nobel Prize for nuclear physics, back in 1967, I doubt that any of us believe in quark theory or it's spawn, Quantum Chromo-Dynamics as a valid nuclear model. Assuming the positive proton to be +1, the slightly, more massive neutral, neutron to be (0 zero) and the minute, more than 1860 times less massive, negative electron to be -1, some of the best and fastest to publish, of two teams of mathematicians, presented Quark Theory. Simply put, (+2/3)+(+2/3)+(-1/3) = +3/3 = +1 = Proton, (+2/3)+(-1/3)+(-1/3) = 0 = Neutron. It is without a doubt, the simplest mathematical solution to the assumed axioms, but, how does it work, what is the mechanism by which these up and down quarks interact and interchange to form protons and neutrons? This brings us to the above mentioned QCD which needs, at least one or possibly up to three more extra complex dimensions, depending upon which quantum particle physicist you ask, an extra fundamental force of nature, (color force) and eight fields of mediating force particles called gluons to explain the mathematics of the "bootstrap" quark hypothesis and nobody knows how it all works, what the mechanism is. As we all know, bootstrap means that the model was created before data was found to substantiate it. Particle physicists have been searching for that data for the last 47 years and still they search on for the elusive quark.

We all believe that the atom or some part of it is a primary oscillator or we wouldn't belong to the Oscillator/Substance (energy-mass) group. Most of us approached this topic from varying directions. I saw an animation in a science film 32 years ago. It nagged at my curiosity. (http://www.protoncosmology.com/new_page_1.htm) What could be causing these peculiar dynamic charge patterns to appear on the proton, as predicted by computer analysis of scattering data? I eventually figured it out,   (http://www.protoncosmology.com/new_page_3.htm), my training in electricity and electronics helped immensely; this led me on to further questions like, what made it work, why did the particles continue to oscillate, what was the spark plug, the mechanism? I eventually arrived at E. = MC2 and the fact that the velocity of the oscillating particles was always slower (sub-luminary) than the spinning charge fields that they generated while in linear motion. It was an extreme Eureka moment! Placing the four particle oscillating model within the confines of a reflective, semi-permeable, sphere, will result in a perpetual primary oscillator, a Proton. (http://www.protoncosmology.com/fundamental_mechanics1.htm.)

I welcome, in fact, urge any member of our group or anyone else who would be good enough to read about it, to comment on this model. If you find fault in this fundamental mechanics hypothesis, please point it out. If you find something pertinent, that can help, PLEASE, point that out too. The only major assumption that I make in the electromagnetic like dynamics of this model is directionality of charge, as mediated by C2, relative to the energy mass spectrum.

Being interested in physics, I'm sure you're all familiar with Occam's razor. "When competing hypotheses are equal in other respects, the principle recommends selection of the hypothesis that introduces the fewest assumptions and postulates the fewest entities while still sufficiently answering the question." – wiki.

Quantum Chromo-Dynamics needs from one up to three extra dimensions, one extra fundamental force of nature, whose force strength works opposite to the four fundamental forces recognized by physics and no valid logical mechanism to describe its workings. The nuclear model, I propose, has a mechanism, operates in just the four perceivable dimensions in which we exist, is in agreement with the accepted Rutherford-Bohr atomic model and can be explained by the mathematics of Classical Hamiltonian Physics. It needs no extra dimensions, or forces. It is a total cosmology, from top to bottom, that can explain what caused the inflationary big bang to occur, what gravity, dark energy and dark flow, quantum foam and the internal mechanics of the atomic nucleus are. For those interested: (http://www.helium.com/items/1304605-how-the-universe-formed).
   Dr. Sinclair, Eski (the old one) as he likes to be called, is the founder of the group. We met through our writings on helium.com and he was kind enough to invite me to join his group. Thanks Doc.
   Doc approaches the oscillator model via various derivations, manipulations and associations of the Planck's constant equation. Max Planck, a pioneering physicist, the first, I know of to propose an oscillating atom. I know you don't like to believe in the neutron, Eski, your cosmological view is, you propose a splitting of energy mass into two separate universes, one matter, and one antimatter. Your nuclear view is a mirror image of that and a neutron, kind of, gets in the way. My model proposes four synchronously oscillating, antimatter universes, whose motions and fields act externally to create our universe and dispense it back to energy.   They've been probing the neutron. It's there: (http://protoncosmology.com/science_news1.htm). In the model I propose, the charge and anti- or opposite charge both occur transitionally, from positive to negative charge during the densest, most compacted, energy mass phase. Can Planck's constant be transposed to accommodate that supposition, Eski, Hugh, anyone? From energy to positively charged mass, to negatively charged mass, to energy. Will that work out on a graph, Hugh? Charles, in your posting you state, "In fact, all of the Planck constants are reduced values as of the elementary charge, 2-pi or the reciprocal of the speed of light." Is there a correlation with the collapsing of two diametrically opposed conical charges upon a point ($V = 2(1/3$ pi c^2 h), where h stands for the cone height, not Planck's constant?
   Next month will bring the Vigier VII symposium in London, Eski, at which you will be afforded 15 minutes to talk about our group. Unless you or someone finds a viable mechanism for your model or someone can demonstrate something vitally wrong with the model I propose, I urge you to please give it a noticeable mention. I guarantee you, Eski, if I win anything. We both won't have to depend on Social Security anymore.
Good Luck and best wishes to all,
hoek

Reply to author   Forward   Report spam

ESKI View profile   More options Jun 4 2010, 4:27 pm

Don't know what is going wrong, get a little way and lose all I've written, Will take other tack and try to post one version, far from complete, (You are not mentioned, Hoek, and I intend to get you a passing comment, at least, as having a somewhat similar theoretical approach, we have a couple of weeks yet, maybe we can get to common ground....) Anyway, that version might clear up some of your misconceptions as to my stance....

Reply to author   Forward

dean sinclair View profile   More options Jun 4 2010, 4:40 pm

This is the first part of a tentative draft of a talk for the Vegier Symposium, it may bear little resemblence to a final draft and anyone is certainly welcome to comment within. I'm posting it in the hopes that it will help to clarify my positions with respect to those taken by you (Hoek) .

A Framework for a Fundamental Theory?

Dean L(eRoy) Sinclair

An idea, "Perhaps one could pretend to be a genius by consciusly trying to 'see the overlooked obvious'," when applied to physical science data--some

using the centimeter-gram-second system of dimensions, at about 4.7i x
10^-19 grams and 4.7x10^-19 centimeters. This 4.7 x 10^-19 g. is some
orders of magnitude larger than the rest mass values for the electron and
proton. The size value turns out to be very close to the 10^-18 cm. value
where Quantum Mechanics is said to fail and the Strings of String Theory
disappear into a 10 dimensional hole.

If the O/S Model be valid, then the Hadron Collider has a very low
probability of ever being able to fulfill the aim of colliding streams of
protons. The problem of moving vortexes which can coordinate, through a
medium countaining other oscillators, some of which may well be separable
into electrons and anti-electrons and deformable into neutrons, is far
different from "pushing a stream of charged particles which will repel each
other through a vacuum...."

 I have covered most of the ideas that were promised to be mentioned in the
"abstract,"

 (see "Sinclair's Abstract, " which can be found by Googleing "Vigier VII
Symposium and checking on "abstracts submitted," please don't be too angry
at the mangled English in the abstract, it wasn't carefully proof read.)

sIt is probably time to try to explain the logic on which all of this is
based. I finally realized, that I might as well go back to an article I
wrote, but have never published, called

The Model That Grew Itself

I hope this will be a bit more understandable and enjoyable than the
previous...

MODEL THAT GREW ITSELF
  A simple, almost naive, Model of Everything, has seemed to almost grow by
itself, logical step by logical step, during the last five years. the
simplest statement of the model is that it postulates that all of existence
is within a "Substance," tending always toward a triple point of equal
distribution of motion, which consists of oscillators or is controlled by
oscillators of a "constant torque family" defined by the equation, "m x r =
h/c," that is mass times radius equals Planck's Constant divided by the
speed of light.

The above is a simple digest of this entire article; however, one may
hope that it is not a total turn-off for readers. Let us back off and start
at the beginning.

About January of 2004, a year before the "Year of Einstein," 2005, which
was exactly a century after Albert Einstein published his work that later
came to be known as "Relativity," an old man working as a part-time janitor
and newspaper inserter, was thinking back to his younger days who, despite
being somewhat learning disabled and slow-thinking he had, somehow,
staggered through college to a rather extensive education in science. HE
realized, however, that despite all the "education," he had not understood
the "Theory of Relativity." even though the term, "Relativity" seemed to be
an obvious concept, the idea that all information is relative to the
observer.   He had a thought, "Maybe Einstein's Genius was the ability to
see things that were so obvious that everyone tended to overlook their
significance. If this be true, then possibly, one could pretend to be a
genius by deliberately seeking to find hidden significance in overlooked
information."

The very first thing that the Oldster thought of looking at was a rather
ironic idea, "Maybe there is something in Einstein's Relativity that is
commonly overlooked?" That started an internal dialog something like this:

 " The most common thing known about Einstein's Relativity is the
mathematical equations containing the relationship, 'v^2/c^2' . Now any
mathematical relationship can be generalized, what would the meaning of 'c'
as a constant be, if we generalized Einstein's Relativity mathematical
equations?"

 " 'c' would be the maximum speed of information transfer in whatever
Universe we are considering whether it be the Universe felt by Whales at

dating back to before 1900--has led to a simple--but inclusive--model
which may provide a framework for a comprehensive theory uniting the fields
of physical science.
Ths model--which could possibly be dubbed "The Research Results of the
Pretend Genius Approach--" is currently called the "Oscillator/Substance
Model." Its basic tenet is "All existence is the result of sequential
'action-reaction-action' interactions within a Substance/Substrate of
undefined basic composition and extent." This all-pervasive basic
substance/substrate can be considered to have the general characteristics of
a liquid substance at its "triple point," where with slight changes it can
act as a gas or solid.

Continued "sequential equilibration" within this results in constant motion
such that the system is composed of/controlled by oscillators.

Some of these oscillators are vortexes which have long term stability.
Their interactions result in "Matter."
Among the results of this view are an explanation for "charges," which are a
result of the orientation, counter-clockwise or clockwise of the
rotation-inversion senses of the vortex oscillators and the related
defiinition of the size, shape and form of electrons and protons and the
corresponding "anti-units." These appear from a reinterperetation of
Planck's Constant and a use of it to define a basic family of oscillators.

A solution to the problem of unifying the "Four Forces of Nature," appears
automatically when it is realized that none of the "Four Forces of Nature"
meet the criterion of a "Force." and that pressures within a substance
does.

Resolving the "Matter of the Missing Anti-matter." becomes a bit more
complicated, but essentially add up to the realization that if one half of
the oscillation of a given oscillator be defined as a matter unit, the
other half would be anti-matter. As the electron and anti-electron can be
considered as the split off halves of a parent oscillator, if the electron
be called a matter unit, the anti-electron is "anti-matter." There is an
extension to this. The electron is a split off "half" of another oscillator,
the neutron. If half an oscillator be matter, and the electron be defined as
matter, then the proton is "Anti-matter." Since it can be shown that if
about 41/42 of the Kinetic Energy of an anti-electron were transformed into
Mass, an anti-electron would apparently become a proton. Similar
calculations show an electron could, theoretically be transformed into an
anti-proton. One may say, therefore, that "There is something the matter
with this whole "matter-anti-matter' idea.

What we call "Matter" is observed in our Universe as a combination of the
"matter unit," the electron and the "anti-matter unit," the proton. A
thorough examination of the idea of oscillators in a medium shows that this
view is probably correct and that the expressiion of the different
rotations of smaller oscillators within a half of a larger oscillator, that
half being our "Universe" which has a rotation of its own, accounts for this
apparent anomaly. In the Anti-verse corresponding to ours, the "electron"
and "proton" of that Universe would have the reversed rotation inversion
orientation of ours, as the Anti-verse has the reversed rotation-inversion
of ours.

Careful examination from nuclear chemistry suggests that the
"proton-neutron" atomic nucleus, hides an interesting clue to the "Lost
Anti-matter" of our universe. If the electrons and protons be considered
"Matter." the "neutron" count may be considered as the instantaneous amount
of the "nucleus" which is being expressed as positrons and anti-protons,
i.e. the "Anti-matter" content. The missing anti-matter, appears to be, by
this model, simply a semantic illusion.

Another situation which may be considered as a semantic illusion is the
missing mass of the universe. This probably occurs because "Mass" is used
in two diffferent ways, as the name of an entity, and as an attribute of an
entity. As an attribute, it is a variable. This variable, we measure,
usually, as a "rest mass" which turns out to be a minimal value. An average
value for the oscillators of our universe can be estimated, along with an
average size of the oscillators of our universe, by noting that if Planck's
Constant be divided by the speed of light, the resulting constant has the
dimensions of mass and distance (radius). As at some instant, mass and
radius would have the same absolute values, each equal to the square root of
this new constant, one can say that this "inversion" point would occur,

1000 ft. in the ocean, or Pony Express Riders carrying the mail in the 1880's. We could even define a 'Perceptual Universe' by the speed of information transfer in it. Einstein's equations, then, describe what happens to information as transmitters and receivers move with relation to one another. The work actually would belong in 'Information Theory.' "

"What then actually is 'c?'" " If we consider the Pony Express Analogy, "c" would be the average speed in any direction from a point that can be expected from averaging over some period the motions of the 'Information Packet Carriers.' The Speed of Light is an average, not an absolute limit, An average of the motions of the carriers, whatever they are, which must exist throughout nature, if nature is consistent."

This view of information transfer being consistent and the speed of light being an average rather than a maximum sets up an entirely different possible view of everything from the conventional picture taught in science classes and the logic which arises leads to a different ...

read more »

Reply to author   Forward

End of messages

« Back to Discussions   « Newer topic   Older topic »

This view of information transfer being consistent and the speed of light being an average rather than a maximum sets up an entirely different possible view of everything from the conventional picture taught in science classes and the logic which arises leads to a different mode. of everything.

Additional questions arose. IF there is a medium, an information carrier, what is its nature? How can it be a "solid/" How does light have both "particle" and "wave" properties? Where do "mass" and "energy" fit in? To be a solid, it must somehow act as if it were a matrix. For information to move there must be some movement, some motion. So the first idea was "Motion in a Matrix." Fitting in Mass and Energy as two aspects of motions, energy as motion along a line and Mas as being related to motion about a point fit well into this. Later it was realized that there would be some slight shifts that would bit better to reality. Mass could be considered as being about a point, but contained within a surface centered on a point. Energy could be considered as a more general term for a total "package of motion," but usually meaning Kinetic Energy, motion associated with motion of an entity along a line, perhaps better called "Translatory Energy."

The focus, however, remaining on where, this started, with the consideration of the importance of the speed of light and hence of electromagnetic phenomena, it was natural to consider what could possibly be learned by examining Planck's relationship between frequency of electromagnetic radiation and energy, the equation, E=hu. Energy equal Planck's Constant, h, times the frequency, "u," {Pronounced "Nu," as the Greek letter.) Analysis of the equation shows that Planck's Constant, h, has the dimensions of Angular Momentum, which is mass times radius (of a spinning body) times velocity. That led to the idea of simply writing, h=mvr and evaluating this at an average value of velocity as "c."

Doing the substitution of "c' for "v" and dividing it out to produce mr=h/c , gave an almost electrifying result when two separate things were realized about this very simple little equation. It was realized, that mass times radius is a "torque.} which can be considered as the effect of one spinning body on another. As "h" and "c" are "Constants of Nature," h/c would be the "Torque Constant of Nature." It was also realized that "mr=h/c is an equation that could be written in the form of xy=K=yx. That is if the "Absolute Numerical Values" of Mass and Radius are switched there will be found the other limit of an oscillator, the balancing values to the observed values.The equation, xy=K=yx, is the form equation for a number of "Balanced Laws of Nature" including the Law of Levers and the Law of Forces. "for each and every force there is an equal and opposite force. " It,, also defines limits for a linear oscillator. Although any natural oscillator would be surely multi-dimensional, such an oscillator, having symmetry, can be analyzed as if it were linear... The mr=h/c equation appears to represent represent such a dimensional analysis applicable to any oscillator of the "family-set" which may represent all of the fundamental oscillators of our natural existence.

Assuming that the oscillator family set, {mr=h/c=rm}, is a valid description of fundamental units of existence, it was postulated that the average value through which all oscillators will vibrate is the balance value of the above equation where m=r=(m/r)^0.5 . That is, the absolute value of mass equals absolute value of radius and each has the absolute value of the square root of the torque.

Checking the above ideas to see whether they would fit with the very important basic units of matter, the electron and the proton, we find that they do indeed fit the set, and in fact, information describing one oscillator limit, the largest radius and the smallest mass set, does occur in the literature. The mass is reported as "rest mass" and the radius is reported as Compton Wavelength. The transposition to determine the other set of limits is not in the literature, nor is there any evidence that the Compton Wavelength is recognized as a radius. A careful examination of the literature definition of the Compton Wavelength showed that it is indeed the same as the "r" defined by our set equation.

Somewhere about this point in the logic that seemed to be leading automatically toward a rather complete "Theory of Everything" starting from very few basic assumptions, there came the realization that what we are looking at as an overall model would fit closely to the idea of a substance which would tend to equalize motion throughout. That is tend to a "triple point," where it could act as any of the three standard states of matter, solid, liquid or gas. This would explain the transverse wave motion of light, the pressure change of the disturbance is enough to cause the medium, "substance," to react as if it were a solid at the instant the disturbance passes through. Electromagnetic radiation would be a transient wave disturbance, "particles" such as the electron would be permanent oscillators/oscillations.

Other ideas began to drop into place. The electron and its reverse, the anti-electron are known to "annihilate" with the release of radiation having a value equal to the "rest the electron" times the square of the speed of light. Conversely, it is known that light at an energy above this value can interact with matter to produce an electron-anti-electron set, "pair Production." It is also known that the electron and anti-electron spend appreciable time in the same vicinity before "annihilation." This implies that annihilation requires a very specific orientation of the two oscillators. As an oscillator spinning counter clockwise and a reversed oscillator spinning clockwise if they became exactly oriented along a common axis could combine to a simple pulsating oscillator with loss of the spinning energy to the milieu, it seems logical to consider that the electron and anti-electron are actually halves of a pulsating parent unit having the same oscillator limits as either of the two. Oscillators in a substance would be expected to fall into three general patterns, full-wave, pulsating spheres, full-wave pseudo-spheres having an axis and an equator with counter-rotating halves, and half-wave vortices which result from the splitting of the second category. The electron, anti-electron and proton would apparently belong to the last category.

Looking at the neutron, it was noted that it does not fit neatly into any of the three above categories as,' having a magnetic moment it is not symmetric. A calculation of what it would take to change an anti-electron into a neutron showed that if about 41/42 of the kinetic energy of the anti-electron were changed to mass, it could become a neutron. As a spherical oscillator could presumably be "squashed" such that half of the unit took the "mass" (it's becoming clear that mass must be a measure of the internal motion within a unit) then it makes sense that the "parent unit" of the electron and anti[electron might, under the right circumstances be distorted to a neutron, which could later rearrange into an electron and proton.

A little consideration showed that the inversion/splitting instant of an oscillator of the kind of oscillator which could produce a "Universe " would certainly release a twisting shock wave in opposite directions, this could distort "parent units," on one side of the split or inversion into neutrons, on the other side, anti-neutrons. If our half-wave oscillators have an average rotational velocity and an inversion rate of "c" the speed of light, their minimum speed is zero and their maximum speed at inversion is 2c.   A pulse trying to move at "2c" against a medium with an average value of "c," in any given direction, would create a shock wave distorting some of the medium.

This not only gives a rationalization for the existence of neutrons, but also gives a possible description of the "Big Bang" postulated as the start of our Universe. The "Big Bang" would simply be the inversion instant of an ultra-low-frequency oscillator in the "Substance of Existence," --or possibly the instant of split, if our Universe actually be an "Electron Analog.""

The conclusions reached at this point give an outline of a theory of everything being a part of a substance of an undefined size, consisting of oscillators--or organized into oscillators--partially or totally belonging to a "family."' defined by the set, $\{m \times r = h/c\}$. This substance is in constant flux due to its tendency to equilibrate motion throughout....

The conclusions that can be reached from this model are interesting. Here are some: We can define mass as a measure of the tension/pressure felt at the surface of an entity balancing the motion within the object under consideration and the rest of Existence. Energy is a general term for a "packet of motion."   Kinetic Energy is a measure of the motion content of an entity moving within the Substance on a given vector. Light, i.e., electromagnetic radiation, is a vibration disturbance within the substance. A disturbance which will dissipate. Electrons and protons are vibrational disturbances which have indefinite life times.

Considering the electron and proton as inverting oscillators of the set, having an average rotational speed of "c" and an inversion rate of once per rotation allows the calculation of probable comparative "sizes" of the two units, their actual motions and the meaning of "positive and negative charges." A "positive charge would be a characteristic associated with a clock-wise spinning unit, the proton. A "negative charge" defines a "counter-clockwise' spinning unit, the electron. (The electron has been shown in cloud chambers to leave a counter-clockwise spiral path in space.)

Since the two entities rotate in opposite senses, they would tend to cancel each others effects in space and hence be "attracted." However, as they have radically different oscillator limits-- therefore, much different inversions frequencies--they cannot "annihilate." Rather than annihilating, they associate in a multitude of ways, resusting in all of the entities which we conider matter.

By this model, we exist behind the shock wave of the expansion phase of an

Esri - July 10, 2010

People, here is my current idea for a script for my talk scheduled in July for the Vigier VII Symposium. I am inviting comment, editing, addition or deletion ideas. Send any ideas to me at deanlsinclair@yahoo.com. As close as I can tell, "reading time" for this as an oral presentation is right on 30 min. I do not know yet how much time will be allowed. Will fill you all in with any details as to the Symposium whenever I get any. Thanks.

THIRD TRY FOR A SCRIPT...

It is my pleasure to welcome the Vigier VII Symposium to the Aberdeen American News here in Aberdeen, South Dakota, U.S. of A. and to be able to thank the News and Dr. Amoroso for this opportunity to introduce to the wider scientific community a Framework for a Comprehensive Theory which seems to be applicable at any scale from sub-atomic to cosmic.

I also wish to thank my friends from two Internet Groups who have contributed information and support over the last three years since the first version of this model was published on Helium.com as "Motion in a Matrix..." The current version has a working title of the Oscillator/Substance Model, as it has become clear that the motions that are involved are primarily, if not totally, oscillatory in nature, and a liquid substance at its triple point where it can also act as a solid or a gas makes a much better model than does a rigid matrix.

One of the groups which I need to thank is the Condensed Matter Nuclear Science Group, cmns, which has furnished much information including the fact that this Symposium existed. Thank you, Jean Pierre

The other group is the Oscillator/Substance Theory Group.

One member of that group, Robert vanderHoek, has a somewhat different version which he feels is superior. That version can be found on line as "Proton Cosmology." He says, ...mention me, and if I win a prize, neither one of us will have to depend on Social Security again." Well, Hoek, you got your mention; however, I have news for you, the chances of either of us winning any prizes are about the same as the chance of the Hadron Collider finding the Higgs Boson. That is about the same as a snow flakes chance of survival in a furnace.

Be that as it may, it is best to get on to the project of using my remaining minutes in trying to convince an undoubtedly, and rightfully. skeptical audience that there is a possibility that an elderly day-laborer may have accidentally accomplished what a man recognized widely as one of the greatest geniuses of all time spent years unsuccessfully trying to do. That is come up with a unifying framework for the physical sciences.

Probably the difference is that this speaker is working with no particular set of preconceptions but is simply following a line of logic based on the thought that he can pretend to be a genius

Gmail Calendar Documents Photos Reader Web more ˅   deanlsinclair@gmail.com ˅

Google groups

« Groups Home

Oscillator/Substance
⊞ Theory

[Search this group] [Search Groups]

View this page "Possible Script for VigierVIITalk"   Options

3 messages - Collapse all - Report discussion as spam

ESKI View profile   More options Jun 10 2010, 3:15 pm

Am posting a possible script as a "page" on this site. Want to know
what the people of this Group think about this version.

I'm finding it more difficult than I thought it would be to condense
the last three years of discoveries, even the most basic, into a
thirty-minute, oral presentation.

I haven't done an oral presentation to a group in almost 30
years,;and, in those days, I always extemporized.

This is my first try with a reading script and I have never done
anything by remote before. I'll need a lot of "fingers crossed." to
pull this one off without making a mess of it. ESKI

Click on http://groups.google.com/group/oscillatorsubstance-theory/web/possibl...
- or copy & paste it into your browser's address bar if that doesn't
work.

Reply to author   Forward

hoek View profile   More options Jun 17 2010, 2:42 pm

Hello Dr. Sinclair,

  Well, it's the middle of the month of June and so far no one, but,
me has responded to your call for assistance in working out a speech
for the upcoming Vigier VII Symposium in July. I want to thank you
for mentioning me and my website in your upcoming speech, but,
mentioning me isn't necessary. My motives are not personal fame or
recognition, but, purely and simply to tell the truth about nature, as
I perceived it. The objective of Vigier VII Symposium is "The Search
For Fundamental Theory". Talking about anything other than a
description of that fundamental theory or where one could be located,
would be off topic. It's interesting how you described your quest
that began in 2004 or 2005, when it seems, you had an epiphany and
realized that there were things relative to Planck's formula that had
been overlooked. My own quest began in 1978. After viewing a science
film, which made me realize that there was valuable data being
overlooked, that didn't agree with the prevailing quark theory, but,
would describe the inner workings of the proton. These are both
interesting stories, but, both are off topic and not what the people
at the Vigier VII symposium want to hear. In my point of view. You
try to oversimplify the universe. It seems you twist substitute and
rearrange Planck's formula trying to get things out of it that just
aren't there, most notably, a mechanism. If your model has no viable
sustaining mechanism, than it is no better than the Quantum Chromo-
Dynamics of "Quark Theory. These people are looking for a FUNDAMENTAL
THEORY! Not the story of a self acclaimed "pretend genius" or day
worker or janitor or old man, who wants to make his mark in the world
before he dies or about the other semi-famous people who are also
named Dean Sinclair. Focus on the topic that they've invited you to
speak about, a FUNDAMENTAL THEORY.
  You've got me by about 14 years Eski. I am the survivor of a liver
transplant, back in 2003. I could also die any time, as any of us
can. I don't complain about it or look for people's sympathy, but, I

thank God the Creator for every extra day that I am allowed to live in this beautiful realm of existence where everything seems to make sense, at least in nature. This pretend genius you speak of, who wants those at the Vigier Symposium to throw out all that they've learned in science from Rutherford's discovery of the nucleus to our present concept of atomic structure. A century of knowledge discovered, not by pretend geniuses, but many real certifiable ones. You ask a lot!!! In her last posting, our group secretary, ka-sala said, " We can go over and over Planck's Theory, and anyone else's. The difference is we are here... now. Today's Theory." I think she's saying you're beating a dead horse, Eski. I noticed in your speech, you mentioned twice about reality, disappearing down the 10 dimension hole of string theory. WHY? Your own hypothesis has nothing to do with other dimensions. It works quite fine in just the three of space, and one of time that we can naturally perceive, as does my own. QCD needs at least one and possibly three, extra dimensions for their mathematical model to work, depending upon which quantum physicist you ask. I've always thought of assumption of extra dimensions as a "fudge factor". A generous assumption factor, a place to hide the sloppy bits that don't quite fit in. When neutrons undergo beta-decay, by emitting an electron and an anti-neutrino and turn into proton's the mathematics of quark theory fails. It seems there's some missing charge energy and mass that needs to be there and isn't. Well, where is it? Oh, I see, it resides in an alternate dimension and pops over just when it's needed. Who would've thought? More dimensions are just more fudge factors and I recommend to you, not to use or acknowledge their existence, that is, unless they are pertinent to your hypothesis. I tell you all this, Eski, not in a mean-spirited or condescending fashion, but, to try and help you and our group make a good impression. I've edited the beginning of your speech in a way that mentions my website twice. It's more important to me that my hypothesis be analyzed, then my name being mentioned. If you mention my website, as well as that of the O/S group, a third time at the end, it's been proven that more people are likely to remember it. You'll notice they do this in almost every TV commercial. Remember to try and focus on the topic, Eski, a fundamental theory and how it works, in describing reality as we know it, at the nuclear and possibly cosmological level as well. By the way, my full name is Robert Kardien Vanderhoek, that's CAR-DEAN VAN DER HOOK. If you do mention it, tell them to google it. It will take them to my Fundamental Mechanics page.

I've taken the liberty to rewrite the intro to your speech. I think it more clearly defines the objectives of all the members of the O/S group. I hope you like it and it helps you, let me know what you think. Thanks again for including me and I wish you and all of us best of luck.
hoek

It is my pleasure to welcome the Vigier VII Symposium to the Aberdeen American News here in Aberdeen, South Dakota, USA and to be able to thank the News and Dr. Amoroso for this opportunity to introduce to the wider scientific community a framework for a Comprehensive Theory which seems to be applicable at any scale from sub-atomic to cosmic.

I also wish to thank my friends from two Internet Groups who have contributed information and support over the last three years since the first version of this model was published on Helium.com as "Motion in a Matrix..." The current version has a working title of the Oscillator/Substance Model, as it has become clear that the motions that are involved are in part primarily, if not totally, oscillatory in nature, and a liquid substance at its triple point where it can also act as a solid or a gas makes a much better model than does a rigid matrix.

One of the groups which I need to thank is the Condensed Matter Nuclear Science Group, cmns, which has furnished much information including the fact that this Symposium existed. Thank you, Jean Pierre

The other group of which I, Dean LeRoy Sinclair, am founder and director of is the Oscillator/Substance Theory Group. At http://groups.google.com/group/oscillatorsubstance-theory

Members, of our group all seek the truth about the nature of the realm in which we find ourselves existing. We all believe that the basic mechanics that govern the universe and the workings of the sub-atomic, quantum, world are oscillatory in nature and mirror each other at the largest and smallest of scales. We come from different backgrounds, with various approaches and models to try to explain this phenomenon. One member of the group, who has a somewhat different version from my own, of which I'll return to in a moment, is at protoncosmology.com. His model agrees with the current Rutherford-Bohr atomic model, but, not the currently accepted quark nucleon hypothesis. His model takes an electro-magnetic like, field dynamics approach based on neglected and recently verified electron-proton scattering data. The model poses the proton as an internally driven spherical oscillator, an electro-dynamic like perpetual motion machine. It is driven by the conversion of charge energy into mass and back again by the function M=E/C2, which is a simple transposition of E=MC2 and the fact that no particle with mass can attain the velocity of light. The full version, which is too lengthy to describe now, can be found on line at " http://www.protoncosmology.com", or on links at: http://groups.google.com/group/oscillatorsubstance-theory

I'll now use my remaining minutes in trying to convince an inquiring audience that we have come up with a unifying framework for the physical sciences.

Probably the difference is that this speaker is working on a model with no particular set of preconceptions, but, is simply following a line of logic examining commonly accepted ideas for hidden or overlooked significances. This is a totally different approach from trying to fit together already accepted viewpoints. This is an open ended journey rather than one focused on some desired destination.

You take it from here Doc

On Jun 10, 4:15 pm, ESKI <deanlsincl...@gmail.com> wrote:

- Show quoted text -

Reply to author   Forward   Report spam

dean sinclair View profile   More options Jun 17 2010, 4:54 pm

- Show quoted text -

Actually, the biggie, was realizing that there was a tremendous amount to be learned by "Looking for the Overlooked Obvious," The second biggie was that The Speed of Light is an AVERAGE. Planck's Constant got only into the act because of logical follow-up on what needed to be happening in a sensible situation where the Speed of Light was an average motion of some sort of information carrier which had to be ubiquitous, it information transfer followed the same principles no matter when or where.

> My own quest began in 1978. After viewing a science
> film, which made me realize that there was valuable data being
> overlooked, that didn't agree with the prevailing quark theory, but,
> would describe the inner workings of the proton. These are both
> interesting stories, but, both are off topic and not what the people
> at the Vigier VII symposium want to hear. In my point of view. You
> try to oversimplify the universe. It seems you twist substitute and
> rearrange Planck's formula trying to get things out of it that just
> aren't there, most notably, a mechanism.

\
Nope, the "mechanism" is inherent to the assumption of a basic substrate in which there is motion, The "Mechanism" is the Law of Forces, or Action, "For each and every action there is an equal and opposite reaction."...W hat needs only to be added to this to get "my version of your perpetual motion machine" is to note that since there is always sequence, this reaction is itself an action and the process of generating motion goes on and on....I only use Planck's Constant and the Speed of Light to give us usable basic information about this Reality in which we exist....I'm not extracting a bit more out than do the QM people and others. The difference is actually that I focus on the Torque, rotation aspects, where as they analyze the same equations from the view of "momentum."

If your model has no viable

> sustaining mechanism, than it is no better than the Quantum Chromo-
> Dynamics of "Quark Theory. These people are looking for a FUNDAMENTAL
> THEORY! Not the story of a self acclaimed "pretend genius" or day
> worker or janitor or old man, who wants to make his mark in the world
> before he dies or about the other semi-famous people who are also
> named Dean Sinclair.

These things were inserted because of the felt need to minimize the "Crack
Pot Factor." of screaming too loudly, "Gee, look how brilliant I am. For the
same reason, I've made no comment about my science background. If the ideas
won't stand on their own from a simple old man, detailing that he actually
had a first rate science education won't help a bitl.

> Focus on the topic that they've invited you to
> speak about, a FUNDAMENTAL THEORY.

Interesting point. The problem, always, in giving a talk is to balance the
situation of some degree of "entertainment" with the information. What you
say about the Group and your material is very correct. As the write up to be
published later can be much "dryer" and fact filled, it will be very
appropriate there. Will think about using it "here."

 You've got me by about 14 years Eski. I am the survivor of a liver

> transplant, back in 2003. I could also die any time, as any of us
> can. I don't complain about it or look for people's sympathy, but, I
> thank God the Creator for every extra day that I am allowed to live in
> this beautiful realm of existence where everything seems to make
> sense, at least in nature. This pretend genius you speak of, who
> wants those at the Vigier Symposium to throw out all that they've
> learned in science from Rutherford's discovery of the nucleus to our
> present concept of atomic structure. A century of knowledge
> discovered, not by pretend geniuses, but many real certifiable ones.
> You ask a lot!!! In her last posting, our group secretary, ka-sala
> said, " We can go over and over Planck's Theory, and anyone else's.
> The difference is we are here... now. Today's Theory."

The point, however, is that our "geniuses," by overlooking what was under
their noses, have grossly over complicated scientific theory. The Universe
and Fundamental Theory, have to be simple!!!

> I think she's
> saying you're beating a dead horse, Eski. I noticed in your speech,
> you mentioned twice about reality, disappearing down the 10 dimension
> hole of string theory. WHY?

You've never seen my explanation of the ten dimensions thing. That is a
mathematical trick, You can triangulate any point from three others, each
of which you assign three arbitrary dimensional axes, hence you end up with
nine dimensions to define your locus, then you give the locus a dimension of
motion, ergo, ten dimensions.   The disappearance into a ten-dimensional
hole is simply because at that size, the mathematics, based essentially
on the two observations of "constants of nature" which I consider
fundamental, i.e.. Speed of Light and Planck's Constant, shows that the
mass and radius scales, (or any other two dimensions based on these two
"Constants") have the same "Absolute Values." We enter into a "reversed"
 dimension, if you wish, of unimaginable smallness and fantastic
mass.....The mathematics used in QM and String Theory, simply fails at this
inversion size and mass.

> Your own hypothesis has nothing to do
> with other dimensions. It works quite fine in just the three of
> space, and one of time that we can naturally perceive, as does my
> own.

Not quite true as to my own hypothesis. It is simply a matter of what one
considers a "Dimension," the term actually simply defines the name of a unit
that we decide to measure by..... In the cgs system, our three dimensions
are centimeters, grams and seconds... The volume below 4.7×10^{-19} cm. in
radius, could be considered an "internal dimension" with respect to our
"outer dimension" larger than that radius.

QCD needs at least one and possibly three, extra dimensions for

> their mathematical model to work, depending upon which quantum
> physicist you ask. I've always thought of assumption of extra
> dimensions as a "fudge factor".

Of course, it is a fudge factor, one can introduce as many dimensions as one pleases. The Standard Model does this, in a sense, if a Quark doesn't, explain it. add a Graviton, if a Graviton doesn't toggle things up add a Gluon, and so on and on....

 A generous assumption factor, a place
> to hide the sloppy bits that don't quite fit in. When neutrons
> undergo beta-decay, by emitting an electron and an anti-neutrino and
> turn into proton's the mathematics of quark theory fails. It seems
> there's some missing charge energy and mass that needs to be there and
> isn't. Well, where is it? Oh, I see, it resides in an alternate
> dimension and pops over just when it's needed. Who would've thought?

This kind of bull is the reason why I try to carefully define "Mass" and "Energy" in most of my work, unfortunately, there really isn't time in a short talk to go into the misconceptions that pervade scientific theory by the loose use of these terms and the fact that scientists seem to go really no farther in defining these terms than the old meaning less and misleading circular definition, "Energy is what moves Mass, and Mass is what is moved by Energy." Lots of hidden mess ups in that bit. Mass as an entity? Or mass as an attribute of an Entity?" Mass as a Constant or Mass as a variable? Energy within or outside a surface? Energy as determined by some balancing process? In any case, what do the words REA:LLY mean. (I finally define mass as a measure of the tension/pressure at the surface of an entity which is a measure of the point centered motions within that entity. I consider "Energy" as a general term for a "package of motion." A term which need in every usage to be much more explicitly defined in context. Both terms, as used in physics, describe variable amounts of motion.

> More dimensions are just more fudge factors and I recommend to you,
> not to use or acknowledge their existence, that is, unless they are
> pertinent to your hypothesis.

Actually, I wasn't acknowledging their existence; but, rather, pointing out that the calculations showed a reason for the failure of the older models.

- Show quoted text -

...

read more »

Reply to author   Forward

End of messages

« Back to Discussions   « Newer topic   Older topic »

Google groups

« Groups Home

Oscillator/Substance
⊞ **Theory**

[Search this group] [Search Groups]

REPEAT of INFINATE VARIABES - Reply to ESKI.   Options

Home

Discussions
+ new post

Members

About this group
Edit my membership
Group settings
Management tasks
Invite members

2 messages - Collapse all - Report discussion as spam

ka-sala View profile   More options Jun 19 2010, 1:49 am

>>>O.K. Ka-Sala,
>>>You are just much too bright for this old man.

* Never believe this of yourself Eski... nor of me!

>>>Or maybe you just think too much.

* It has been no secret between us that I have almost completed a
Report/Book. I try to share 'snippets' but cannot give it all away...
yet. OK ?

>>>It seems to me that you are missing a lot of my points, particularly my point about
defining what we are talking about.

* If I am, I am sorry. I personally know what I am talking about and
the link is to you all and the Oscillation Substance Theory. You know
that! But I never claimed to speak your theoretical language,
knowing another of my own, and that is hard to translate.

>>>Just what do you mean by 'unified weak and electromagnetic interaction' ?

* To save you the confusion of my language, let me just explain it in
more depth via Wikipedia.
A/ Particle Physics.
1/ Elementary subatomic constituents of matter and radiation, and the
interactive relationship between them. Also called high energy
physics
Read more... Look it up.

B/ The unified description of two of the four fundamental interactions
of nature: electromagnetism and the weak interaction. Though these
two forces appear very different at everyday low energies, the theory
models them as two different aspects of the same force. Above the
unification energy on the order of 100GeV, they would merge into a
single electroweak force. If the Universe is hot enough (approx. 10 to
the power of 15 K,) a temperature reached after the Big Bang, then
the electromagnetic force and weak force will merge into a combined
electroweak force.

>>>"By my thinking both the "weak" and "electromagnetic" interactions are terms from
>>>conventional physics which are meaningless if we consider the "four forces" as all being
>>>observational aspects of one "force--" pressures within a "substance-substrate" having
the >>>basic characteristics of a liquid at its triple point.
>>>Could you define "atemporality" for me, hopefully in terms of my postulated "Aether?"

* I cannot see that you have not already formulated the timelessness
of this Aether/Ether but if you are moving into the dimentions in
which Alchemy operates, then you are looking at Life's Elixir. So
naturally your 'substance substrate' is very illuminating. The very
catalyst you want.

>>>Also, can you give me a good definition of "Entropy?"

* Which one do you want Eski ? For now, just stay with the catalyst
but...! Once you have your energy interaction and in balance -
therefore equal - what you term 'repair' becomes in quantum
thermodynamics a propellent beyond any shock waves, and in absolute
silence the interacting forces within your term, 'Aether' wraps itself
around the electromagnetic forces of your Oscillation, as if in a safe
cocoon. (Inclusive of pragraph below's answer.)

>>>I've never seen a really good one, and I must admit that I'm not sure how to fit the
concept >>>in. Does it fit in with the concept of Mass as I define it, with the problems of
Kinetic Energy, >>>or is it a measure of some sort of the "degree to which some action-
reaction sequence >>>somehow repairs the chaos behind a shock wave in which we exist?
>>>"Infinity," to me, has a practical definition of "the point, instant,or number just beyond
where >>>we stop, quit counting, quit examining, or our means of information gathering or
transfer >>>fail...." Eternity would have a similar connotation. An extent just beyond the
extent of >>>our "Perceptual Universe."

* Who can measure Infinity Eski. But while you are working on this O/
S I believe the 'google' will be big enough to measure what you want.

>>>As long as we've been co-workers in this, I don't think that you have listened closely to
old >>>Eski...

* You wait this long to tell me ??? Then forgive my efforts to
contribute.

>>>It could, of course, be that Old Eski, "the Pretend Genius," a person of admittedly slow
and >>>limited intellectual abilities, who can only try to be intelligent by looking for different
ways to >>>interpret what others take for granted, is totally lost in comprehending the ideas
of those who >>>have true genius level capabilities. I only vaguely discern what either you
or O'Sullivan is >>>saying a great deal of the time.

* That was the problem I had when trying to explain the incident of my
Profile. No-one seemed to understand, until it was found that I had
what was 5 months prior to the announcement of the 1979 Nobel Prize.
At least even some-one like me had it! Your earth language is not
easy !!! Facts are I cannot 'pretend'... only extend. And to you
Eski my friend, it was a helping hand. I proclaim no qualities of a
genius. I'm just a bit alien to many thoughts and ideas which have
been boxed in by their theoretical masters.

>>>I do hope that some how, Eski can put together something that is at least rational for a
30 >>>min. presentation at the Vigier Symposium so we don't end up appearing a bunch
>>>of "nincompoops."

* I am sure you know what it is you want to say and have to offer. Do
it your way as we do offer ours. One cannot be a ' nincompoop ' to
make such a presentation. Just be sure of what it is you have!

>>>If you were to suggest one point that you think he should make in that presentation,
what >>>would it be?

* Don't ever think it's too late, nor that what you have is not
presentable. Anyone with enough insight will see through your efforts,
and if I – little me – can come up with Nobel material. Anyone can!
One just has to know what it is that you are presenting, and that it
is fact..

Would we be behind you if we were not with you ?

>>>Allaha asmaladik...DLS

' Ve size' dear Dr Dean L(eRoy) Sinclair
Kind regards Eski,
Ka-sala

PS.. Have just been requested to submit material re. radio-active wate
issue to Federal Minister for Enviroment; Australian Parliment !

Reply to author   Forward   Report spam

ka-sala View profile   More options Jun 19 2010, 1:52 am

PS> Correction typo. Radio-active waste

On Jun 19, 4:49 pm, ka-sala <irrir...@gmail.com> wrote:

- Show quoted text -

Reply to author   Forward   Report spam

End of messages

« Back to Discussions   « Newer topic   Older topic »

Google groups

« Groups Home

Oscillator/Substance
⊞ Theory

Home

Discussions
+ new post

Members

ADDENDUM TO TALK   Options

3 messages - Collapse all - Report discussion as spam

ESKI View profile   More options Jun 22 2010, 3:56 pm

In thinking about it, particularly realizing that the criticism, "You
are asking an awful lot!!!" (In a justifiable criticism of the
temerity of an admittedly average person operating rather jokingly as
a "pretend genius" daring to say, in effect, "some of the geniuses
were possibly wrong, Eski, has decided that he should probably go
ahead and point out some of the very definite differences in what he
is saying and what the conventional wisdom says.... Get some of it out
in the open and "let the brick bats fly."

So here is a possible addendum for near the end of the "London Talk."
Criticism and comment are, of course, invited. How about some of you
who haven't had much to say chiming in??

Addendum to talk...

This is called a "Framework" as it is a simple start toward new
construction. The model, developed essentially independent of
consideration of theoretical models from a thought that communication
should be consistent whether the information be carried by by Pony
Express Riders or Electromagnetic Waves, turns out to be definitely in
contrast to most scientific attitudes. In fact, we might say that it
is usually anywhere from 90 to 180 degrees out of phase.

Where the standard view seems to be that what is needed is somehow a
theory to unify many diverse parts, this model takes the view that
there is a "unity," " a Substance/substrate of undefined extent and
undefined basic unit." This may be paraphrased, " There is a Fact of
Existence which we may never be able to totally understand or define,
let us accept that and move on to what we can do.l"

Where the general concensus is that there is nothing in a "vacuum,"
this model postulates that there is an all pervasive substance, even
in "vacuums" from which the vortex aggregates which we call "matter,"
have been removed.

Matter and "void" are considered, in this model, as being composed of
different "arrangements" within the same basic "substance/substrate."

Where the usual view of electrons is as some sort of probability
cloud, this model sees them as rotating, inverting vortex
oscillators...similarly, reality of size and shape are given to other
subatomic units.

Where the conventional picture is that electrons and positrons combine
to annihilate converting totally to "electromagnetic radiation," this
model says that they combine to another type of oscillator with
dissipation of half of their total motion in the form of
"radiation." Whereas, conventionally, "pair-production" is some sort
of a mysterious conversion of "Energy" into "particles" in the
presence of matter, in this model, pair production is simply the
splitting of the "parent oscillator" when supplied with enough excess
motion...

Much of conventional physics theory is based on an idea similar to

About this group

Edit my membership

Group settings

Management tasks

Invite members

View this group in the
new Google Groups

Einstein's supposed comment, " Mathematics is the reality." It is even assumed by many that if theoretical ideas are not expressed in differential equations, they have no validity. The view here is that math. is a tool, and that it is probably best to work with the simplest tools possible. Although advanced math. could have been used, this entire presentation has used nothing that was beyond grade school level.

Where Mass and Energy are accepted as being somehow fundamental and inter convertible, without truly defining either; this model defines both with respect to motions relative to a point.

Where the "Unification of the Four Forces of Nature" is considered conventionally as a major theoretical problem, this model almost cavalierly dismisses the situation by pointing out that none of the "Four So-Called Forces" meets the definition of a force, whereas pressure does.

The Matter of the Missing Anti-matter has been discussed earlier. Clearly the view is very different from the conventional.

The problems of the "Missing Mass" of our Universe, Dark Energy, and some of the other related concepts may turn out to be due to several factors, one could be the semantic confusion between the use of the term, "Mass," as describing a "physical body," and "mass" as a scientific term describing an attribute of that body. This differentiation clearly shows in this model.

Where, conventionally, there are many constants of nature, this model implies that there should be few, and those will be not absolute limits of any sort but are more likely to be statistical averages. Furthermore, combinations, multiples, and roots of "constants" are logically also "constants" which may furnish information. For instance, the square root of the speed of light, $(c)^{0.5}$, about 173 Kc/sec., might be a very interesting frequency, as it is the value at which frequency and wave length will have the same "Absolute value."

This model does not consider positive and negative charges as mysterious, accepted things of nature, but as manifestations of the rotation, inversion senses of vortex oscillators. As such, they are not constant values....
\
Where the Standard Model considers the units found as results of atom-smashing experiments as somehow being fundamental particles released by the experiments, this model would imply them to be different, alternative states of matter created in the experiments. That is artifacts, rather than fundamentals.

This is by no means a conclusive listing of the differences in philosophy and attitude of this model from the conventional situations.

In presenting this framework, this person, is not asking that all the ideas and information collected by all the geniuses who have contributed in the past be discarded. He is simply suggesting that this model may be a frame work into which profitable re-examination of data and ideas could be fitted,

Remember the Oscillator/Substance Google Group is an open membership group where you can post "Kudos" or "Boos." I hope that this presentation will receive feed back.

Reply to author   Forward

Hugh V View profile   More options Jun 28 2010, 10:26 am

Yep, Doc,
You certainly do end up very much out of step with the conventional situation. It's about time someone did. What a confused mess scientific theory is in!

Have you noticed that the two experimental results that you focus in

on, "MM" and Planck are where Einstein's Space Time, Quantum
Mechanics, and String Theory all have their start, except that you're
interpreting from the opposite side from where they are?

In an old copy of Astronomy Magazine there is a comment about
something called the space ship mystery or some such thing. Seems
there is some mysterious force that is braking space ships that are
going away from Terra in opposite directions by the same amount. Seems
this would fit, the ships if they don't get destroyed in the meantime
by collision with something else would be eventually turned around and
sent back to their origin?? Seems to fit with the idea of a substance
which tries to balance. Maybe?
Good luck with the talk. Who knows, someone might actually listen.
HAV

On Jun 22, 3:56 pm, ESKI <deanlsincl...@gmail.com> wrote:

- Show quoted text -

Reply to author   Forward   Report spam

ESKI View profile   More options Jun 30 2010, 11:29 am

Hugh,
You point out an interesting fact that I had sort of overlooked. The
currently fashionable, or at least for the last few years,
fashionable, theories, all have at their base information from the
Michelson-Morely and Planck's work. It would appear that they, and we,
are agreed that there are basic answers hidden there, However, the
others all interpret the results in one way, we use an almost opposite
version. \

 They were trying to invent a theory while we were just following up a
string of logic to see where it would go.

It could be that there is a definite point in this that needs to be
somehow gotten across,,,the answers are probably in front of our
noses, if we are looking in the right direction, in the right way.
Maybe that "Theory of Now" is somewhere around what we are talking.
It seems to me that what has developed so far seems to explain more
things and make more sense than what the "real geniuses" have come up
with in the last 100 years or so.

The idea from Astronomy Mag.is interesting. We could use some good
astronomers in this group , or an astrophysicist or two.!!

 I just sent in for a "Great Courses" copy of "The Joy of Mathematics"
which is a popularization of math. which takes one up through Calc.
I'm hoping that I can go thru it and get some of my old math. training
back in focus. I was quite good with Calculus for about 6-7 years
after I took the course, but that course was about 1952-53. I..don't
know if any of that will help with fitting the work to the cosmos, but
it can'[t hurt....

On Jun 28, 10:26 am, Hugh V <hughv...@gmail.com> wrote:

- Show quoted text -

Reply to author   Forward

End of messages

Google groups

« Groups Home

Oscillator/Substance
⊞ Theory

[   ] [Search this group] [Search Groups]

Vigier VII Symposium   Options

5 messages - Collapse all - Report discussion as spam

hoek View profile   More options Jul 2 2010, 1:30 pm

Hello again, Dr. Sinclair,
    In your prompt, inconsiderate, dismissive and unacceptable
response to the below paragraph, which I ask that you read in your
upcoming Vigier VII Symposium address. An address that you asked all
members to help you with as I did below:

It is my pleasure to welcome the Vigier VII Symposium to the Aberdeen
American News here in Aberdeen, South Dakota, USA and to be able to
thank the News and Dr. Amoroso for this opportunity to introduce to
the wider scientific community a framework for a Comprehensive Theory
which seems to be applicable at any scale from sub-atomic to cosmic.
I also wish to thank my friends from two Internet Groups who have
contributed information and support over the last three years since
the first version of this model was published on Helium.com as
"Motion
in a Matrix..." The current version has a working title of the
Oscillator/Substance Model, as it has become clear that the motions
that are involved are in part primarily, if not totally, oscillatory
in nature. One of the groups which I need to thank is the Condensed
Matter
Nuclear Science Group, cmns, which has furnished much information
including the fact that this Symposium existed. Thank you, Jean
Pierre
The other group of which I, Dean LeRoy Sinclair, am founder and
director of is the Oscillator/Substance Theory Group. At
http://groups.google.com/group/oscillatorsubstance-theory
  Members, of our group all seek the truth about the nature of the
realm in which we find ourselves existing. We all believe that the
basic mechanics that govern the universe and the workings of the sub-
atomic, quantum, world are oscillatory in nature and mirror each
other
at the largest and smallest of scales. We come from different
backgrounds, with various approaches and models to try to explain
this
phenomenon. >One member of the group, who has a somewhat different

> version from my own, of which I'll return to in a moment, is at
> protoncosmology.com. His model agrees with the current Rutherford-
> Bohr atomic model, but, not the currently accepted quark nucleon
> hypothesis. His model takes an electro-magnetic like, field dynamics
> approach based on neglected and recently verified electron-proton
> scattering data. The model poses the proton as an internally driven
> spherical oscillator, an electro-dynamic like perpetual motion
> machine. It is driven by the conversion of charge energy into mass and
> back again by the function M=E/C2, which is a simple transposition of
> E=MC2 and the fact that no particle with mass can attain the velocity
> of light. Thus, the field deflections always precede the oscillating particles. The full
version, which is too lengthy to describe now, can
> be found on line at " http://www.protoncosmology.com".

You Respond,
"This I will probably find some way to use in the "write up" for
publication,
its too technical a coverage for here, I think that it is better to
note
that you have a different version which may possibly be preferable,

give
your name and the references."
    "Too technical a coverage for here", Sure Eski , much too technical
for a group of scientists to understand. Dean, I really think they'll
be able to handle it and probably they'll be accompanied by someone
who'll explain it to them, if they can't.

  Anyway, I realize that it cost you 75 Euros to enter the Vigier VII
Symposium and I will contribute (pay to you) 25 Euros if you'll just
read the paragraph from, >One member, to " http://www.protoncosmology.com".
above, the way I've written it. I timed it 5 times. I read it aloud,
moderately, concisely, annunciating each word and it took on average,
one minute and 16 seconds for all of it, the longest reading being one
minute and 22 seconds. Send me your mailing address or paypal or
however you would like me to pay you for 1/3 of your submission cost
and I'll pay you the day I received it. By the way, I and I'm sure
the other members of the O/S group, would like to know the exact time
you'll be broadcasting to the scientists attending the symposium?
What are the call letters and frequency of the station you'll be
broadcasting from?
I hope you'll find it in your heart to be a fair, responsible and just
group leader. Remember, I'll pay you for a third of your time for
just under a minute and a half of it.
Stay well,
hoek

Reply to author   Forward   Report spam

hoek View profile   More options Jul 9 2010, 4:11 pm

    ATTENTION, ATTENTION!!!!! Where is Eski????
It's been a week since my last posting in which I asked our group
leader to post the time, date and broadcast frequency of his address
to the Vigier VII Symposium. There has been no response and the
symposium starts in only three days!!!! Is he alright, does anybody
know??? It's not like him to be so long in responding to a post,
especially one as important as this. In the past, he's usually
responded to my postings within in a few hours. DOES ANYONE KNOW IF
HE'S ALRIGHT, HUGH, KA-SALA ????

Reply to author   Forward   Report spam

dean sinclair View profile   More options Jul 10 2010, 4:08 pm

Sorry, Hoek. Haven't had a chance to check on things. Was out of town on
family business for a week, no computer available... We just, finally got
a date and time, 1500 hrs.. London Time. on Wed. the 14th. I think that that
is 9 AM. here in SD.

 Check the Vigier VII Symposium site, As to "braodcast frequency" and
address at/to the Symposium, that probablly won't be available until
Nonday. Seems that my contact there is doing everything at the last minute,
like everyone else.....Will try to keep you people informed. " Eski"

- Show quoted text -

Reply to author   Forward

ESKI View profile   More options Jul 15 2010, 9:05 am

Well, Folks
,The VigierVIISymposium has come and gone, and, as we might have
expected, if anything can go wrong it will.

  It did.

  I don't know what happened; but, on July 13, I got word :
"Sorry ...can't get our act together this time (Internal
problems),,,," I wrote back, sending a copy of my final short script,
asking if someone else could give the talk for me so that the time
slot, 1500 hrs., Wed, July 14, 2010, London time, would not be empty
and told my correspondent, in a separate communication, "Como jefe
de este symposio tu tienes la responibilidad ultima por qualquier

'chingandos de patos' que pasan...sea que tu puedes dar la lectura
para mi en person?" In blunter English, this adds up to "You were the
boss, so you bear the ultimate blame for any foul ups (I used the
Spanish translation for an certain English vulgar, slang expression
for messing up thoroughly) so how about you giving the talk?"

Answer this morning July 15, was, "Sorry, we couldn't accommodate....
time to concentrate on (the write-up for the 'Proceedings') ....make
sure it is well referenced...."

Well, there it is.

We got a little exposure as the "abstract" is still posted on the
Vigier VII site, and, possibly in part from there, we seem to be
getting consistent hits here on the site of a hit or so an hour.
When I signed in it said, "100 web views in the last 4 days." Now I
have the problem of getting the final paper done and in by Nov. 1. I
am going to try to put a personal deadline of Sept. I on it,

Incidentally, have not yet received an address to which to send the
final draft. The paper is supposed to be "in Microsoft Word with a
pdf file for comparison" what ever that means? Can someone
translate?

Socio-economics of science... If there had been any way to have a
representative of our group there, we could have probably done a
couple of papers.... As a research group, we are definitely
underfunded.

Well, " back to the drawing board. "

Of course, if I were being a bit paranoid about it, I'd think that we
ran into some scientific politics and got deliberately stalled. There
is more and more information that is showing up that says that O/S is
probably very close to "right on." Some part of the "Establishment"
could possibly have done a sabotage job to stall this set of ideas
giving more time for damage control on the "current frameworks."
After all, no one wants to have some upstarts point out that a lot of
the theory of the last 100 years is "somewhat off base." (A toggle up
of Quantum Mechanics to cover what we are saying might well earn
someone a Nobel Prize and the ever-lasting gratitude of the
"Scientific Establishment." There are rumours of a Russian Group that
seems to be on somewhat of this course.)

An instance of more information showing up is the following bit of
information: Some work that had a preliminary publishing in May says
that certain units, which are shown to oscillate between "Matter and
Anti-matter" states, spend less time in the Anti-matter state. We
can predict this by saying, "In our Universe the expression of the
proton-electron, anti-proton-positron states which we call matter, is
read as 'protons and electrons.' The expression of smaller vortex
oscillators, in the larger vortex oscillator which we call "the
Universe" will be affected by the rotational/inversional sense of that
oscillator..."

It is a bit disgusting to have spent so much time in trying to get a
short talk done. At least 40 hrs. ;or more of typing and retyping,
writing and rewriting. The result probably wasn't that great, but it
is frustrating!!! Eski

On Jul 10, 4:08 pm, dean sinclair <deanlsincl...@gmail.com> wrote:

- Show quoted text -

Reply to author   Forward

dean sinclair View profile   More options Jul 15 2010, 9:45 am

Nope, Ka-Sala, we beat that deadline. We were supposedly "on" to give the
talk up until almost the last moment: I could post the interchanges back and

forth for the several days as we were trying to set up the teleconferencing net, but consider e-mails as semi-confidential.

We were OK from this end; but. apparently, either they couldn't get a link into the Conference room in London or someone said, "We have some inconvenience in setting this up so, why should we waste our time and effort on what clearly is some Crackpot nonsense." I hope that the comment, "internal problems," meant the former. The fact that we did have--as far as I know, still do have--an abstract posted and had a time slot reserved (those can be checked on the VigierVIISymposium site) tends to indicate that we did everything we could from this end, and, at least at some point, we were scheduled in.

. The collapse of the "talk" was some problem at their end, not ours. We honestly tried. DS

- Show quoted text -

Reply to author   Forward

End of messages

« Back to Discussions   « Newer topic   Older topic »

Gmail Calendar Documents Photos Reader Web more ˅   deanlsinclair@gmail.com ˅

Google groups

« Groups Home

Oscillator/Substance
⊞ # Theory

[Search this group] [Search Groups]

Home

Discussions
 + new post

Members

Fwd: Can someone else read a script into the 1500 slot schedule so that there isn't a "blank" space where I'm supposed to be?
Options

About this group

Edit my membership

Group settings

Management tasks

Invite members

1 message - Collapse all - Report discussion as spam

 View this group in the new Google Groups

dean sinclair View profile   More options Jul 15 2010, 10:01 am

Sponsored links

People, here is the e-mail that I sent to the Symposium director when he wrote, "Sorry to disappoint after...." DS

Maid Service Software
Easy-to-Use. Web-based. Most Widely
Used by Maid Services - Free Trial
ServiceCEO.insightdirect.com/Trial

- Hide quoted text -

---------- Forwarded message ----------
From: dean sinclair <deanlsincl...@gmail.com>
Date: Tue, Jul 13, 2010 at 10:29 AM
Subject: Can someone else read a script into the 1500 slot schedule so that

Collaboration Software
Simple and secure collaboration
for leadership teams
www.boardvantage.com

there isn't a "blank" space where I'm supposed to be?
To: noet...@mindspring.com

Project Management Tools
Manage Projects from Anywhere.
Register Now For a Free Demo!
www.Tenrox.com/ProjectManagement

Dr. Amoroso,: Assuming that the "Internal Problems" of which you speak are technical problems of scheduling and getting a Teleconferencing set up, . rather than some sort of Scientific Politics problem, is it possible that some one could read the following into the record at the scheduled time? This would not be considered an endorsement of the ideas but merely an endorsement of the principle that ideas are more important than the speaker.

See your message here...

Cheers, Dean Sinclair

This is the final script that was intended to be used for the oral version of the paper that was scheduled for the 1500 slot on Wed. July 14, 2010, by Dean L. Sinclair speaking from the Aberdeen American News in Aberdeen, South Dakota, US of A to the Vigier Symposium in London. Difficulties at the London end prevented Dr. Sinclair from being able to present this material.

It is my honor to present some information about a model called. the Oscillator Substance Model which may provide a framework for a comprehensive theory uniting the fields of physical science.

The basic tenet is, " All existence si the result of sequential action-reaction-action interactions within a Substance/Substrate of undefined extent and undefined basic composition. A Substance/Substrate which may be considered as if it be a liquid at the triple point, able to respond to slight pressure differences as any of the three basic phases of solid, liquid or gas.

The continuous, sequential equilibration within the substance results in constant motion such that the system is composed of/controlled by oscillators.

Some of the oscillators are vortexes having long term stability. these vortexes, the electron and proton and their mirror units, the positron and anti-proton, interact to form what we know as matter.

This view produces valuable insights that often differ from the conventional viewpoint by 180 degrees.

Positive and negative charges are seen as the result of reverses rotation/inversion senses of vortex oscillators. As these vortex units have mass and radius limits--and, corresponding frequency limits, charges will vary from a maximum value to zero and back, Charges are not fixed values, but have limits and an an average.

As the vortexes responsible for charges have determinable limits, their sizes and shapes can be estimated, The results of these determinations have interesting results for the theories of atomic structure.

As the equilibration process in a substance can be considered to result in constant pressure adjustment. Pressure fits the criterion for a true force; therefore, the various Forces of Nature can be seen to be a result of interpretations of pressure adjustments.

The "Missing Anti-matter" Matter has a double explanation. The separation--or inversion instant --of an oscillator, which we know as the "Big Bang" resulted in the definition of two oscillator halves, having reversed rotation/inversion orientations. Smaller oscillators, within these halves will be influenced by the larger oscillator, the orientation of one half will tend to be stretched, the other compressed, so that one rotation/inversion will tend to be expressed differently than the other. In our Universe, it appears that the stretched form which appears most obviously is the electron. Its "almost identical' mirror, the "positron," appears to be somewhat suppressed.

If one half of a separable oscillator be considered "Matter," and the other half be "Anti-matter," then, as the electron is always considered matter, the positron is anti-matter. The two units are logically halves of a separable oscillator, to which they rejoin in the "annihilation" process.

The electron and proton are halves of another separable oscillator, the neutron. The electron is still "Matter," hence the proton, as the other half of a separable oscillator, is ANTI-MATTER.!
Since the neutron is "neutral" it may be considered as either Matter or Anti-Matter and, like the B sub s Meson, could probably be shown to invert between the two states.

From the foregoing, we see that what we call "Matter," combinations of electrons and protons, are actually combinations of the "Matter" electrons, and the "Anti-matter" protons.

Logically, there is, somewhere, an "anti-Verse" where the rotation inversion dominant expressions are the opposite of ours, but, also, we have no truly "Missing Anti-Matter," we only have semantic confusion. The "neutron count" of an atom can be considered simply as the number of nucleons that , at any given instant, are in anti-electron, anti-proton states.

Considering the electron and proton as vortex oscillators which can associate gives a clue as to why the Hadron Collider apparently breaks down soon after starting up. Vortex oscillators can not only associate with different vortexes, but also self-associate. Electron-electron association has been long known. However, no one seems to have realized the same to be true for protons. Additionally, conventional science gives no hint of the possible existence in "vacuums" of pulsator-oscillators some of which may be separable into electrons and positrons and deformable into neutrons.

Pushing a stream of mutually repulsive "Charged Particles" through a void, is very different from trying to control vortexes whch can self-associate through a possibly-reactive medium.

If the O/S --Oscillator/Substance--view be correct, the Hadron Collider, designed to be a sophisticated particle accelerator, may well be acting for a short time as a rather primitive fusion reactor before feed-back causes a break down.

What the basics of this framework are have been stated, and a few implications covered. A few words about the start of the ideas leading to the O/S Model and the basic reason for its almost reversed view from the conventional may help.

This model started to develop quite innocently in the Spring of 2004 with
the realization that basic ideas of Einstein's Special Relativity fit into
communication theory, where they would apply to any Perceptual Universe
defined by a maximum, practical velocity of information transfer, whether
that velocity be determined by Pony Express Riders or Electromagnetic Waves.

Since, in every case, practical maximum velocity of information transfer is
going to be a bit less than the average speed of the packet carriers, the
Speed of Light, is logically an average which acts as a practical maximum
velocity of information transfer.

By the Summer of 2008, when the Oscillator/Substance Google Group was set
up, follow-ups on the initial insight noted above, had led to the
realization that there was a "T.O.E" available, as outlined at the start of
this talk, which would have been seen a Century ago had the Mickelson-Morley
Experiment which determined the Speed of Light been reversed in
interpretation from ruling out an "Aether," to partially defining an Aether.
If then, a few years later, Planck"s Constant had been considered a Constant
of Angular Momentum and used to define characteristics , of that Aether,
this model could have come into existence 100 years ago.

Equating Planck's Constant,"h," to its definition as an angular momentum
and evaluating the resulting equation at the Speed of Light, "c." leads to
the equation, $m \times r = h/c = r \times m$., This arises from the fact that one
definition of angular momentum is the resultant of a mass , m, rotating at
a radius, "r," from a point, with a tangential velocity, "v.;" As Planck's
Constant applies at the Speed of Light, it makes sense to evaluate at the
speed of light and to simply by dividing out that speed from the left side
of the equation to form a ratio constant, "h/c." The resulting equation, m
$x r = h/c = r \times m$ is an example of a common, very valuable relationship in
physics, the law of levers, the balance law used in weighing, the law of
conservation of momentum, the law of conservation of energy. Here it can be
used to determine the oscillator limits for a family of constant torque
oscillators , defined by the set,$\{m \times r = h/c = r \times m \}$, with a torque of
h/c and inversion at the state where $r = m = $ square root of h/c. In the cgs
system, this value is about 4.7×10 ^-19 grams at 4.7×10^-19 cm.

This implies a hidden half of any basic oscillator which is smaller than 4.7
$x 10$ i^-19 cm.
Coming to these two basic sets of data from the opposite view of the more
standard theoretical approaches such as Space-Time, Quantum Mechanics and
String Theory, this model has a reversed orientation on maniy issues. Very
heretical, it asks for re-examination of the accepted percepts of modern
physical theory.

It may, however, turn out that this model will be complementary to much
theory rather than contradictory. In its definition of Mass as a measure of
the tension-pressure at a surface of the point-centered motions within that
surface, a characteristic of entities that is measured by comparison, and
suggesting that the term, "Energy," usually means a measurement of a
package of motion which includes a point and its associated motions along a
line, a unit whose effects are usually observed as the results of
collisions, that is. "Kinetic Energy," it appears that this model tends to
focus on the "Mass" aspect , whereas most theoretical approaches focus on
"Energy" for the most part, and considering " Mass" as generally a
constant value of some sort.

There is far too much too much information developed from this model and
clolsely associated ideas which cling easily to it as a "Framework." to
even begin to cover in this short presentation.
I refer you to the web site of the group previously mentioned,
Groups.Google.com/Oscillatorsubstance (written as one word) hyphen theory.
Where most of the extant material has been collected as "pages" which vary
in size from a half-page to 23 pages and counting.

There, also, you can meet some interesting people including the Canadian,
Al Zeeper, who has an analysis of of the various mathematical energy
expressions. There are the two ladies of the group, "Nish Laverz," a young
writer from the North of England, and "Ka-Sala." the most philosophical of
the group, and Group Manager, who is from the South of Australia. There
is.also, Robert Kardien Vanderhoek, "Hoek," who has another version of an

Oscillator theory which he considers superior to the version presented here, it can be examined on the Internet as "Proton Cosmology. "

I hope that the rest of the crew do not take offence that they are not mentioned here. They, each and every one, are separate and interesting personalities. !

The site is open membership, anyone can join-- any of you who'd like to pat me on the back or kick some other part of my anatomy is welcome to do so there. I hope to get some feedback, and "see" some of you there.

I'd be remiss, also, if I did not thank Jean-Paul from the cmns, Condensed Matter Nuclear Science Group, for the "heads up" about this Symposium which led to this paper. The cmns group , of which I have the honor of being a member, is a widely diverse, international group of experimentalists, theoreticians and others who are interested in the area which I personally think of as "sub-atomic chemistry." I expect that, in the not too distant future, out of that group and its associates will come some of the most valuable scientific information of our time.

Although, in a sense, I am a representative of the two groups mentioned above, I need to emphasize that the ideas and opinions expressed are my own and any errors and misinterpretations are strictly my own responsibility.

Thank you, Dr.Amoroso, for all your help and for inviting me to participate in this Symposium. I regret that I could not attend in person.

Thanks, everyone for taking the time to stay with me in this brief examination of the Oscillator/Substance Model as a possible Framework on which to Build a Fundamental Theory.

"

Well, there it is.

We got a little exposure as the "abstract" is still posted on the
Vigier VII site, and, possibly in part from there, we seem to be
getting consistent hits here on the site of a hit or so an hour.
When I signed in it said, "100 web views in the last 4 days." Now I
have the problem of getting the final paper done and in by Nov. 1. I
am going to try to put a personal deadline of Sept. I on it,

Incidentally, have not yet received an address to which to send the
final draft. The paper is supposed to be "in Microsoft Word with a
pdf file for comparison" what ever that means? Can someone
translate?

Socio-economics of science... If there had been any way to have a
representative of our group there, we could have probably done a
couple of papers.... As a research group, we are definitely
underfunded.

Well, " back to the drawing board. "

Of course, if I were being a bit paranoid about it, I'd think that we
ran into some scientific politics and got deliberately stalled. There
is more and more information that is showing up that says that O/S is
probably very close to "right on." Some part of the "Establishment"
could possibly have done a sabotage job to stall this set of ideas
giving more time for damage control on the "current frameworks."
After all, no one wants to have some upstarts point out that a lot of
the theory of the last 100 years is "somewhat off base." (A toggle up
of Quantum Mechanics to cover what we are saying might well earn
someone a Nobel Prize and the ever-lasting gratitude of the
"Scientific Establishment." There are rumours of a Russian Group that
seems to be on somewhat of this course.)

An instance of more information showing up is the following bit of
information: Some work that had a preliminary publishing in May says
that certain units, which are shown to oscillate between "Matter and
Anti-matter" states, spend less time in the Anti-matter state. We
can predict this by saying, "In our Universe the expression of the
proton-electron, anti-proton-positron states which we call matter, is
read as 'protons and electrons.' The expression of smaller vortex
oscillators, in the larger vortex oscillator which we call "the
Universe" will be affected by the rotational/inversional sense of that
oscillator..."

It is a bit disgusting to have spent so much time in trying to get a
short talk done. At least 40 hrs. ;or more of typing and retyping,
writing and rewriting. The result probably wasn't that great, but it
is frustrating!!! Eski

On Jul 10, 4:08 pm, dean sinclair <deanlsincl...@gmail.com> wrote:

- Show quoted text -

Reply to author   Forward

dean sinclair View profile   More options Jul 15 2010, 9:45 am

Nope, Ka-Sala, we beat that deadline. We were supposedly "on" to give the
talk up until almost the last moment: I could post the interchanges back and
forth for the several days as we were trying to set up the teleconferencing
net, but consider e-mails as semi-confidential.

We were OK from this end; but. apparently, either they couldn't get a link
into the Conference room in London or someone said, "We have some
inconvenience in setting this up so, why should we waste our time and
effort on what clearly is some Crackpot nonsense." I hope that the
comment, "internal problems," meant the former. The fact that we did
have--as far as I know, still do have--an abstract posted and had a time
slot reserved (those can be checked on the VigierVIISymposium site) tends to
indicate that we did everything we could from this end, and, at least at

some point, we were scheduled in.

. The collapse of the "talk" was some problem at their end, not ours. We honestly tried. DS

- Show quoted text -

Reply to author   Forward

End of messages

« Back to Discussions   « Newer topic   Older topic »

Google groups

« Groups Home

Oscillator/Substance
⊞ # Theory

Home

Discussions
 + new post

Members

About this group

Edit my membership

Group settings

Management tasks

Invite members

 View this group in the new Google Groups

Matter-Anti-matter Annihilation   Options

7 messages - Collapse all - Report discussion as spam

ESKI View profile   More options Jul 27 2010, 3:32 pm

Hi, everybody,
I have a question that I need an answer for, It is:. IS THERE
ACTUALLY, ANYWHERE, A VERIFIABLE CASE OF MATTER-ANTI-
MATTER
ANNIHILATION AT THE LEVEL OF THE hYDROGEN ATOM OR
ABOVE?

It seems to be dogma that "Matter" and "Anti-matter" annihilate on
contact to revert to pure "Energy."

Our work suggests that this is nonsense, and in fact, "Matter,as we
know it, is made up to a great extent of a unit, the electron, and
another unit, the proton, which could be considered as, respectively,
"matter" and "anti-matter" units by a logic which I have outlined
elsewhere. In fact, there seems to be evidence developing that the
"nuetron count" of an atom is actually a count of the number of
nucleons which are at any given instant in a
state of "anti-electron, anti-proton" i.e. an "Anti-matter state."
By this view, everything is made up of a balance of what we consider
as "Matter," (electrons and protons) and Anti-matter (anti-electrons
and anti-protons.)

The point is that I can't find a thing in the literature of a true
case of matter-anti-matter annihihaation other than the known electron-
anti-electron situation which , aat least by my reasoning, is a
combination "reaction" to form another ocsillator rather than a
distruction to wave motion (electromagnetic radiation.)

IF SOME ONE HAS THE TIME, COULD THEY DO A LITERATURE
SEARCH ON ANTI-
MATTTER AND SEE IF, ANWHERE, THERE IS A VERIFIABLE CASE OF
ACTUAL
"MATTER-ANTI-MATTER ANNIHILATION' FOR ANYTHING OTHER
THAN THE ELECTRON-
ANTI-ELECTRON CASE? On a quick scan, I could find nothing....

seems to me this might make a basis for a good science article for
some aspiring wirter, with a title something like "Anti-Matter
Myths..."

Thanks, Dean (Eski)

Reply to author   Forward

ESKI View profile   More options Jul 27 2010, 3:33 pm

Hi, everybody,
I have a question that I need an answer for, It is:. IS THERE
ACTUALLY, ANYWHERE, A VERIFIABLE CASE OF MATTER-ANTI-
MATTER
ANNIHILATION AT THE LEVEL OF THE hYDROGEN ATOM OR
ABOVE?

It seems to be dogma that "Matter" and "Anti-matter" annihilate on
contact to revert to pure "Energy."

Our work suggests that this is nonsense, and in fact, "Matter{,as we
know it, is made up to a great extent of a unit, the electron, and
another unit, the proton, which could be considered as, respectively,
"matter" and "anti-matter" units by a logic which I have outlined
elsewhere. In fact, there seems to be evidence developing that the
"nuetron count" of an atom is actually a count of the number of
nucleons which are at any given instant in a
state of "anti-electron, anti-proton" i.e. an "Anti-matter state."
By this view, everything is made up of a balance of what we consider
as "Matter," (electrons and protons) and Anti-matter (anti-electrons
and anti-protons.)

The point is that I can't find a thing in the literature of a true
case of matter-anti-matter annihihaation other than the known electron-
anti-electron situation which , aat least by my reasoning, is a
combination "reaction" to form another ocsillator rather than a
distruction to wave motion (electromagnetic radiation.)

IF SOME ONE HAS THE TIME, COULD THEY DO A LITERATURE
SEARCH ON ANTI-
MATTTER AND SEE IF, ANWHERE, THERE IS A VERIFIABLE CASE OF
ACTUAL
"MATTER-ANTI-MATTER ANNIHILATION' FOR ANYTHING OTHER
THAN THE ELECTRON-
ANTI-ELECTRON CASE? On a quick scan, I could find nothing....

seems to me this might make a basis for a good science article for
some aspiring wirter, with a title something like "Anti-Matter
Myths..."

Thanks, Dean (Eski)

Reply to author   Forward

Discussion subject changed to "Micro-Black-Holes & Matter-Anti-matter A

Jack O Suileabhain View profile   More options Jul 28 2010, 6:23 pm

DEAN you KILLJOY(not)~;-) Hey; If we blow the 'anti-matter' cover then
Star-Trek get's CANCELED. . .

AND: Aren't the super-colliders racing a micro-thimble's worth of Anti-
Matter around the track so that a virtual ARMY of
ping-pong playing & tenured physicists can maintain their credentials and
PAYCHECKS? Mostly this is an elaborate & intentional 'campaign of
disinformation; methinks.'

HEISENBURG'S 'Uncertainty' is having big-play in this global CHARADE;
now you see the wee lil' Anti-Matter(s) & Now you DON'T.

MOSTLY-------> It's never been anything but DON'T.

When Galaxies (Gyro-Toroidal energy-mass bodies with Singularity Centres-balanced 'grey-holes') reach light-speed at the Universe-bubble border
their respective hub-grey-hole singularity centres dialate to BLACK-HOLE status at the AexoDarkEnergy-Hyperspace frontier and to the fanfare of MEGA-GAMMA-RAY-BURSTERS
are 'eaten' back into parallel-adjacent AexoDE-Hyperspace.

Hadron-Cern copies the above process using super-energized-accelerated Protons aka (MICRO Gyro-Toroidal energy-mass with Singularity Centres-balanced (micro-grey-holes').
At the appropriate quasi-light-speed acceleration point the Proton-grey-hole-Centre dialates to MICRO-Black-Hole Status and SUCKS itself-shut to the fanfare of a MICRO-GAMMA-RAY-BURSTER(notable) energy release. . . . and this process is that which was touted as the FICTIONAL 'anti-matter-annihilation energy-release' phenomenon. But MUCH energy IS being released but NOT from fictional anti-matter-annihilation.

?Why does not said Micro-Black-Hole-Singularity EXPAND dangerously.
 EASY. The hugely-overwhelming proximal PLANETARY MASS-PROTON dynamic equilibrium simply SQUASHES-SHUT the Einstein-Rosen Micro-Black-Hole access to parallel-adjacent AexoDE-Hyperspace.
 This could be the LAW of AETHYR-M-BRANE PROXIMAL ELECTRO-GRAVIONIC FIELD HYPER-TENSOR EQUILIBRIUM(especially near High-mass Bodies; like planets AND even being near a 'free-black-hole' the PROXIMAL M-BRANE TENSOR FIELD would 'preclude' any 'other' relatively 'near-by' singularity formation.)

EVEN the same parallel macro-phenomenon of a Fission-Bomb's Plasma-Gyro-Toroidal creates the 'micro-singularity-centre' which ingresses the classic INGRESS MUSHROOM PILLAR, which initiates a FUSION PILLAR within the FISSION RELEASE GYRO-TOROIDAL MAELSTROM.
 And even this notable (miniscule but relatively large/to-us) ingress of parallel-AexoDE-Hyperspace plasma is QUICKLY SQUASHED-SHUT by the Solar-System/Planetary Electro-Gravionic Field tension upon our immediately surround 'space-M-brane.'

LOS ALAMOS: Manhattan's 'Teller etc.' thought that the 'bomb's' reaction would 'cascade-open-exponentially' and consume the ENTIRE MASS OF THE EARTH and maybe the solar system.

But 'Singularity' is the MODEL & RULE of all 'Mass-Matter-Gyro-Toriodal' constructs in the Bubble-Universe which accounts for pretty-much EVERYTHING. And the natural M-Brane/Aether Electro-Gravionic super-tensor field-state STABILIZES this COSMIC/AEXOCOSMIC fluid-dynamic energy circulation SYSTEM.

So obviously the above HIGHLY FUNCTIONAL MODEL pretty much excludes the so-called anti-matter & thusly anti-matter annihilation-model.

And because of the above stated LAW of M-Brane FIELD-TENSOR Equilibrium the notion of MORONS that Cern-Hadron is 'creating' so-called DEADLY BLACK-HOLE is mostly BS.

HOWEVER: Being able to FIRE-DIRECT a stream of BLACK-HOLE STATE ACCELERATED PROTONS aka 'Micro-Black-Holes' is probably DOABLE & would make ONE HECK OF A POTENT WEAPON>(maybe already in existence)< Cheers Jack Harbach O'Sullivan~:-)

- Show quoted text -

Reply to author   Forward   Report spam

dean sinclair View profile   More options Jul 28 2010, 7:43 pm

Hi Jack,
 As usual, we seem pretty much in agreement, although, again. as usual, I
don't quite understand your DeutsberTypeWordSprache.
                              Deutscher

The problem with the Hadron super-collider racing a pico-thimble full of
anti-matter around is that the darn thing just can't "get off the ground."
Odds are that it is broke down again, right now.

 The idea that protons can be accelerated in the same way as the more
massive units can is an error. Protons can, and do, associate with one
another; and, to make it worse, the proton is an inverting oscillator which
probably, under the right circumstances, can convert to an
Anti-proton--think conversion of the Hydrogen Molecular Cation to the
Deuteron--there are lots of things that can happen that the designers had
no
idea of..

.In the talk that I didn't get to give, i was going to note that there is a
possibility that the Hadron Collider, while designed as a sophisticated
particle accelerator may well be acting as a crude fusion reactor before
feed-back causes it to break down....

 The tenured professors may end up sweating a bit, if their jobs depend on
the Hadron Collider finding the Higgs Boson....'Tain't gonna happen....

Apparently, you have no better luck finding any authentic Matter-Anti-
matter
Annihilation material than I did.   Cheers, Dean

On Wed, Jul 28, 2010 at 6:23 PM, Jack O Suileabhain <

braghgoerin...@hotmail.com> wrote:
> DEAN you KILLJOY(not)~;-) Hey; If we blow the 'anti-matter' cover then
> Star-Trek get's CANCELED. . .

Horrors. But didn't Star Trek us DiLithium.....???

 - Show quoted text -

Reply to author   Forward

Discussion subject changed to "Matter-Anti-matter Annihilation" by ka-sala

ka-sala View profile   More options Aug 1 2010, 1:26 am
 QUOTE

> seems to me this might make a basis for a good science article for
> some aspiring writer, with a title something like "Anti-Matter
> Myths..."

> Thanks, Dean (Eski)

Hello Eski,
I know of one writer who wrote about all of this; and I have to agree
with Jacko here, despite you say you find some of his language
'DeutscerTypeWordSprache'. a bit hard to understand. Yet he is right
when he says...
QUOTE

>probably DOABLE & would make ONE HECK OF A POTENT WEAPON
(maybe already in existence)<

Plus as you have said, some of mine is #@!!***(....) which maybe of
interest to you that in the beginning, '70's, was done in a form of
hieroglyphics for protection. So breaking this down into English, I
am the first to admit much of what I say goes over the heads, due to
reading it according to what only the theorists claim. When in fact,
it it given with a wisp of insight into whatever subject matter is
being discussed. Perhaps it is this 'wisps' you need?

This is one reason why this section of the already written book, is
written in such a way, that hopefully, the energy within this
potential you seek, will only be used for peaceful purposes. It seems
to be the disregard of a lot in science, to keep things that way, once
anything seen can be used otherwise.

There is no intention of mine in wanting anything out of this - except
an intelligent mind to understand - and to put it into practice. It is
simply being held back due to the unstable factor of greed, intentions
of self worth, and warfare. It is no myth Eski, far from it! It
would be an absolute waste of time spending a life-time of blood,
sweat and tears on a myth; only to find the warring nature of many
turn it back on their fellow man.

So it remains... always oscillating, until the receiver is worthy of
this knowledge. Being a receiver itself, it naturally transmits this
energy, and the whole intention is for the benefit of mankind to
utilize and not annihilate one another in their race to the top
without full knowledge of how.

It would be like letting loose the knowledge of what created the known
atomic bomb - despite all the good which has come out of nuclear
medicine - the facts remain they did not first know the potential
energy of what they still so ignorantly call 'waste'. Instead now the
damage done is too late in understanding, what could have been... in
place of it's radio-active harm. One should not open the doors to some
energies, without first understanding it is all worth-while when used
correctly. And when no harm can come upon the heads of any human
being.

NASA seeks to make their Lightships - prototypes - but they have a way
to go before understanding enough in the direction in which we speak.

Getting back to where we were at the time of the ' VigierVIITalk' -
and what didn't happen - what are you wanting further to bring this
together the way you wanted it to be? It appears you are still
searching for more ? Have you personally had any outside feedback re
as far as you got prepared for this? Is this the reason you now need
more, and why you need an answer by further searching...
>A VERIFIABLE CASE OF MATTER-ANTI-MATTER

ANNIHILATION AT THE LEVEL OF THE hYDROGEN ATOM OR
ABOVE? <

Where are your intentions of this O/S leading to? I guess I am
asking, since you had it all ready to put forward?
Regards always,
ka-sala

PS. Don't overlook the WISP in Science Eski... I don't have the
Pleiades avatar for nothing. But NASA won't let me in! As they read
my report in the 80's... wouldn't it be an irony ? I have no proof
they copied nothing... when they had the lot!

On Jul 28, 6:32 am, ESKI <deanlsincl...@gmail.com> wrote:

- Show quoted text -

Reply to author   Forward   Report spam

ESKI View profile   More options Aug 3 2010, 12:13 pm

Ka-sala,
As always, you are apparently several steps--or is it light years ?--
ahead of this slow old man.Have to admit to being lost--again. Don't
recognize the acronym, WISP, and do not know about your Pleiades
Avatar and, somehow, what was your 80"s report slips my mind. Would
like to see what has been written about the myth of Anti-matter other
than my own little short article on Helium. I should check there for
the other articles on the same subject. Regards, "Eski"

On Aug 1, 1:26 am, ka-sala <irrir...@gmail.com> wrote:

- Show quoted text -

Reply to author   Forward

ka-sala View profile   More options Aug 4 2010, 7:22 am

http://www.sal.wisc.edu/WISP/
You might get through NASA's Link here...
https://dns.l4x.org/cddis.gsfc.nasa.gov

http://www.kevin.harkess.btinternet.co.uk/wisp_ch_7/wisp_ch_7.html

http://www.kevin.harkess.btinternet.co.uk/wisp_ch_5/wisp_ch_5.html

http://ffden-2.phys.uaf.edu/212_fall2003.web.dir/tyler_freeman/modern...

http://www.hypergeometricaluniverse.com/?p=2893

Don't know if you can bring up this next
one!
Wisp Unification Theory - almost the theory of everything

http://www-public.jcu.edu.au/news/JCUPRD1_060264

Not intending to lose you Eski... Just a little food for thought to
add to the subjects within the Light which you seek. Or should I say
energy ?

Don't worry 'bout Pleiades... Just a starry Cosmological Link.

But I am serious...

Many have gone down in history as having 'lost it'... but there is
anti matter as equally as there is matter. It is no Myth! Just as
everything has an opposite. In O/S you are looking for this force.
Right from the start, I brought this in. Nothing can change this, but
minds can confuse it.

It's like there is Light and dark/black Light energies. Just because
it reaches into nanos and quantum's it's still there, as even

annihilation leaves something in it's wake...   This is why I
mentioned the WISP (wisp) so lightly.

This link ties in with just the title alone.
'Antimatter Atoms could shed light on the universe.'
http://www.usatoday.com/news/science/2002-10-30-antimatter_x.htm

Draw from anything what you will Eski. I have never been one to
believe all theorists – just because it is written – or claimed (until
proven wrong.) I guess I have enough understanding to simply hold to
that... as I have all my life. And the same here is just as
applicable... Until proven wrong, I will just carry on.

It's very probable you have seen these sites, but then again, there
maybe just one small word which triggers your own search here.
Anything I say would not change where you are heading. I have been
there – in my way – and believe I have given it all in a nut shell,
and cannot take you further than you are, if you are not open to new
avenues.

I hope somewhere you can find more peace of mind within your
understanding. Not feel nor see yourself as some 'old man'... and
most of all, even think that some-one who may simply understand things
in a different yet realistic light, as not being light years ahead of
you, or any others. I just haven't waited this long for an intelligent
enough mind to grasp what I'm saying....

Quote. 'Truth is ever to be found in the simplicity, and not in the
multiplicity and confusion of things.'
*Izaac Newton
Regards until next time,
ka-sala

On Aug 4, 3:13 am, ESKI <deanlsincl...@gmail.com> wrote:

- Show quoted text -

Reply to author   Forward   Report spam

End of messages

« Back to Discussions   « Newer topic   Older topic »

Gmail Calendar Documents Photos Reader Web more   deanlsinclair@gmail.com

Google groups

« Groups Home

Oscillator/Substance
⊞ Theory

[Search this group] [Search Groups]

Matter-Anti-matter and the possible HH+ --> D+ transform. Options

4 messages - Collapse all - Report discussion as spam

ESKI View profile   More options Sep 23 2010, 1:41 pm

Fairly recently it has been shown that certain entities known as Beta
sub s in Standard Model Parlance undergo Matter-to-Anti-matter
oscillation....

It is not inconceivable that this type of oscillation is general
rather than confined to this type of special case.I

The consequences of such a supposition are interesting. Let us look at
one of the possible "beginning reactions" of the build up of the
atomic nuclides . This is the possible transformation of the Molecular
Hydrogen Cation to the Deuterium Cation. That is HH+ (p+, e-, p+) ---

>D+.

HH+, in our Universe would be expected to be a clockwise rotor with a
wobble.... If, in addition to the inversion-rotations of the two
protons in this dual centered unit there were also the matter/anti-
matter complication so that at any instant there might be the
situation not only , p+,e-, p+, but also, p-, e+, p- . and the other
possible combinations....

With continuing rotation and loss of vibrational and translational
motion to the milieu, it is conceivable that at some point the three
units eventually drop into a complete coordination of all the modes,
inversion, rotation and "rotation/inversion" (i.e. matter-anti-
matter "switch") so as to become one coordinated entity , the
Deuterium Nucleus, a basic unit of many other nuclides..

Is this Science Fiction Fantasy or Legitimate Scientific Speculation?
Who knows at this point.   Dean (Eski)

Reply to author   Forward

ka-sala View profile   More options Sep 27 2010, 7:24 am

Hello Eski,

In attempting to bring the stellar energy to earth, and in dealing
with matter or antimatter, the fact of fusion lies yet in the
inability to master it. With fragments of theory - instead of dealing
with the cold fusion facts - the obvious eludes due to the simplicity
of what really is.

OUOTE
< in addition to the inversion-rotations of the two
protons in this dual centered unit... >

Remember Hoek, and the Proton ?

QUOTE
< "(The point our Universe is at, given that size and time are
infinitely relative). The U particle has already been confirmed by
particle physicist, only they call it the "strange quark".>

Home

Discussions
+ new post

Members

About this group
Edit my membership
Group settings
Management tasks
Invite members

Whatever the Elements name, Deuterium or anti-deuterium, Hydrogen or anti-hydrogen, any matter or antimatter automatically meets in the middle, creating a 3rd. Call it the Third Law, or even a Muon Catalyst of the 3rd.degree. One could even say a 3 in 1.

When performed correctly, it is not Science Fiction Fantasy but Legitimate Scientific Speculation. I repeat your term of speculation, as you have, until it is mastered here. In all, there are 7 Principles which rule these Laws - and like the harmony of a scale - one could liken them to the Octave. The 1st. is the completion of all, when the 8th - the bridge - is the frequency of all combined. [As in the Octave.]

As you say...
QUOTE.

< With continuing rotation and loss of vibrational and translational motion to the milieu, it is conceivable that at some point the three units eventually drop into a complete coordination of all the modes, inversion, rotation and "rotation/inversion" (i.e. matter-anti-matter
"switch") so as to become one coordinated entity... >

Oscillation will always be, whether Matter or Anti-matter, or we would have nothing.
It is...

QUOTE
< one of the possible "beginning reactions" of the build up of the atomic nuclides...>

No-one need a big bang to safely work with cold fire. We only need a Light Ship to get back there! Nor do we have to blow ourselves up in the efort. And though I maybe the most philosophical of you all, which came first. Philosophy or Science ? In truth they are linked just as we are to the Cosmos.

In closing... did you check the 'WISP'?

Best regards,
Ka-sala

On Sep 24, 4:41 am, ESKI <deanlsincl...@gmail.com> wrote:

- Show quoted text -

Reply to author   Forward   Report spam

dean sinclair View profile   More options Sep 28 2010, 2:10 pm

KaSala,
Yes, a lot of what Hoek says makes sense...   Nope, i didn't check the WISP?
i do not know the WISP....D

- Show quoted text -

Reply to author   Forward

ka-sala View profile   More options Sep 30 2010, 8:01 pm

It is all linked Eski... WISP and all!
I gave you some links earlier to see... In the Matter-Antimatter
Annihilation if you want to know more to bring you up to date?

Your QUOTE
< Is this Science Fiction Fantasy or Legitimate Scientific
Speculation?
Who knows at this point.   Dean (Eski) >

Remember... what was once Science Fiction are now facts in many
fields.
Regards,
Ka-sala

On Sep 29, 5:10 am, dean sinclair <deanlsincl...@gmail.com> wrote:

- Show quoted text -

Reply to author   Forward   Report spam

End of messages

« Back to Discussions   « Newer topic   Older topic »

Gmail Calendar Documents Photos Reader Web more ·   deanlsinclair@gmail.com ·

Google groups

« Groups Home

Oscillator/Substance
⊞ **Theory**

[Search this group] [Search Groups]

FW: [Vo]:FZ-Quantum Transistion-LENR-Podkletnov-Casimir
Options

2 messages - Collapse all - Report discussion as spam

Jack O Suileabhain View profile   More options Oct 6 2010, 1:16 pm

Dean: The following is another 'short' that I 'woke-up' with this morning. . . Cheers; have a
good Autumn!~Jack~ aka Jack Harbach O'Sullivan/O'Suileabhain

* * *FRANK ZNIDARSIC & QUANTUM TRANSITION re: LENR/Podkletnov corollary* * *

*Quantum Transition root-ingress-Plasma as original pre-atomic energy state a la Frank
Znidarsic*

I've said this all before, but the data is beginning to agree with the posit of Protons' within
Electro-Valient Capacitance shells as micro-Singularity systems. This is an extension of
Einstein's Solar Voltaic electron-field as 'Capacitors'-model and the Casimir atomic
postulates dove-tailing with the original posit.

The corollary of "One Million Meters per second'/Quantum-Transition re. LENR results and
also Frank Znidarsic's finding the
same within the extended Super-Fluid-Toroid Podkletnov model->re. the Quantum-
Transition Velocity is 'huge.'

POSIT: Turn this around 'not' as back-tracking the phenomenon but rather as positing
Quantum-Transition speed-density(velocity) as a new insight rather as QUANTUM-
TRANSITION-ROOT-PLASMA that is formative energy state of 'ingress-plasma' into the
'eye' of the Proton(all protons) as balanced-worm-hole connected to said proton-eye as a
'each atom is a balanced singularity system.'

Corollary: The Einstein Solar Voltaic modeling of the Electron-shell(s) as an energy
quantum-whole-field CAPACITOR state of graduated familiar electro-valent energy multple
levels in onion-layer-like configuration etc.

Case in Point: This Einstein Solar Voltaic Electron shell(s)as Capacitor-Field(s) WORKS.
 And the Solar Votaic effect is proof of the pudding of this model by producing effective
technologies. And in the negative; the 'old model' of the electron-as-particle orbiting a fixed
proton making the atom mostly relatively 'empty-space' is a non starter. The whole-atom is
a circulating whole-energy 'field-system.'

A co-posit is that 'electron-flow' aka 'electrical current' is exactly like 'light-Photonic 'current'
& is a 'wave-form' that I have referred to as 'helicoid-wave-string' of Quantum-Electron
current via 'Quantum-Electron' Velocity-Density Momentum. Much is semantic, an
admittadly Frank Znidarsic tends to state the matter more succinctly. Hydrogens Electro-
Valent Capacitance 'shell' equals 'One Quantum Electron' and work up from there. And
Quantum-Electron wave-current helicoid-string flow would quantify as One Quantum
Electron would equal=> one Q-Elec. Helicoid-spiral-Wave from wave-crest-to-wave-crest;
ad Planck-Dirac angular momentum calculations etc.

But specifically I posit that the Frank Znidarsic observation of the 'unity' of the various
gradient(spectrum) offshoots of 'forces'(EM-Nuclear-Photonic etc) from the convergent 'One
Million Meters per Second,' indeed indicates here in is a Unified Source Quantum-energy
state of INGRESS-PLASMA from adjacent Dark-Energy HyperSpace. This Source-Parent-
Hyperspace is posited as a hyper-velocity/hyper-fluidic while hyper-dense quasi-infinite
Super-M-Brane state. And this is what we have been poking toward under the names of
Zero-Point-Energy &/or Quantum Vacuum which tend somewhat to misnomer because
Hyperspace is hardly a 'vacuum.'

'Singularity' is the model for all interflow fluid dynamic balance from the atomic micro-singularity level to the macro-cosmological level. This is including the original Hyperdimensional singularity that 'big-banged' our Bubble-Universe into inflation as well as a virtual-infinite-myriad of other universii-sister-bubblels more or less similar to our own. These sister-universii-bubbles are inflated via routine Hyperspace current dynamics that form hypervelocity-hyperdense-hyperfluidic swirling current super-eddies which are the birth torus's of bubble-universes like bubbles within a virtually infinie Hyperspace Champagne.

AND SO: When Jeremy posits the confinguration of Super-Fluid 'bagel/torus' reactors to achieve the Podkletnov-effect, he is definitely on the right track. And if the above model is correct this Hi-Density-EM generated Bagel/Torus/Toroid Reactor which is functioning like a 'SUPER ATOM' according to Znidarsic-&-Jeremy should get much-more than they are looking for.

The Pot-o-Gold of this model; should be that cross-spectrum/transdimensional field viscosty should also intitiate a parallel-hyperspace adjacent torus-field which should act like a virtual cross-dimensional clutch & pressure plate. This should tend to create a common-Einstein-Rosen incipient-worm-hole quasi-singularity connecting the Reactor Torus Eye with the kick-started parallel Hyperspace Torus. And this should elicit a super-Podkletnov effect as the 'bleed-through' Hyperspace super-velocity-density field while also producing hyper-gravity effects. At this point the bleed-through effect will reverse the bagel-EM-input into the 'reciever-inductee' reactor phase and becoming virtually a perpetually inducted EM-energy-field reciever from Hyperspace ingressed-QUANTUM-TRANSITION-ROOT-PLASMA. * * * And this is the 'point.' * * *

CASIMIR: Casimir indicates that the Proton-eye-singularity ingressed Quantum-Trans-Root-Plasma creates a Torus within a specific gyro-gravionic-centrific hypercompressed energy-shell-wall. The Proton's 'eye' as a balance-singularity connecting to adjacent Hyperspace would thusly create a micro-Dyson-Sphere-live gyro-centric-hi-gravionc 'shell' & likewise would be illiciting the Casimir-cavitation effects that housing a transdimensional-micro-singularity would indicate.

The axial-flow of this micro-energy-gyrotorus is indeed the flow-feed of the circulating-electro-valent shell(s)f outer Capacitor-acting-like onion-layered field. Within the Proton-gyro-torus-wall are correctly posited Casimir trans-temporal geometric effects considering that Hyperspace is a Virtual-No-Time/Virtual-No-Distance hypervelocity medium. Also this indicates that the quasi-Hyperspace-ingress of Quantum-Trans-Root-Plasma creates a High-Gravitic-axial-lobe-effect mirrors in parallel both the atomic-level as well as with our macro-level Super-Fluid Bagel/Torus Reactor.

A short jump should lead us to surmise that the hyper-gravitic Quantum-Trans-Root-Plasma around our Bagel-Reactor would tend toward trans-temporal manipulation a la' Casimir-Effect. And if this Super-Fluid Bagel Reactor were within a 'craft,' then we might expect it's flight characteristics to elicit quasi-Casimir-transtemporal distortion effects that would exactly mirror likewise posited Casimir effects at the atomic-proton level in parallel with the Bagel-Field's(quasi-macro-electron) shell-field-bubble of a Quantum-Transition-Root-Plasma 'gate' Reactor.

I'm thinking that this above outlines the ultimate destination-conclusions and R&D futures of what Frank Znidarsic has labeled the Quantum Transition Effect. And I'm calling it the Quantum-Transition-Root-Plasma Effect.

The Znidarsic LENR cross-corollary of 'One Million Meters per Second' Quantum-Transition velocity is indicating really cool possibilities for the further developement of inter related Cold-Fusion future strides.

Case in Point: The Super-Fluid-Bagel/Torus Reactors promise virtually-infinite access to cross-dimensional inducted power for global & regional powergrids. And for mass-global-transit such will also welcomely obviate high Carbon/CO^2 generating, JP-4/Jet Fuel guzzling Airline-traffic etc. Thus simply 'not-fouling' our precious atmosphere is worth the price of admission exponentially.

However: Practically for day-to-day, up-close-&-personal uses Cold-Fusion is the related technology of choice. We don't need 'large-transtemporal-effects' occurring under the bonnets of our personal vehicles for instance. . . . Super-Fluid Reactors &/or Cryo-Super-Conductor forms of same do not lend themselves to Personal Vehicle &/or home-dwelling functions etc. & compact-sophisticated functions. But Cold Fusion most certainly is the most promising avenue of application of these extended Quantum-Transition-root-Plasma principals for practical energy system(s) that are vital to our overall global energy uses for these types of applications relative to humanities' day to day needs & crucial compact-energy quasi-overunity technologies. In short; batteries will 'not' cut it for the future.

Methinks that this future is already happening and not 'Blarney.'~:-)-Cheers; Jack Harbach
O'Sullivan

dean sinclair View profile   More options Oct 6 2010, 3:32 pm

On Wed, Oct 6, 2010 at 1:16 PM, Jack O Suileabhain <

braghgoerin...@hotmail.com> wrote:

> Dean: The following is another 'short' that I 'woke-up' with this morning.
> . . Cheers; have a good Autumn!~Jack~ aka Jack Harbach
> O'Sullivan/O'Suileabhain

> * * *FRANK ZNIDARSIC & QUANTUM TRANSITION re: LENR/Podkletnov corollary* *
> *

> *Quantum Transition root-ingress-Plasma as original pre-atomic energy
> state a la Frank Znidarsic*

I'm going to have to try to make some translations into my vocabulary.
"pre atomic energy state swounds like my subsance/substrate.

> I've said this all before, but the data is beginning to agree with the
> posit of Protons' within Electro-Valient Capacitance shells as
> micro-Singularity systems. This is an extension of Einstein's Solar Voltaic
> electron-field as 'Capacitors'-model and the Casimir atomic postulates
> dove-tailing with the original posit

Yeah, so the " little proton" moving back and forth only a bit from the 4.7
x 10^-19 cm. radius is easily buried in the ce3nter of the much greater
volume electron.   Is that what all this verbiage says?

> The corollary of "One Million Meters per second'/Quantum-Transition re.
> LENR results and also Frank Znidarsic's finding the
> same within the extended Super-Fluid-Toroid Podkletnov model->re. the
> Quantum-Transition Velocity is 'huge.'

Are you talking here about th same thing as my " matter-anti-matter "
transition frequency,?

 Turn this around 'not' as back-tracking the phenomenon but rather as
positing Quantum-Transition speed-density(velocity) as a new insight rather
as QUANTUM-TRANSITION-ROOT-PLASMA that is formative energy state of
'ingress-plasma' into the 'eye' of the Proton(all protons) as
balanced-worm-hole connected to said proton-eye as a 'each atom is a
balanced singularity system.'
Would agree that each atom is a balanced system which has a , mensurable
lifetime.

Seems like about three years ago, "Hugh" came up with a suggestion that
every thing cduld be considered a composed of black holes...(That'd be
something that had an event-horizon (read inversion sphere or circle) and
singularity, (read inner limit)

> Corollary: The Einstein Solar Voltaic modeling of the
> Electron-shell(s) as an energy quantum-whole-field CAPACITOR state of
> graduated familiar electro-valent energy multple levels in onion-layer-like
> configuration etc.

This view sounds simplistic, although useful, it is also very reminiscent
of Mills' neutrino shell modelling of the Hydrogen atom.

> Case in Point: This Einstein Solar Voltaic Electron shell(s)as
> Capacitor-Field(s) WORKS. And the Solar Votaic effect is proof of the
> pudding of this model by producing effective technologies. And in the
> negative; the 'old model' of the electron-as-particle orbiting a fixed
> proton making the atom mostly relatively 'empty-space' is a non starter.

> The whole-atom is a circulating whole-energy 'field-system.'

I'd sort of go along with that. However, I am not comfortable with the
terms, "Energy" and "Energy Field s" I wouLd prefer motion volume
systems.   This is in part because I feel the term, "Energy" is too loosely
defined and I can , starting with the fact that momentum is the
instantaneous change of "Energy" with respect to time, derive three
different equations which would repre4sent "Energy. [$mv^2/2$, $m^2v/2$, and
$m^2v^2/2$, }

> A co-posit is that 'electron-flow' aka 'electrical current' is exactly like
> 'light-Photonic 'current' & is a 'wave-form' that I have referred to as
> 'helicoid-wave-string' of Quantum-Electron current via 'Quantum-Electron'
> Velocity-Density Momentum. Much is semantic, an admittadly Frank Znidarsic
> tends to state the matter more succinctly. Hydrogens Electro-Valent
> Capacitance 'shell' equals 'One Quantum Electron' and work up from there.
> And Quantum-Electron wave-current helicoid-string flow would quantify as One
> Quantum Electron would equal=> one Q-Elec. Helicoid-spiral-Wave from
> wave-crest-to-wave-crest; ad Planck-Dirac angular momentum calculations
> etc.

 My theorizing would suggest that the above would need a lot of

> dissection, Sounds like several concepts are running into each other. also
> you and I are not yet using the same concepts of the electron and proton....

> But specifically I posit that the Frank Znidarsic observation of the
> 'unity' of the various gradient(spectrum) offshoots of
> 'forces'(EM-Nuclear-Photonic etc) from the convergent 'One Million Meters
> per Second,' indeed indicates here in is a Unified Source Quantum-energy
> state of INGRESS-PLASMA from adjacent Dark-Energy HyperSpace.   This
> Source-Parent-Hyperspace is posited as a hyper-velocity/hyper-fluidic while
> hyper-dense quasi-infinite Super-M-Brane state. And this is what we have
> been poking toward under the names of Zero-Point-Energy &/or Quantum
> Vacuum which tend somewhat to misnomer because Hyperspace is hardly a
> 'vacuum.'

Again, by my thinking, these Zero-Point Energy Quantum Vacuum Ideas are a
mishmash of confused theory which mixes point-centric motion (essentially
what is known as "Mass") and the relative motions measured along collision
vectors which result in what is usually called "Energy."

> 'Singularity' is the model for all interflow fluid dynamic balance from the
> atomic micro-singularity level to the macro-cosmological level. This is
> including the original Hyperdimensional singularity that 'big-banged' our
> Bubble-Universe into inflation as well as a virtual-infinite-myriad of other
> universii-sister-bubblels more or less similar to our own. These
> sister-universii-bubbles are inflated via routine Hyperspace current
> dynamics that form hypervelocity-hyperdense-hyperfluidic swirling current
> super-eddies which are the birth torus's of bubble-universes like bubbles
> within a virtually infinie Hyperspace Champagne.

Yes, I'd say we are part of one half of an oscillator which could be called
a "bubble Universe," there would be an unknown number of such and they may
well overlap and intertwine. Bubbles in Hyperspace Champagne? Lot more
romantic than my prosaic half oscillatorsw is a Subsance-Substrate.

> AND SO: When Jeremy posits the confinguration of Super-Fluid 'bagel/torus'
> reactors to aohieve the Podkletnov-effeci, he is definitely on the right
> track. And if the above model is correct this Hi-Density-EM generated
> Bagel/Torus/Toroid Reactor which is functioning like a 'SUPER ATOM'
> according to Znidarsic-&-Jeremy should get much-more than they are looking
> for.

\
I don't know what the Podkletnov effect is. unless it is possibly something
like the motion of a Bagel winding in space for "no apparent reason."
Incidentally, I'm not acquainted, at this point with Z---- and Jeremy or
their "Super atom Toroid" ......

- Show quoted text -

Google groups

« Groups Home

Oscillator/Substance
⊞ **Theory**

Fwd: Vigier AIP Paper Submission. Correspondence, I give up on this one. I blew it...	**Home**
	Discussions
Options	+ new post
	Members

2 messages - Collapse all - Report discussion as spam

dean sinclair View profile   More options Oct 12 2010, 3:42 pm

Geez, people, I hate to admit being a "Quitter" but this one I give up on .

Will post my current version of the paper that I'm giving up on because of
publication technicalities and time constraints. I'ts been an expensive
learning experience for an old man.ESKi

Here is my latest correspondence re the paper with the Symposium
coordinator.

- Hide quoted text -

---------- Forwarded message ----------
From: dean sinclair <deanlsincl...@gmail.com>
Date: Tue, Oct 12, 2010 at 3:26 PM
Subject: Re: Vigier AIP Paper Submission.
To: noet...@mindspring.com

Richard,
Going through the guidelines and instructions, I see that I am deep in over
my head, I'd have to have had this information months ago, and probably then
have to have gone to some of my friends at the newspaper to sort it out for
me. I don't understand any of the technical jargon of publication and don't
know how to do a pdf file....

I'll never be able to beat the paper into shape in three days with all the
other things that are going on.;.In any case, if I could beat it into
"publishable technical format," AIP would be unhappy with the lack of an
institutional address, to say nothing of the fact that the paper has at
least thirty points of disagreement with orthodox dogma... To make it worse,
it is written in straight American English rather than Physics Jargonese,
so--Heaven Forbid--any reasonably literate person could probably understand
it! Seems as if that would be about the last thing that AIP would want for
a paper they published......

In a sense, the paper is getting lost in the same way the talk did. At the
last minute, it is becoming clear that the technical details are just too
complicated to overcome.   Sorry to have wasted so much of your time. DS
On Wed, Sep 29, 2010 at 8:07 PM, <noet...@mindspring.com> wrote:

> Hi Dean

> Attached are the guidelines for the submission
> Please follow precisely

> Need a copy in MS Word and Pdf

> Papers are due by 15 October

> if trouble with this email also send to

> amor...@noeticadvancedstudies.us

> This email you sent got through so send a similar without attachment to
> notify it, then if I see it I will know paper didnt follow

> rla

> >Need specific details whom to submit paper to for the API publication of
> >the
> >Vigier VII Symposium. E-mail submission or Hard Copy? Address for
> >submission. E-mails don't seem to be making it to this address. Thanks,
> >Dean L. Sinclair

> Richard L. Amoroso -Director
> Noetic Advanced Studies Institute
> www.mindspring.com/~noetic.advanced.studies
> noet...@mindspring.com
> (+) 510 435 1013

Reply to author   Forward

ka-sala View profile   More options Oct 26 2010, 8:08 pm

Hello Eski,
I cannot blame you for how you feel here.
I went over everything you've said, and even trid this
www.mindspring.com/~noetic.advanced.studies myself, but only to have
mail returned.

I'm sure I have seen this Richard L. Amoroso -Director, Noetic
Advanced Studies Institute on some TV Programs??? I also feel his
approach is more from the metaphysics point of view and may answer for
my own approach to things? Just guessing, but he does not appear
'appoachable' without 'joining' his site?

There must be a way round or another address??? Yet you got through
this one time?
Sorry all this ended like this, for now anyway. Just don't give up on
your group who are with you, despite our different approaches.

Till again, take care,
Ka-sala

On Oct 13, 7:42 am, dean sinclair <deanlsincl...@gmail.com> wrote:

- Show quoted text -

Reply to author   Forward   Report spam

End of messages

« Back to Discussions   « Newer topic   Older topic »

Gmail   Calendar   Documents   Photos   Reader   Web   more ▾

deanlsinclair@gmail.com ▾

Google groups

« Groups Home

Oscillator/Substance
⊞ Theory

[Search this group]   [Search Groups]

Google is closing our "Pages" section. SjAVE the "Pages_{Options} Library?!"

Options

4 messages - Collapse all - Report discussion as spam

ESKI View profile   More options Oct 26 2010, 4:56 pm

Google is going to close down access to "Pages" as of Feb. 1 That will
destroy our "Library." I suggest that you-all might want to transfer
the "Pages" from this site, en masse, to a file on your home
computers.

I'm going to try to move the material to a companion
OscillatorSubstance Site, but am not sure how to do it. Anyone want
to volunteer to take care of this for me?

Way things are going here, I'm only finding about 6 hrs. a week to do
any computer work and I don't get any where near what all needs to be
done... DLS

Reply to author   Forward

ka-sala View profile   More options Oct 26 2010, 7:53 pm

Wish I knew this one Eski, but I don't now how to. I wasn't sure after
your last post if we were even continuing with this work ? I'm sure
there is someone amonst us who can help here.

Take care, and what is meant to be, will be.
Look after yourself also!
Ka-sala

On Oct 27, 8:56 am, ESKI <deanlsincl...@gmail.com> wrote:

- Show quoted text -

Reply to author   Forward   Report spam

dean sinclair View profile   More options Oct 27 2010, 3:26 pm

In the information that Google posted about closing the Pages, they also
said something about a Zip drive and apparently gave a way to move the whole
file to a "Zip drive," I tried moving the material to this public computer,
apparently successfully, but couldn't get it back off here over to the new
site.   Cheers, Dean

- Show quoted text -

Reply to author   Forward

ka-sala View profile   More options Oct 27 2010, 3:54 pm

I will be greatful if anyone can save this site to ZIP to forward a
copy to those unable to, as myself.
Thank you,
Ka-sala

Home

Discussions
+ new post

Members

About this group
Edit my membership
Group settings
Management tasks
Invite members

 View this group in the
new Google Groups

Gmail Calendar Documents Photos Reader Sites Web more ⁓   deanlsinclair@gmail.com ⁓

Google groups

« Groups Home

Oscillator/Substance
⊞ **Theory**

[   ] [Search this group] [Search Groups]

Message from discussion <u>Fwd: Copy of Framework II</u>

ESKI <u>View profile</u>   <u>More options</u> Nov 17 2010, 4:51 pm

This is the paper that I didn't get correctly formatted to send in for
publication in the AIP Publication of the Proceedings of the Vigier
VII Symposium.... It is not in their format. The format used here is
more like that used in papers on Helium.com.

For some unknown reason, a couple of other attempts to post this have
failed, as I promised to post this about a month ago. DS

On Nov 17, 3:43 pm, dean sinclair <deanlsincl...@gmail.com> wrote:
> --------- Forwarded message ----------
> From: <deanlsincl...@gmail.com>
> Date: Wed, Oct 6, 2010 at 3:45 PM
> Subject: Copy of Framework II
> To: deanlsincl...@gmail.com

>   I've shared Copy of Framework
> II<https://docs.google.com/document/edit?id=1Razysz5QR2j5zJ6k1_JX2CKE6Tg...>
> Message from deanlsincl...@gmail.com:Click to open:

>   - Copy of Framework
> II<https://docs.google.com/document/edit?id=1Razysz5QR2j5zJ6k1_JX2CKE6Tg...>

> A Framework for a Fundamental Theory?

> Dean L. Sinclair

> Email: deanlsincl...@gmail.com

> Abstract. A model called the "Oscillator/Substance Model" may provide a
> framework for a comprehensive theory uniting the fields of physical science.
> Its basic tenet is "All existence is the result of sequential
> 'action-reaction-action' interactions within a Substance/Substrate of
> undefined basic composition and extent." This continued "sequential
> equilibration" results in constant motion such that the system is composed
> of/controlled by oscillators. Some of these oscillators are vortexes which
> have long term stability. Their interactions result in "Matter." Among the
> results of this view are an explanation for "charges," and the related
> definition of the size, shape and form of electrons and protons; a solution
> to the problem of the "Four Forces of Nature," and an explanation of the
> "Matter of the Missing Anti-matter." More details of the developmental
> process which led to this model, some definitions which arise from it and
> the relationship to some of the other theoretical approaches is also
> covered, as well as implications of the model in various areas. For one
> example: What may be a very basic reason the Hadron Colider has a rather low
> probability of ever being able to fulfill its original mission is mentioned.
> Much of the information covered in the paper is available in a number of
> short "pages" athttp://www.Groups.Google.com/group/oscillatorsubstance-
> theory<http://www.groups.google.com/group/oscillatorsubstance-theory>

> This is an edited and annotated expansion of a talk prepared for the
> VigierVII Symposium. London, England, July 12-14, 2010, which. because of
> certain systemic failures, was not presented.

> It is my honor to present some information about a model called. the
> Oscillator Substance Model which may provide a framework for a comprehensive
> theory uniting the fields of physical science.

> The basic tenet is, " All existence is the result of sequential

> action-reaction-action interactions within a Substance/Substrate of
> undefined extent and undefined basic composition. A Substance/Substrate
> which may be considered as if it be a liquid at the triple point, able to
> respond to slight pressure differences as any of the three basic phases of
> solid, liquid or gas.

> The continuous, sequential equilibration within the substance results in
> constant motion such that the system is composed of/controlled by
> oscillators.

> Some of the oscillators are vortexes having long term stability. these
> vortexes, the electron and proton and their mirror units, the positron and
> anti-proton, interact to form what we know as matter.

> This view produces valuable insights that often differ from the conventional
> viewpoint by 180 degrees. One reason being that the model, by shifting the
> generally accepted interpretation of the Michelson-Morley Experiment from
> proving the non-existence of an "Aether," to showing some characteristics
> necessary to an all-pervasive Aether, and thence changing the idea of the
> Speed of Light from an absolute maximum velocity of anything, to an average
> velocity which is a limit of information transfer, changes the viewpoint
> considerably. When, in addition, Planck's Constant be reinterpreted from a
> Constant of Action (Energy times Time) to a more prosaic, Constant of
> Angular Momentum, there arises a form which can be combined with the Speed
> of Light as an average tangential velocity to produce an equation, mass
> times radius equals Planck's Constant divided by the Speed of Light. This
> is an equation which can be used to define a family of oscillators. This
> series of shifts produces a totally different view of reality from the same
> basic, century-old data which underlies Space-Time and Quantum Mechanics.

> [A good coverage of the Michelson Morley Experiment with annotation is given
> in the

> Wikipedia Article, "Michelson Morley Experiment "

> Discussion of the development and usual interpretations of Planck's
> Constant, is given in the Wikipedia Article, "Planck's Constant."]

> In the model which arises positive and negative charges are seen as the
> result of reversed rotation/inversion senses of vortex oscillators. As these
> vortex units have mass and radius limits--and, corresponding frequency
> limits, charges will vary from a maximum value to zero and back, Charges
> are not fixed values, but have limits and an average.

> As the vortexes responsible for charges have determinable limits, their
> sizes and shapes can be estimated, The results of these determinations have
> interesting results for the theories of atomic structure.

> The liberty is taken to insert, here, most of the contents of a "page"
> previously published on the Internet Oscillator/Substance Google Site under
> the title, "The Electron and Proton as Oscillators."

> http://www.groups.google.com/group/oscillatorsubstance-theory/web/the...

> If it be taken that the relationship between energy and electromagnetic
> radiation be a fundamental relationship of our universe, then examination of
> that relationship should furnish clues as to the nature of our universe.

> If Planck's Constant--the constant which relates energy to electromagnetic
> radiation--be equated to its definition as an angular momentum, one obtains
> the equation, $m \times r \times v = h$. Evaluating this at "c," the speed of light,
> we obtain, $mrc=h$ which can be rearranged to $mr=h/c$.

> As any equation of the form $xy=K$ can be taken to describe an oscillator, by
> writing it in the form, $xy=K=yx$, to emphasize the interchangeability of the
> values of the two variables, we can see that the equation, $mr=h/c$, can be
> taken to define a family of oscillators of constant torque, h/c. With an
> "average" value of mass and radius where $m = r = (h/c)^{0.5}$.

> When the electron is checked to see if it fits into this family, it is found
> to fit with one limit set with the "rest mass" as the mass, "m", and the

> "Compton Wavelength" as the radius,"r." For an oscillator the absolute
> values can be switched to determine a "reciprocal limit." . For this
> particular oscillator, the other oscillatory limit would be absolute value
> of Compton Wavelength as mass and the absolute value of the "rest mass" as
> the value of the radius, "r."

> The set of values best known to this writer is the centimeter-grams-second,
> "cgs," system, which will be used throughout this discussion in analysis of
> oscillatory motion.

> Noted in our Universe as "Rest Mass" and "Compton Wave Length" for the
> electron are 9.10953 x 10^-28 g. for the mass, correlated to 2.42631 x
> 10^-10 cm. for the Compton Wavelength which corresponds to the radius.
> Switching the absolute values of the units, the other oscillatory limit
> would be 2.43631 x 10^-10 g. correlated to 9.10953 x 10^-28 cm. (The rest
> mass and Compton Wavelength values are taken from the Chemical Rubber
> Company, Handbook of Chemistry and Physics, 78th Edition.)

> We can analyze the proton in the same way to obtain the values:

> Minimal mass, maximal radius:1.67264 x 10^-24 g. and 1.321401 x10^-13 cm. ;
> balanced by the other limit of maximal mass, minimal radius-- 1.321401 x
> 10^-13 g. and 1.67264 x 10^-24 cm.

> It is to be noted that the electron is both heavier and lighter than the
> proton and larger and smaller, depending upon where the observation be
> taken. At the average values they are the same, as would be expected of all
> oscillators of this family.

> Often oscillators are defined by the frequency limits within which they
> operate. If we do this for the electron, using the above radius figures as
> expected wavelengths,we find the maximum frequency limit to be about 3.3 x
> 10^39 cycles per second and the minimal frequency to be about 1.25 x 10^20
> cps. The inversion frequency would be about 6.4 x 10^30 cps. For the
> proton the corresponding frequencies would be about 1.8 x 10^34 cps. and 2.3
> x 10^23 cps. The inversion frequency--which corresponds to the average or
> inversion situation, where m = r = (h/c)^0.5, which is about 4.7 x 10^-19
> grams at 4.7 x 10^-19 cm., would be the same as for the electron, about 6.4
> x 10^30 cps. These figures would indicate the electron would be operating
> over a band width some eight orders of magnitude greater than that of the
> proton.

> If the mathematics above accurately reflect the "real world," observations
> are of limit situations of minimal masses and maximal sizes of the electron
> and proton when measured in "Our Reality," which is "balanced" in an
> alternative, and equally valid, alternate reality by a maximal mass and
> minimal size limit.

> It is possible that both electrons and protons may be considered as
> combination oscillators with one oscillator operating between minimal size
> at maximal mass and the 4.7 x 10^-19 limits and the other operating between
> the maximal size at minimal mass limits and the "4.7 inversion situation."

> Both the electron and proton would show "nodes" at the oscillation limits
> and at the central inversion "equator." This 2/1 ratio of limit to
> inversion point has an interesting coincidence to the idea that "Quarks"
> occur in pairs, one with "2/3 of a charge" the other with "1/3 of a
> charge." Possibly the Quarks, considered fundamental particles in the
> Standard Model of Particle Physics are observational phenomena due to this
> nodal characteristic of oscillators.The Interaction of this "internal
> structure" of electrons with that of the proton and other "particles." may
> be the reason that Quarks are supposedly confirmed by scattering data.

> [The Internet Wikipedia Article, "Quark
Model,"http://en.wikipedia.org/wiki/Quark model gives an idea of the stunning
> complexity of the Quark approach to atomic theory.. It would seem that
> there should be a much simpler set of explanations for the "alternate states
> of matter" which arise from "atom-smashing."]

> The equilibration process in a substance can be considered to result in
> constant pressure adjustment. Pressure fits the criterion for a true Force;
> therefore, the various Forces of Nature can be seen to be a result of

> interpretations of pressure adjustments.The " Four Forces of Nature" are
> usually said to be Gravity, Electromagnetism and the Strong and Weak Nuclear
> Forces. Gravity can be explained as differential pressure between vortexes
> and/or vortex aggregates. It is a set observational phenomenon which has
> been explained as an "Attractive Force." Electromagnetism is another set of
> observational phenomena due to interactions of vortexes in the medium.
> Again, these interactions are describable mathematically but do not fit the
> criterion for a Force, "For every Force there is an equal and opposite, for
> every action a reaction." (1)

> The "Strong and Weak Nuclear Forces" may be considered to arise from the
> idea of there being neutrons, as such in nuclei. The proton-neutron nucleus
> model, used since the 1930's, is good for "book-keeping" purposes;' but, it
> requires a good deal of mental agility to try to justify it logically.
> Particularly when--on occasion--it requires transformation of an electron
> and proton into a neutron, a reaction which is the reverse of the known
> "spontaneous," exo-energetic transformation of the neutron into an electron
> and a proton.

> .

> The "Missing Anti-matter" Matter can be given a double explanation. The
> first part of an explanation is that the separation (or inversion) instant
> of an oscillator--an event which we know as the "Big Bang--" resulted in the
> definition of two oscillator halves, having reversed rotation/inversion
> orientations. Smaller oscillators, within these halves, will be influenced
> by the larger oscillator. The orientation of one half of these smaller
> oscillators will tend to be stretched, the other compressed, so that one
> rotation/inversion will tend to be expressed differently than the other. In
> our Universe, it appears that the stretched form which appears most
> obviously is the electron. Its "almost identical" mirror, the "positron,"
> appears to be somewhat suppressed.

> If one half of a separable oscillator be considered "Matter," and the other
> half be "Anti-matter," then--as the electron is always considered matter--
> the positron is anti-matter. The two units are logically halves of a
> separable oscillator, to which they rejoin in the "annihilation" process.(3,
> 4)

> The electron and proton are halves of another separable oscillator, the
> neutron. The electron is still "Matter," hence the proton, as the other
> half of a separable oscillator, is "Anti-matter!"

> Since the neutron is-- like the B sub s Meson which has been shown to
> oscillate between matter and anti-matter states--"neutral," it may be
> considered as either Matter or Anti-Matter. It may also be possible that all
> basic units, including the electron and proton, invert between "matter" and
> "anti-matter" states as does the B-sub-s Meson. (5, 6, 7)

> From the foregoing, we see that what we call "Matter," thought of as
> combinations of electrons and protons, might be more accurately considered
> as combinations of the "Matter" electrons, and the "Anti-matter" protons.

> Logically, there is, somewhere, an "anti-Verse" where the rotation inversion
> dominant expressions are the opposite of ours; but, also, we apparently have
> no truly "Missing Anti-Matter." We only have semantic confusion. (8) The
> "neutron count" of an atom can be considered simply as the number of
> nucleons that, at any given instant, are in anti-electron, anti-proton
> states. It is possible to do an accounting process as follows: Assume that
> the "atomic weight" to the nearest whole number, represents a sum of the
> total "heavy nucleons, " considered as protons and anti-protons. Assume
> also that the atomic number represents the number of electrons and the
> number of protons. Now, we assume that there is an exact balance of matter
> and anti-matter units. According to the discussion above that would mean
> there would be the same total number of protons and anti-electrons as there
> are electrons and anti-protons. ((If instead of neutrons in the nucleus, we
> guess there to be an anti-proton-anti-electron association where we have
> always said, "Neutron.") Both of these summations add up to the atomic
> weight.   We could also look at this as an electrical neutralization
> balance, as the number of electrons plus the number of anti-protons
> balances the number of protons and positrons.

> It. also, may be noted that the postulated proton-anti-electron pairs may be

> considered as having energy levels parallel to those that are written for
> the "outer electrons," that is. the proton-electron pairs, and,correlations
> may be made to a set of energy levels including both sets. Taking a simple
> example: Li7, atomic number 3, atomic weight 7, would have the standard
> "outer electron distribution" of "1s2, 2p1, " which we can say would cover
> the energy levels for electrons and protons, the distribution for the
> anti-electrons and anti-protons would be "1s2, 2s2; " while a combination
> distribution would have the structure "1s2 , 2s2, 2p3." The
> electron-anti-proton energy level postulate represents what has long been
> known as a stable set ,and the combination number represents a stable set
> plus a "half-filled, sub-shell" which is known also to be a situation of
> some stability. Considerations of this type lead to interesting
> correlations within the "Periodic Chart..(9)

> In addition to the "halves of oscillators" argument above, there is a
> mathematical argument that, given that both are "positively charged units,"
> the positron and the proton belong to the same category. If a positron of a
> certain kinetic energy were slowed down enough, with the "lost" Kinetic
> Energy all being converted to Mass, the Positron would be convertible to a
> Proton.

> If we let "m " and "v" be the masses of a Positron and "M" and "V" be the
> masses of the Proton, equate the Kinetic Energy expressions for the two
> units, and forget about the meaningless--for our purposes--one-half value
> which is in both, we can write $mv^2 = MV^2$ and rearrange this to

> $m/M = V^2/v^2$. Inserting the "rest masses" for the Positron and the Proton
> in this equation, we see that were a Positron slowed to about 1/42 of some
> initial velocity, it could be converted--at least theoretically--to a
> Proton. (10)

> The logical combination of the electron and positron to form a combination
> oscillator with release of "Energy" as "Annular Radiation" does not seem
> to have been published anywhere prior to the Sci-Scoop article, "Negatron
> plus Positron Equals Zerotron?" (3) There are several logical arguments
> for this model for "annihilation" and "pair-production" being reciprocal
> processes involving a previously unsuspected combination oscillator.

> One of these arguments involves another often neglected aspect of science
> theory. This is the fact that both mass and velocity are variables. When
> the momentum expression, "mv," is integrated to form an "Energy"
> expression, mass is usually considered to be constant and only velocity to
> vary, producing the well known "Energy" expression, $E= (mv^2)/2$. This
> expression possibly should be said to apply only to "Kinetic Energy," the
> "motion package" associated with moving a point and its associated,
> "point-centric" motions along a vector.

> If it be considered there could be a situation wherein the velocity cannot
> change, then mass would be the variable. Doing the integration under these
> circumstances, produces another equation, $"E" = (vm^2)/2$. (11)
> Considering that, in reality, both mass and velocity will vary, it might
> be better to simply integrate, "p", momentum, itself, as a variable, to
> obtain the expression, $(p^2)/2$ as a more accurate picture of the "total
> Energy" that is, of a "total motion-package." This leads to $(m^2v^2)2$, the
> expression obtained by reinserting "mv" for "p."(12)

> In the "annihilation," it is noted that the Energy release is "mc^2,"
> where "m" is the "rest mass" of each particle and "c" is the speed of
> light. This is the "Energy" release expected from the dissipation of the
> Kinetic Energy of two units meeting "head-on" on the same vector at
> velocity,

> "c." It is usually considered that the electron and positron are
> "destroyed" rather than combining.

> It doesn't seem to be generally realized that when objects meet "head on,"
> the Kinetic Energy is dissipated, the objects may be changed, but they are
> not converted totally to "Energy." In the usual handling of the case of the
> Electron and Positron, the "Second Energy Expression," $(vm^2)/2$, is not
> considered, and all of the "Motion" which would be described by (m^2v^2) is
> also not considered.

> . As it may be argued that the "Collision Energy" would more properly be
> represented by m^2c^2, than mc^2, it can be seen that there is definitely
> an "m" value which is not accounted for.... It makes sense that the two
> units, when they finally become oriented on the same vector, after some
> time existing as one or another of the forms of "Positronium," the
> short-lived (in our time scale) "near-zero-mass" analog of Hydrogen. [This
> is a unit having some characteristics of both atomic and molecular Hydrogen.
> (13)] When the two "halves" reach the proper orientation, Kinetic
> Energy--possibly actually vibrational Energy--is lost as a wave
> disturbance, "Annular Electromagnetic Radiation," and the two vortexes
> coalesce into a pulsator.

> [This would be somewhat analogous to a molecule collapsing to an atom, or a
> molecular cation collapsing to an atomic cation. These may be the processes
> whereby Deuterium may arise from the Hydrogen Molecular Cation and Helium 4
> from the Deuterium Molecular Cation. The latter may be an explanation for
> the observation of the formation of "Helium 4," along with "excess heat" in
> certain electrolysis experiments with Palladium electrodes. This is the most
> studied type of reaction in the ongoing research in the field once called
> "Cold Fusion," and now known as "Condensed Matter Nuclear Science," or
> "cmns" for short.

> The definitive history of this field, as of about 2006, can be found in the
> book, "The Secrets of Low Energy Nuclear Reactions " by Dr. Edmund Storms.
> A web site maintained by Dr. Ludvik Kowalski contains information which
> has developed since Storms' book was written. (14)]

> Considering the electron and proton as vortex oscillators which can
> associate may give a clue as to why the Hadron Collider apparently has had
> difficulties. Vortex oscillators can not only associate with different
> vortexes, but also self-associate. Electron-electron association has been
> long known. However, no one seems to have realized the same to be true for
> protons. Additionally, conventional science gives no hint of the possible
> existence in "vacuums" of pulsator-oscillators, e.g., the "Zerotron"
> mentioned previously, some of which may be separable into electrons and
> positrons and deformable into neutrons or even other neutron-like entities.

> Pushing a stream of mutually repulsive "Charged Particles" through a void,
> would be very different from trying to control vortexes, which can
> self-associate, through a possibly-reactive medium. If the O/S
> --Oscillator/Substance--view be correct, the Hadron Collider, when trying
> to accelerate protons might have been acting for a short time as a fusion
> reactor before feed-back caused breakdowns.

> [If the "Zerotron" unit idea be valid, and, also, the idea that the neutron
> may result from shock-wave distortion of this entity, then there is a
> possibility that the attempts to accelerate protons can cause shock wave
> effects in the Substance-Substrate, creating neutrons, this soon would be a
> source of electrons and additional protons. It is not too much of a
> stretch to see this situation resulting in the HH+ --Hydrogen molecular
> cation--which, very possibly can convert to the "He4+" unit, the Helium 4
> Cation.... The overall effect would be a release of "Energy" into the
> system, from a very unexpected source. A system that was supposed to be
> absorbing "Energy" would instead be emitting it...]

> The basics of this "Framework" have been stated, and some implications have
> been covered. However, a few words about the start of the ideas leading to
> the O/S Model and the reasons for its quite different view from the
> conventional may be in order.

> This model started to develop quite innocently in the Spring of 2004 with
> the realization that basic ideas of Einstein's Special Relativity fit into
> communication theory, where they would apply to any Perceptual Universe
> defined by a maximum, practical velocity of information transfer, whether
> that velocity be determined by Pony Express Riders or Electromagnetic Waves.
> (15)

> Since, in every case, practical maximum velocity of information transfer is
> going to be a bit less than the average speed of the packet carriers, the
> Speed of Light, is logically an average which acts as a practical maximum
> velocity of information transfer. This idea, also, would rise quite
> naturally from the use of the term, "Speed," in the name, "Speed of Light,"
> as a "Speed" may be considered either as an average over time in a given

> direction of an number of velocity vectors, or an instantaneous "velocity"
> of undefined direction.

> A line of logic stemming from the above consideration. led to two 2007
> papers published on Helium com under the title, "Motion in a matrix as a new
> model of the physical universe." (16, 17) These papers outlined the
> reasoning involved up to that time. . These papers were the progenitors of
> the present theoretical form. It was later realized that the idea of a
> "Matrix." as a true solid, would be an error. and the model of the "Matrix"
> was modified to "A medium having the general characteristics of a
> substance at its triple point."

> By the Summer of 2008, when the Oscillator/Substance Google Group was set
> up, follow-ups on the initial insights noted above, had led to the
> realization that there was a "T.O.E." available, as outlined at the start of
> this talk, which could have been seen a Century ago. Had the
> Michelson-Morley Experiment been reversed in interpretation from ruling out
> an "Aether," to partially defining an Aether; and then, a few years later,
> Planck"s Constant had been considered a Constant of Angular Momentum and
> used to define characteristics , of that Aether, this model might well have
> come into existence 100 years ago.

> Equating Planck's Constant,"h," to its definition as an angular momentum
> and evaluating the resulting equation at the Speed of Light, "c." leads to
> the equation, m x r = h/c = r x m., This arises from the fact that one
> definition of angular momentum states that angular momentum is the
> resultant of a mass , "m," rotating at a radius, "r," from a point, with a
> tangential velocity, "v."

> As Planck's Constant applies at the Speed of Light, it makes sense to
> evaluate at the speed of light and to simply by dividing out that speed from
> the left side of the equation to form a ratio constant, "h/c." The resulting
> equation, m x r = h/c = r x m is an example of a common, very valuable
> relationship in physics, which occurs in the law of levers, the balance law
> used in weighing, the law of conservation of momentum, the law of
> conservation of energy....

> Here, this relationship can be used to determine the oscillator limits for
> a family of constant torque oscillators , defined by the set,{m x r = h/c =
> r x m } , with a torque of h/c and inversion at the state where r = m =
> square root of h/c. In the cgs system, this value is about 4.7 x 10 ^-19
> grams at 4.7x 10^-19 cm. This implies a hidden half of any basic oscillator
> which is smaller than 4.7 x 10 i^-19 cm.

> Coming to the work of Michelson, Morley and Planck from a somewhat
> opposite view of the more standard theoretical approaches such as
> Space-Time, Quantum Mechanics and String Theory, this model has a quite
> reversed orientation. As such, it asks for re-examination of many of the
> accepted percepts of modern physical theory.

> It may, however, turn out that this model will be complementary to much
> theory rather than contradictory. In its definition of Mass as a measure of
> the tension-pressure at a surface of the point-centered motions within that
> surface, a characteristic of entities that is measured by comparison; and ,
> in suggesting that the term, "Energy," usually means a measurement of a
> package of motion which includes a point and its associated motions along a
> line--a unit whose effects are usually observed as the results of
> collisions--that is. "Kinetic Energy," it appears that this model tends to
> focus on the "Mass" aspect , whereas most theoretical approaches focus on
> "Energy," for the most part, and consider "Mass" as generally a constant
> value of some sort.

> There is far too much information developed from this model--and closely
> associated ideas --which cling easily to it as a "Framework," to cover in
> this short presentation. Therefore, I refer you to the web site of the
> group previously mentioned, http://groups.google.com/group/oscillatorsubstance-theory .
 In this site
> most of the extant material has been collected as "pages," which vary in
> size from a half-page to 23 pages and counting..

> This model is called a "Framework," by this writer who considers it as a
> simple start toward new construction. The model developed from a thought
> that communication should be consistent whether the information be carried

> by by Pony Express Riders or Electromagnetic Waves. Since it almost "grew
> itself--" essentially independent of consideration of other theoretical
> models--it turns out to be definitely in contrast to many attitudes current
> in the scientific community. In fact, we might say that it is "usually
> anywhere from 90 to 180 degrees out of phase." Here are some examples, some
> of which may not have been explicitly mentioned before.

> Where the standard view seems to be that what is needed is a theory to unify
> many diverse parts, this model takes the view that there is a "unity, a
> Substance/substrate of undefined extent and undefined basic unit." This
> idea might possibly be interpreted to mean, "There is a Fact of Existence
> which we may never be able to totally understand or define. Let us accept
> that and move on to what we can do."

> Where the general consensus is that there is nothing in a "vacuum," this
> model postulates that there is an all-pervasive substance, even in "vacuums"
> from which the vortex aggregates which we call "matter," have been removed.

> Where the usual view of electrons is as some sort of probability cloud, this
> model sees them as rotating, inverting, vortex oscillators.... Similarly,
> reality of size and shape are given to other subatomic units. (The
> probability cloud idea arises from an average positioning of the electron as
> an entity in space, but is generally taken as a representation of the
> structure of the electron.)

> The conventional picture is that electrons and positrons combine to
> "annihilate," converting totally to "electromagnetic radiation." This model
> says that they combine to another type of oscillator with dissipation of
> half of their total motion in the form of "radiation." Whereas,
> conventionally, "pair-production" is some sort of a mysterious conversion of
> "Energy" into "particles" in the presence of matter-- in this model, pair
> production is simply the splitting of the "parent oscillator," when supplied
> with enough excess motion That is, pair-production is considered the
> reverse process of "annihilation."

> Much of conventional physics theory is based on an idea similar to
> Einstein's supposed comment, " Mathematics is the reality." It is even
> assumed by many that if theoretical ideas are not expressed in differential
> equations, they have no validity. The view here is that mathematics is a
> tool, and that it is probably best to work with the simplest tools possible.
> This entire presentation has used nothing beyond grade school level, except
> for reference to Integration of the Momentum equation to form Energy
> equations.

> Where Mass and Energy seem generally accepted as being fundamental and

> inter-convertible--without truly defining either--this model defines both
> with respect to motions relative to points. (Mass is considered a measure
> of the tension/pressure at a surface, a measure of the motions concentrated
> about points within that surface. Energy specifically refers to Kinetic
> Energy, A measure of motion of an entity--a surface having a center of
> mass--along a vector. This is a value which is determined by velocity with
> respect to a point on that vector...)

> Where the "Unification of the Four Forces of Nature" is considered
> conventionally as a major theoretical problem, this model dismisses the
> situation by pointing out that none of the "Four Forces" meets the
> definition of a Force, whereas Pressure does.

> The problems of the "Missing Mass" of our Universe, Dark Energy, and some of
> the other related concepts may turn out to be due to several factors. One
> could be the semantic confusion between the use of the term, "Mass," as
> describing a "physical body," and "mass" as a scientific term describing an
> attribute of that body. This differentiation clearly shows in this model.

> Where, conventionally, there are many constants of nature, this model
> implies that there should be few, and those will be not absolute limits of
> any sort but are more likely to be statistical averages. Furthermore,
> combinations, multiples, and roots of "constants" are logically also
> "constants" which may furnish information. For instance, the square root of
> the speed of light, $(c)^{0.5}$, which is about 173 Kc/sec. (and $1.73 \times 10\,^5$
> cm) may be a very interesting frequency as it is the value at which

> frequency and wave length will have the same "Absolute value." The use of
> the square root as two separate units which may have different titles but
> have the same "absolute values" is related to two short papers published on
> SciScoop. (18, 19)

> This model does not consider positive and negative charges as mysterious,
> accepted things of nature, but as manifestations of the rotation, inversion
> senses of vortex oscillators. As such, they are not constant values....

> Where the Standard Model of Particle Physics originally considered the units
> found as results of atom-smashing experiments as being fundamental particles
> released by the experiments, this model would imply them to be different,
> alternative states of matter created in the experiments. That is, artifacts
> rather than fundamentals.

> This is by no means a conclusive listing of the differences in philosophy
> and attitude of this model from the more conventional situations. It is
> simply a listing of some of the differences that come immediately to this
> writer's mind.

> In presenting this "Framework," this person is most certainly not asking
> that the ideas and information collected by all the workers who have
> contributed in the past be discarded, he is simply suggesting that this
> model may be a framework into which profitable re-examination of data and
> ideas may be fitted. This is a framework which appears, at least to this
> writer, to furnish a simpler, more easily understandable view of Existence
> than is currently available elsewhere.

> _____

> References:

> 1. deanlsinclair, "Four Forces or One Substance?," SciScoop.com

> 2. ibid.

> 3.deanlsinclair, Positron plus Negatron equals Zerotron?.SciScoop.com

> 4.Sinclair, Dean L.,, ON THE MATTER OF ANTI-MATTER, What is hidden where?

> deanlsinclair.blogspot.com

> 5.Perricone, Mike, It might be...it could be...it is!!! Fermilab Press
> Release, Sept 25, 2006

> 6.Jamieson, Valerie, Flipping particle could explain missing anti-matter,
> New Scientist, 18 March 2008

> 7.Overbye, Dennis , A New Clue to Explain Existence, The New York Times,
> May 17, 2010

> 8. Sinclair, Dean L., Anti-matter, the basics. Helium.com

> 9. William Harrington, Charles William Johnson. and Dean L. Sinclair,
> private correspondence dealing with continuing work.

> 10, Sinclair, Dean L., Could protons be "reformed" anti-electrons?
> Helium.com

> 11.deanlsinclair, Two Energy Expressions Interact? SciScoop.com

> 12.ibid

> 13._____, Positronium, Wikipedia.org

> 14. Kowalski, Ludvik, Index to Cold Fusion Items, montclair.edu

> 15,Sinclair, Dean L.,Emulating Einstein in 2004, a privately circulated, 7

> pg. report.

> 16.Sinclair, Dean L., Motion in a matrix as a new model of the physical
> universe. Helium,com

> .17. Vreeland, Hugh, Motion in a matrix as a new model of the physical
> universe.Helium,com

> 18.deanlsinclair, Roots and Directed Numbers, SciScoop.com.,

> 19. deanlsinclair, Problems in Mathematics-Signs and Signed Numbers,
> SciScoop.com , _____

> Dean L. Sinclair, B.A., M.S., Ph.D.

> Aberdeen, SD, USofA

> Oct. 8, 2010

> Google Docs makes it easy to create, store and share online documents,
> spreadsheets and presentations.
> [image: Logo for Google Docs] <http://docs.google.com/>- Hide quoted text -

> - Show quoted text -

Reply to author   Forward

Gmail Calendar Documents Photos Reader Web more ⌄   deanlsinclair@gmail.com ⌄

Google groups

« Groups Home

Oscillator/Substance
⊞ **Theory**

[Search this group] [Search Groups]

Elements from "Empty Space," yep, looks like it's accidentally been done; and they darn near killed themslves doing it. ....

Options

Home

Discussions
+ new post

Members

About this group
Edit my membership
Group settings
Management tasks
Invite members

 View this group in the new Google Groups

1 message - Collapse all - Report discussion as spam

dean sinclair View profile   More options Dec 1 2010, 4:35 pm

Researchers,posting on a closed Internet site have reported that, in a
Cavitation Collapse experiment, they produced--among other things--a melange
of elements, consistent with what they called a "Nova Event." They also
report that they became very ill. The symptoms are of radiation poisoning.
They also note that later checking showed a "neutron burst."

O/S work indicates that there may be as many as 10^{54} --plus or minus a few
million orders of magnitude--of neutral oscillators in any centimeter of any
space. It also indicates that some--possibly even most--of these are
possibly "Shock Wave Convertible" to neutrons or "neutron-like" units.

In consideration of the above, the results reported seem creditable, and
raise a red flag for anyone working with experiments which could create an
intense shock wave.

Dean Sinclair (Eski)

Reply to author   Forward
End of messages

« Back to Discussions   « Newer topic   Older topic »

Gmail Calendar Documents Photos Reader Web more ⁊   deanlsinclair@gmail.com ⁊

Google groups

« Groups Home

Oscillator/Substance
⊞ **Theory**

[Search this group] [Search Groups]

A bit of update from Eski.   Options

1 message - Collapse all - Report discussion as spam

ESKI View profile   More options Dec 10 2010, 5:39 pm

Apparently Google has backed off trom blocking access to current
content on our "pages," that is good news.

There is one bit of personal news. Although i do not yet have
Internet access except through public computers, at least not useable
access, i do now have ove of those little toys that Americans seem to
now feel indispensible. a "Cell" phone, thanks to a daughter and step-
son who got together and got me one.

The phone no., of the "Net 10 Phone" is a United States of America
number, 1-605-290-2154 , in case anyone gets a wonderful idea that
they can't wait to give me a "heads up" on... International calls only
cost me 15cents a minute which isn't bad consiering the usual rates!
Cheers, Eski

P.S. Merry Christmas, Happy Hannukah, Joyeux Noel, Feliz Navidad,
Froeliges Weinachtszeit, and Welcome Back to the Sun for all of you
who remember the original reason for celebrating a few days after the
Winter Solstice!

Reply to author   Forward

End of messages

« Back to Discussions   « Newer topic   Older topic »

Home

Discussions
+ new post

Members

About this group

Edit my membership

Group settings

Management tasks

Invite members

   View this group in the
new Google Groups

Sponsored links

Video Chat on Facebook
Video Chat & Connect With Friends
On Facebook. Sign Up Today!
www.Facebook.com

Small is the New Big
HP® Veer: Intuitive Software &
Apps
in a Powerfully Small Design. Shop!
www.hp.com/veer

Download Google Chrome
A free browser that lets you
do more of what you like on the web
www.google.com/chrome

See your message here...

Google groups

« Groups Home

Oscillator/Substance
⊕ **Theory**

♜Comment on Schroedinger Equatiƒon, Also. Happy New Year

Options

1 message - Collapse all - Report discussion as spam

dean sinclair View profile   More options Dec 30 2010, 3:21 pm

Here is my " take" on the Schroedinger Equation, the basis of Quantum Mechanics.

The first element of the Equation notes a wave function involving the second
derivative with respect to time , apparently of Kinetic Energy. The first
derivative of Kinetic Energy is Momentum,, and the second derivative, the
differential of momentum, would be either velocity or mass, whichever one
wished to consider as being a constant. The integrations which would
result
from working with the second derivatives along the x, y , z coordinates of
the basic "Energy Wave " considered in the mathematical conventions of
positive numbers with Cartesian Coordinates, when followed logically,
describe only one type of unit, albeit in ten dimensions;
♯Each double integration adds three dimensions, doing this three times
gives nine dimensions, and the combination of the right, up, forward,
"positive numbers" conventions adds a counter clockwise twist to the
whole
thing which adds a tenth dimension of counter clockwise spin. :This could
be   disappear
the general basis for the String theorists idea that the strings slapper
into a then d-dimensional hole at 10I^-18 cm. which would be about the
4.7 x
10^-19 radius which our speculations suggests is the average diameter of
the
oscillators of our universe..}

This creates a "picture of a model" which is rotating counter clockwise into
the octant to the right, up and forward from the chosen origin, So, right at
the start, The Wave Function, in this view, is covering only 10/80 of the
dimensions of reality, we can go on doubling up, clockwise instead of
counter-clockwise, in rather than out, integrating with velocity constant
rather than mass, I think I'm up to ten out of 640 at this point, and,I
think that I could double up a couple of more times, .In other words, If
Schroedinger's Equation be valid as a descriptor, it is of a very limited
view of the possibilities.

Anybody out there who is a QM expert, I'd like to hear your comments.

Some ideas for projects, resolutions for this group for 20ll....
1. Get this thing some publicity so it will get checked out more widely...
Write a short book?

2. Fit the "baryons" etc. for the "Standard Model, Atom Smashing work in
...
3. Extend model beyond Chem. and Physics into Cosmology....
4..Accomplish what we can ala the Serenity Prayer. as applied to this
stuff!

Happy New Year, Prospero Ano Novo, etc.   Eski

Reply to author   Forward

End of messages

« Back to Discussions   « Newer topic   Older topic »

Gmail Calendar Documents Photos Reader Web more ⌄   deanlsinclair@gmail.com ⌄

Google groups

« Groups Home

Oscillator/Substance
⊞ Theory

[Search this group] [Search Groups]

Rumoured Soviet Weapons, Death Ray and Hammer   Options

Home

Discussions
+ new post

1 message - Collapse all - Report discussion as spam

Members

ESKI View profile   More options Mar 1, 5:28 pm

If as I suspect, the matter-anti-matter duality noted in the Bsubs
units is actually common throughout existence even in electrons and
protons, then it may well be that the rumoured weapons could have been
developed by technologists who, unhampered by theoretical
considerations,
simply applied greater and greater accelerative forces to electron
streams, alternately accelerating and compressing the units toward the
"Sin-Vree" unit (4.7 x 10^-19 g. at 4.7x10^-19 cm.) with a relative
velocity of "c."
This resulted product, as a continuous emission stream would be
expected to be a very effective "death ray" and, as a pulsed emission,
would appear to be some thing which would indeed have a hammer
effect....
.I'm simply speculating, noting that there is no reason that, since
proton streams can be created and accelerated, ala the Hadron
Collider, there really is no reason that the much easier to control
and create electron streams could not also have been accelerated to
form beam "tools," i.e. weapons.... DS

About this group
Edit my membership
Group settings
Management tasks
Invite members

 View this group in the
new Google Groups

Reply to author   Forward

End of messages

« Back to Discussions   « Newer topic   Older topic »

Create a group - Google Groups - Google Home - Terms of Service - Privacy Policy
©2011 Google

Gmail   Calendar   Documents   Photos   Reader   Sites   Web   more ⌄   deanlsinclair@gmail.com

Google groups

« Groups Home

Oscillator/Substance
⊞ **Theory**

[Search this group] [Search Groups]

Reality and math. short comment   Options

1 message - Collapse all - Report discussion as spam

dean sinclair View profile   More options Mar 14, 5:00 pm

Reality and Mathematical Definitions

We consider mathematics as a model for reality. Perhaps it would be
worth while to consiee4 how certain mathematical definitions fit as we
look at the physical world.

First, let us look at the concept of Zero, usually considered as the
symbol of nothingness. However, is this true? Zero is the starting
point for counting, the starting point for any journey, the crossing
for the Cartesian axes of most conventional graphs. It might be
better to say that the symbol, "Zero," is actually the symbol for the
fact of existence. In the real world, the starting "point" may be of
any size, any shape. Zero, then. is not without existence, without
dimension, rather it is the symbol of the very first dimension, the
Dimension of Existence.

What then of the number, "One? " That's simple, its the 'counting
number." However, isn't it a lot more than that? It is the symbol of
wholeness. it represents a whole starting point, a whole line, which
actually has to be made up of two starting points, a whole surtace,
made up of at least three starting points, etc. Hence the number one
may represent many things. If we atttach a sign to the number,
implying a motion, then +l, represents the motion of a whole starting
unit one unit to the right, or possibly up, or forward. We say that
one times one times one equals one, but we always assume that there is
a positive value atttached to the one, so if the first one represents
a motion of one space to the right, the next one reprensents the
motion of the first "one" upward, and the third one represents the
motion of the second generated one a unit forward. Therefore, if we
attach the positive notation to "one" which, by convention we do, one
times one times one actually means one cube generated to the right,
above and forward from the origin of a set of axes by a set of
motions whch are actualy counter clockwise,

Looking at "One" as the symbol of wholeness has many uses. One
interesting one arises is one looks, for instance, at figuring a
maximum frequency for our particular Universe.   The equation for
the movement of Energy by electromagnet radiation is $E=hu$, where "h"
is Planck's constant and "u" is cycles per some unit of time. If we
place E equal to one Energy unit in any set of units, the maximum
frequency, expressed in that set of units, will be seen to be "1/h"
. This presumably would be the "hlgh frequency cut off" for
communication."

At the other end of the scale is the symbol of "Infinity. the Number
Beyond All Numbers." For mathematicians this is a perfectly good
definiion; but in the real world we have to use more rational
definitions. Does it really make sense to say that we can measure mass
of something moving with relation to us up to a velocity as near the
speed of light as we care to but say that the mass will become
"Infiinte, meaning "without limit" at the speed of light? Is the
darkness just beyond the flashlight beam a void? Couldn't that
darkness be considered an Infinity? We can't see into it.   In the
practical world we probably should conisder the concept of "Infinity"
as representing simply the point just beyond the last point that we

Home

Discussions
+ new post

Members

About this group
Edit my membership
Group settings
Management tasks
Invite members

 View this group in the
new Google Groups

can measure with the instruments at hand, the number beyond where we
stopped countig, for whatever reason.

Reply to author   Forward

End of messages

« Back to Discussions   « Newer topic   Older topic »

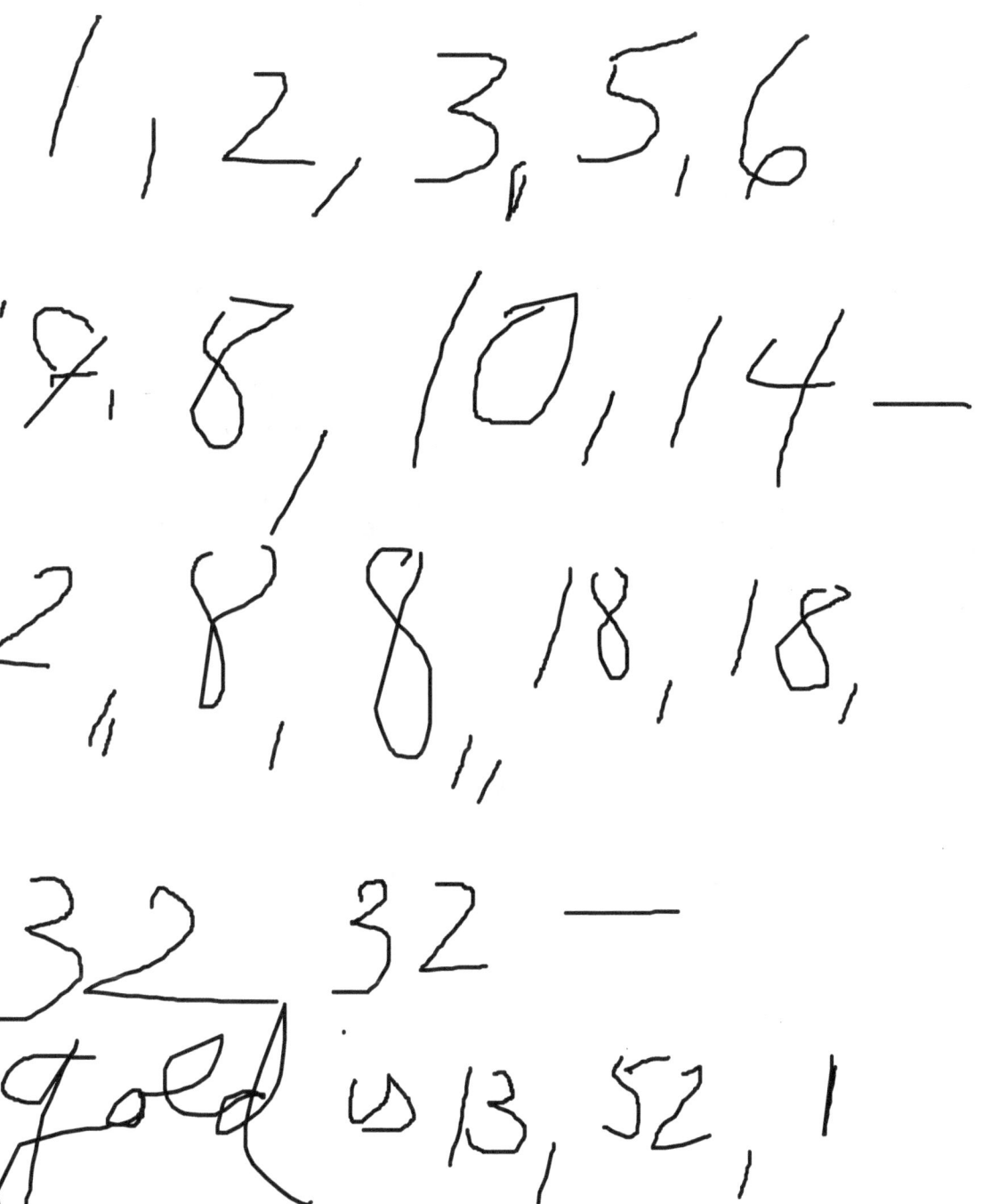

Google groups

« Groups Home

Oscillator/Substance
⊞ **Theory**

[Search this group] [Search Groups]

Home

Discussions
+ new post

Members

About this group

Edit my membership

Group settings

Management tasks

Invite members

 View this group in the
new Google Groups

VALUE OF INFORMATION ON THIS SITE   Options

2 messages - Collapse all - Report discussion as spam

ESKI (deanlsinclair@gmail.com) View profile   More options May 10, 2:10 pm

Dear Reader.:
Google has removed the welcome message and the pages fro this group,
inadvertingly destroying its usefulness to most readers.

To get maximum value, rather quickly, firnd the paper, "Essentials of
O/S" lsted in the discussions, I t is on about the second page of
"Discussions.

Another site at deanlsinclair.blogspot.com has most of the very
essential papers.

SORRY! Eski

Reply to author   Forward

ESKI (deanlsinclair@gmail.com) View profile   More options May 25, 11:49 am

Note the comment in red on the top of the page, I'd recommend any
visitor who is working off their own computer to down load the
"zipped" pages to be perused at leosure, They are the "historical
library of the site," and loaded with information including some
information about other theories

,On May 10, 2:10 pm, "ESKI (deanlsincl...@gmail.com)"

- Show quoted text -

Reply to author   Forward

End of messages

« Back to Discussions   « Newer topic   Older topic »

Create a group - Google Groups - Google Home - Terms of Service - Privacy Policy
©2011 Google

Go gle groups

« Groups Home

Oscillator/Substance
⊞ **Theory**

A "Rant " I posted to another site that someone may find Options indesting.

1 message - Collapse all - Report discussion as spam

dean sinclair View profile   More options May 25, 12:25 pm

GmailCalendarDocumentsPhotosReaderWebmoreSitesOn a Coherent Theory, a "Rant."

Let me introduce myself, for those who do not personally know me, my name is Dean LeRoy Sinclair. I am a 79 year old, former science teacher whose background includes a Ph.D. in Organic Chemistry, Mathematics through Differential Equations and Non-Euclidian Geometry and Electronics Background which includes Radar Repair Training at Ft. Monmouth, a Radio Telephone General Class License and an Amateur Radio Technician"s Class License.

In the last few years. it has become clear, not only to me but to many others, that the current theories and beliefs in the physical sciences can not account for the phenomena which are being discovered every day.

This is quite understandable, considering that perhaps the latest of the commonly accepted ideas is the "Standard Model of Particle Physics" which received a Nobel Prize in the 1970's. . and, in my opinion, bears about as much relationship to "reality" as the Geocentric Model did to the Solar System.

The sad state of modern physics theorizing is evident in the fact that the 2010 Nobel Prize in Physics can be said to have been won by a chemical structure, graphene, which is an extended mono-layer of graphite.

It has been proposed on one Internet site, of which I am honored to be allowed to be a posting member, that a new basic theoretical approach is needed. I am totally in agreement with this, However, some have said that this must be, "....in the common language of science and considering the tried and true principles that have served us so well." This last I totally disagree with. To make progress. I feel that we must start over carefully reexamining what we have so long accepted as truths.

The "tried and true principles" have to have errors and misconceptions or there would not be so many conflicting concepts which simply do not fit together, such that people seem to say with straight faces things like, "The electrons in a deflated state of a naught orbital are so highly relativistic that they will cause the quark flavour to penetrate the Coulomb Barrier such as to allow fusion by means of the Strong Force." The above statement, is. of course, pure nonsense, but theoreticians seem to string together pieces of different approaches which have no connection to one another into pronouncements which actually have no more meaning than the above string or fragments.

I propose that we start over.examining ideas from "Point Zero."

Zero, the first number, which in mathematics is considered as representing "Nothing," or more usefully, the starting point from which to travel, or measure. In mathematical conventions it can be the center of a circle or the "origin point of a set of Cartesian Coordinates." In reality, then, as   compared to mathematical abstraction, Zero, the starting point will have a size and shape, and may be called the "First Dimension," the Dimension of Existence. The number, One, has, in reality, a number of meanings, it may represent an entire object, or it may represent a line, which has to be made up of two points, hence, a line can be considered as The Second Dimension ' Line. Iff we say "one times one equals one," we are saying that we have done a second mathematical operation, so our new "one" represents one surface, a third dimension. One times one times one adds another dimension, one volume. Since we cannot easily visualize the next form we may consider that we start over with a "Zero Dot" which is now a Volume and go on defining a line of volumes,, a surface of volumes, and a volume of volumes. Etc...

Let us note here that. if we use the convention that any "one" with out a sign attached is the signed number, +1, "Positive One," use Cartesian Coordinates , and the conventional "positive directions" of right, up and forward, on the "x, y and z axes.," we might consider that we are describing a potential model which expands counter-clockwise right, up, and forward into the "Totally Positive Number Octant" of the "Cartesian Field."

 Having noted that an automatically accepted mathematical convention may well be excluding at least ⅞ of "Reality" from a model, let us look at another mathematical number, "Infinity," defined as the number beyond all numbers.... Infinity can be consider in a practical sense as the point, position, number, or whatever, which is just beyond where we stopped measuring or counting, either by choice or because our "tool," is no longer usable. This should be kept in mind wen we look at things such as the statement, "At the speed of light, mass goes to infinity...."

Having mentioned the Speed of Light, brings us to the idea of "Constants of Nature." Since most of our information about nature seems to have been transmitted by means of "Electromagnetic Waves. " two constants of nature which deal with Electromagnetic Waves may possibly be considered as fundamentals for any theory. These are The Speed of Light in a "Vacuum" and Planck's Constant. Let us realize that a constant is not a limit, it is more likely to be a statistical average. In fact, the term, "Speed" implies an averaged velocity between two points. We may take a logical view that the "Speed of Light" is an averaged velocity of some sort. Likewise, Planck's Constant has a possible interpretation as an angular momentum. probably, also, an average value.

If we go one step further and guess that these velocity and angular momentum constants are somehow related and related to something about whatever the "Substance of Existence" might be, we may set Planck's Constant. "h," equal to a definition of angular momentum and get the equation, "$h=mrv$. " Assuming that the average "v" involved is "c," the speed of light, we can write "$h=mcr$," and rearrange this to "$mr=h/c$."   Since mass times radius is called "torque" the push or pull on a spinning body, we have combined these two constants into another constant, a "torque constant of nature." We can even go a step further and say that at some point there would be a situation where the value of m equals the value of r equals the square root of "h/c."

We have come to an expression which can be evaluated to have possible physical reality, in cgs units this "square root set" is about 4.7×10^{-19} grams at 4.7×10^{-19} cm. (In terms of certain "accepted theories" this is interesting, Quantum Mechanics is said to fail at below 10^{-18} cm. and String Theory Strings vanish into a 10 dimensional hole at the same value..)

There is another interesting factor, the equation, $mr=h/c$ is an example of am equation which seeems to be the very commonest mathematical relationship in nature, the "balancing equation," $xy=K=yx$, which shows up in many guises, and in this case could be taken to possibly define the limits of an oscillator. That is is there has been determined a limiting mass, the corresponding radius could be found from tis equation, and by reversing the coefficients, the corresponding, "balancing" other limit be determined. As "rest masses" may well be limiting values and are known for such things as electrons and protons, this little equation could lead to many insights.

 Planck and others have noted that harmonic oscillators make excellent models for natural phenomena. These gentlemen do not seem to have taken the next step to the possibility that may natural phenomena are harmonic oscillators.

Another speculative leap may be taken about the ubiquity of the balance equation in nature and ask the question, "Since this balancing seems to be everywhere, yet the only thing that seems constant is change, couldn't a rational theory be developed on the basis or a "Substance of Existence" which was constantly in flux because of its tendency to "regress toward the mean," to balance all motions throughout. with, however, each action creating a reaction which was in turn is another action--this "Basic Substance" acting something like a physical substance at its Triple Point?

If anyone reading this far sees a theory developing above, which is totally independent of Relativity, Quantum Mechanics, and the Standard Model of Particle Physics, yet one which might at some point give insights into all of these, that person is absolutely correct, over the past few years, starting in 2004, the writer followed a similar line of reasoning and developed what was first called , "Motion in a Matrix," and later, the "Oscillator/Substance Model" A Little scouting around on the Internet, will find information on these if anyone is interested, However, I'd like to issue a challenge to intelligent, creative people who have read this far, "Try to ditch your preconceptions as much as possible about what is 'True' in physical science, reduce to as few simple basics as you can, and try to put together for yourself a coherent model, one where you will be satisfied that you have an understanding of all those things we've taken for granted and never really questioned, mass, energy, charge, gravitation, the proton-neutron nuclear atom, Relativity, Quantum Mechanics, ever, perhaps, Elements. See what you can come up with for a model."   Most, if not all, of you are far more intelligent than I, it should take you far less than seven years to put something together. Of course, I expect you to come out about the same place I have, but. also, hope there will be some pleasant surprises....

It is intended to post this several places on the Internet, as a result there is a potential audience of several hundred people. It is hoped that a few will take seriously what I have said and take up the challenge.

Dean L. Sinclar   deanlsincl...@gmail.com

Reply to author   Forward

End of messages

« Back to Discussions   Older topic »

$$3^{++}, 3^{+}, 2^{+}, 1^{+}$$

$$Au \quad 13, 52, 1, \bigcirc$$

Gmail Calendar Documents Photos Reader Web more ⁔   deanlsinclair@gmail.com ⁔

Go⁚gle groups

« Groups Home

Oscillator/Substance
⊞ Theory

[Search this group] [Search Groups]

Members

[Search members]

Sort: nickname▽   membership type   join date   15 members « Previous | Next »

 eskia...@mail.com
Member - joined Jun 1 2010

 Brad Guth
Member - joined Aug 30 2010
   Name: Brad Guth
Location: Olalla Washington
      Bio: 1-253-8576061 or 459-9790
        email: bradguth@gmail.com
        Blog and Google document pages:
        http://groups.google.com/group/guth-usenet?hl=en
        http://bradguth.blogspot.com/
        http://docs.google.com/View?id=ddsdxhv_0hrm5bdfj
        https://docs.google.com/File?id=ddsdxhv_4fdgd46df_b

 ESKI (deanlsinclair@gmail.com) (you)
Group owner - joined Jul 26 2008
   Name: Dean L. Sinclair
Location: Aberdeen, SD, USA
      Bio: Born Jan.3, 1932, MIdland, SD Lived on hard scrabble ranch/farm West
        River South Dakota to 1944, when the grasshoppers finally ate us out....
        Graduated ;Murdo SD HS, class of '49. , valedictorian ;Yankton College,
        BA, cum laiude, Chem., Math. 1953 ; Army, Pfc., Radar Repairman, '54-
        56, M.S. Chem., Okla. State, 1959, PhD, Organic Chem.,Kansas State,
        1967. Some 375 SH of total credit.with major hours in Math., Science,
        Languages, Psychology , Education, Counseling

        Interests: Science theory. Languages, Reading, Sketching....
        My current major project is"Oscillator/Substance Theory." which I hope is
        logic based and not just my "Cracked Pottery."

 heycollin
Member - joined May 7 2010

 hoek
Member - joined Oct 13 2008
   Name: Bob Vanderhoek
Location: Northeast US

Home

Discussions

Members
+ invite new members

About this group
Edit my membership
Group settings
Management tasks
Invite members

 View this group in the
new Google Groups

Group info
Members: 15
Language: English
Group categories:
Science and Technology
> Physics
Science and Technology
> Chemistry
Science and Technology
change categories
More group info »

Hugh V
Member - joined Aug 7 2008

JackOS
Member - joined Oct 25 2009

ka-sala
Manager - joined Oct 16 2008

Name: ka-sala

Location: Cosmos

Bio:

The load of all my concerns was of the nuclear waste issue, which grew into a new Energy Source.

By 1979, needing confirmation of a small scientific subject I was working on, it took me to a Professor Cunningham, at the Tech. for Higher Education in the Department of Physics. Cape Town SA.

Five months later... he phoned.
"Three scientists, have just won the Nobel Prize for what you were explaining to me. It's all connected to the Cosmic Rays!" (The previous year had been the dicovery of the Cosmic Microwave Background Radiation.)
"Oh..." was my only answer. Truth was, I knew it within me, and this was my confirmation. I certainly wasn't looking for a Nobel Prize... but I could at least thank him. *

*** These 3 Scientists can be found on Nobelprize.Org.

The year =1979

'for their contribution to the unified weak and electromagnetic interaction between elementary particles, including inta alia, the prediction of weak neutral current.'

*** No... there is no mention of me. All I was doing was making report of 'My Reason'.

I will be happy to share some, as it was recognized... This may not be in exact English Science Jargon.

magickid
Member - joined Oct 5 2008

Major Ray
Member - joined Jan 12 2009

Name: Major Ray

Location: Dedham, Massachusetts

Bio: I may have several degrees in chemistry, including a Ph.D with honors, but I have no respect for the bias prevalent throughout organized science. Racism and religious bias are the reasons I created BIAS, Inc. (Boston Institute for the Advancement of Science, Inc.) On my site I address a number of issues related to the MATRIX. I believe in the theory of everything.

NewYearGreetings
Member - joined Jan 1 2010

nishlaverz
Member - joined Apr 1 2010

ollin
Member - joined Jan 18 2010
Name: Jorge Luna
Location: New Orleans, USA

Ronald Jennings
Member - joined Nov 4 2010

SamisadMark
Member - joined Nov 2 2010

+ Invite new members   15 members   « Previous | Next »

Edit member list and member permissions >

Create a group - Google Groups - Google Home - Terms of Service - Privacy Policy
©2011 Google